Principles of Wet End Chemistry

William E. Scott, Ph.D.

TAPPI

TAPPI PRESS

Atlanta, GA © 1996

TAPPI PRESS
Technology Park/Atlanta
P.O. Box 105113
Atlanta, GA 30348-5113 U.S.A.

ISBN 0-89852-286-2
TP 0101R241
Printed in the United States of America

Library of Congress Cataloging-in-Publication Data

Scott, William E., 1940-
Principles of wet end chemistry / William E. Scott.
p. cm.
Includes bibliographical references.
ISBN 0-89852-286-2
1. Papermaking -- Chemistry. I. Title.
TS1120.S15 1996
676'.23--dc20 95-52282
CIP

Preface to the First Edition

For much of its history, machine papermaking has been primarily considered to be a mechanical process. This is understandable, since the mechanical aspects of stock preparation, sheet forming, pressing and drying are readily observable, or at least apparently so. Whereas, the chemical interactions that occur in these processes are not readily visible and understanding them requires a knowledge of basic general chemistry principles and a capacity to think abstractly about "invisible" chemical processes. The combination of an invisible collection of chemical processes and the necessity to do abstract reasoning has produced a slower pace of technological development in the wet end chemistry arena than in the mechanical papermaking arena.

As is often the case, the development of one area of technology brings about pressures that force the development of related areas. This has been occurring in the area of wet end chemistry for the past 25 years as a string of mechanical advancements, such as high speed papermaking, hydraulic headboxes, twin wire forming, high intensity pressing, reduced basis weights, increased white water system closure, have made it clear that the papermaker must gain improved control over the chemical processes in papermaking. More recently, the trends toward alkaline papermaking and increased use of recycled fibers have accelerated the emphasis on wet end chemistry control. Concomitant with the increased awareness of the importance of wet end chemistry has been increased activity by instrumentation and process control companies in the development of on-line sensors applicable to wet end chemistry process control. Thus, the pace of wet end chemistry advancement has accelerated and the future of the discipline is very promising.

There is general agreement among wet end chemists that much remains to be learned before the major wet end chemistry processes can be placed under real-time, closed-loop process control. Consequently, it will be necessary for production personnel and process engineers to be intimately involved in process control decision-making in the foreseeable future, which brings us back to the need for an understanding of basic wet end chemistry phenomena. TAPPI has recognized this need and produces a number of seminars, short courses, and technical publications dealing with the various aspects of wet end chemistry. This book is intended to be part of these offerings.

The primary objective of this project was to produce an introductory overview of the subject of wet end chemistry. Most of the topic areas associated with the discipline are treated. The intent is to introduce the jargon and concepts and provide entry to the key reference literature in special topic areas. It is assumed that the reader has had chemistry training at the college freshman level. This should make the text accessible to a wide readership within the paper and allied industries.

Preface to the Second Edition

This edition of "Wet End Chemistry: An Introduction" represents a major rewrite of the original book. All of the chapters have been revised and new chapters dealing with paper properties, water chemistry, charge measurement, fillers, recycled fibers and wet end chemistry process control have been added. Every attempt has been made to retain the original introductory nature of the treatment and the same audience is addressed as in the first edition.

When asked to describe wet end chemistry, papermakers and others working in the field have no problem coming up with appropriate adjectives. Words such as "difficult," "complex," and "impossible" immediately come to mind. What is it about the subject that produces these responses? A moment's reflection will reveal that there are many reasons for this. For example:

- A papermaking system has multiple, uncontrolled inputs and feedback loops that affect wet end chemistry phenomena.
- Most important wet end chemistry reactions are irreversible.
- There are interfering and interacting mechanical factors to be taken into account.

- Most wet end chemistry reactions and phenomena are dynamic (i.e. time-dependent).
- Wet end chemists must work with competing aggregation and dispersion chemistries — often at the same time.
- The wet end chemist's real-time process knowledge is limited due to a paucity of available on-line data.

This is a daunting list, to be sure. But, the fact that all of the above factors are recognized and, to a great extent, understood, represents a major step forward. Once the "enemy" is known, it becomes easier to deal with.

Another aspect that must be taken into account is that there are two views of the "wet end chemistry world." One view is held by paper machine operators, production managers, process engineers, and chemical technical sales and service people. These are the practitioners whose wet end chemistry world has the following attributes:

- <u>It is empirical and experience-driven.</u>
  Practitioners collect vast amounts of technical information in the course of doing their job. Some of it is experimental data from trials and other experiments. Some of it is more subjective or of an anecdotal nature. Cause-and-effect relationships and rules-of-thumb are developed over time and used to support decision-making.

- <u>It is "practical."</u>
  This is an obvious characteristic of the practitioner's world. The bottom line rules — however it is defined. Cost-effectiveness, productivity, quality and sales are utmost in one's thinking at all times.

- <u>Absolute results are most important.</u>
  The papermaker and supplier are interested in achieving specific numerical targets or goals in their work. Paper specifications, production rates and costs are all expressed as quantitative absolutes. It is another aspect of the "practical" nature of the job.

- <u>There is much "noise" in experimental data.</u>
  Experimental data collected from papermaking systems has much random variation associated with it. Consequently, real process changes are impossible to discern in many instances because they are obscured by random data fluctuations. This is a fact of life in the papermaker's world.

- <u>Industrial processes have multiple, interacting variables.</u>
  By its very nature, the papermaking process is influenced by a number of variables - many of which interact with one another. This means that effects resulting from changes in a specific variable (e.g. filler level, sizing level, machine speed) may depend upon the level of another variable (e.g. pH, conductivity, temperature, coagulant level, dye level). Sometimes these interactions involve more than two variables. Such occurrences make it very difficult or impossible to develop cause-and-effect relationships among process and product parameters.

- <u>"Models" are difficult or impossible to derive from mill data.</u>
  Operators, process engineers, technical service people, production managers and scientists all think about their processes and products in terms of "models." These models come in many different forms. Some involve mathematical equations. Others are conceptual. They can be either quantitative or qualitative. Whatever their form, they are necessary to effective problem solving and decision-making. Unfortunately, the factors mentioned above (process noise, multiple variables, interacting variables) all mitigate against the development of reliable models from process data.

The second "world view" of wet end chemistry is that of academia and the industrial wet end chemistry researcher. What are the important attributes of their world?

- The researcher undertakes laboratory experiments or "controlled" mill experiments.
  The researcher either works in a laboratory, whether it is on-site at a mill or off-site in a research center, or collects data from mill trials that are controlled as closely as possible. This is the most essential requirement of a researcher's work environment.

- The researcher designs experiments with fewer and more controlled variables.
  This is self-explanatory. By focusing on a limited number of controlled variables, the chances of learning how they affect the process or product under investigation are increased. The laboratory environment has less experimental noise (i.e. fewer, random, uncontrolled variables) and the noise that does exist can be taken into account by rigorous, statistical experimental design procedures.

- Models can be derived from laboratory data.
  The development of models is really what research is all about. In other words, a researcher tries to produce a description of a process (used in the general sense) that can be extended or used as the basis for future decisionmaking The description can consist of mathematical equations, graphs, narrative, or be of any other form appropriate to the situation. The underlying theme is that the model must describe a process in terms of the independent variables of interest.

- Relative results are more important than absolute results.
  This is a key distinction between the practitioner and the researcher. Researchers recognize that, no matter how hard they try, it is practically impossible to reproduce the "real world" in the laboratory. Consequently, experimental data collected in laboratory experiments will not duplicate mill data. Therefore, researchers must be content to develop experiments that produce relative data aimed at elucidating relationships and effects. The underlying assumption is that even though the absolute values of the data will be different, many interrelationships among variables hold equally in the laboratory and in the mill.

So, where do the worlds of the researcher and practitioner intersect? The answer to that question lies in a common feature that exists between the two groups — both rely strongly on models to do their jobs. Researchers develop the data from which models can be constructed. Practitioners use the models to facilitate the thinking processes required in process and product trouble-shooting and optimization. This is true in all areas — not just wet end chemistry.

And where does this book fit into all this? The answer again lies in models. The book is full of them. For example, in describing the function of retention aids, we rely upon "bridging" and "patch" models. An understanding of how fillers enhance opacity depends upon models for light scattering from solid surfaces. To explain sizing, we invoke a model of sizing molecules that has hydrophobic and hydrophilic regions. And so on and on. Models are invoked to describe and explain practically every concept included in the book. It is hoped that these devices will enhance the teaching effectiveness of the manuscript and serve as reasoning aids in your future wet end chemistry trouble-shooting and optimization endeavors.

Wet End Chemistry "Themes"

When studying a broad subject like wet end chemistry, it is helpful to identify the major themes that tie the material together. The following four examples are particularly relevant to this book.

- The importance of chemical equilibria
  This most elementary of chemistry subjects is often overlooked by papermakers when they are faced with the complexity of a typical paper machine system. However, many wet end chemistry phenomena are based on chemical equilibria and so the papermaker needs to be well-grounded in these fundamental concepts.

- The importance of particle specific surface area
  Surface interactions such as adsorption, aggregation and ion exchange are extremely important in wet end chemistry. All of these phenomena are affected by the specific surface area of participating fibers, fines, and fillers. Since the specific surface area of the fines and filler fraction of a papermaking suspension is usually much greater than that of the fiber fraction, it is not surprising that fines often influence wet end chemistry to a far greater extent than their weight proportion would suggest. This fact is referred to many times throughout the book.

- The role of surface charge and dissolved charge
  Practically all papermaking furnish components have associated electrostatic charges that play important roles in adsorption and aggregation processes. The papermaker often strives to control these processes by manipulating the charge on one or more furnish components. A good understanding of the basic phenomena involved is required in order to do this effectively and efficiently.

- The adverse effects of interfering substances
  Interfering substances are introduced early in the book. Their very name indicates that they are not "good" things to have in a papermaking system. In fact, they can interfere with most functional additives and some processing aids. Unfortunately, current trends such as closed white water systems and increased secondary fiber use point to increasing levels of these materials in the future. Consequently, a papermaker must be aware of their existence and be knowledgeable about their properties and understand how to counteract them in a given situation. Their effects are described throughout the book.

Organization of the Book

In revising the book, a new order of presentation was selected. Chapter 1 briefly reviews the impact of wet end chemistry on basic paper properties. Chapter 2 deals with the basic properties of water, the medium within which all wet end chemistry phenomena occur. Next, fibers and fines are introduced and the characteristics that affect wet end chemistry are described. Interfering substances are the subject of Chapter 4 At this point, the basic materials present in most papermaking systems will have been introduced.

Chapter 5 then discusses the fundamental colloid and surface chemistry concepts that govern the interactions among system components and Chapter 6 reviews the very important topics of dissolved and surface charge measurement. Chapters 7-9 cover the basics of dyes, dry strength agents and wet strength agents. These materials all interact with fibers molecularly through surface adsorption mechanisms. The complexity of their reactions increases in the order in which they are presented.

Next, we move to additives that occur as particles or droplets: fillers, alum-rosin sizing agents, and synthetic sizing agents. Chapters 10-14 are involved.

Chapter 15 deals with retention, which is the desirable aggregation of fine particles during papermaking. Chapter 16 synthesizes much of what has gone before from the process control standpoint. Finally, a glossary of important terms and definitions has been added.

The author recognizes that the book is an unfinished work because deposit control, foam, and other adverse wet end chemistry interactions are not included. This was an unfortunate, but necessary, omission in order to publish the manuscript in a timely fashion.

The book is organized so that relatively simple systems are treated first, followed by more complex areas. It is hoped that this organization will help the reader to see the connections from one topic to the next and facilitate comprehension of the material. Feedback regarding whether or not this objective has been met would be very much appreciated.

About the Author

William E. Scott is a professor in the Miami University Paper Science and Engineering Department, Oxford, Ohio. He teaches and carries out research in various areas related to papermaking chemistry. In addition to his academic experience, Dr. Scott has worked for Champion International and served as a consultant to many companies. He is active in various TAPPI committees and is a former TAPPI Director. He has published over 50 technical papers, books, and chapters. He is a TAPPI Fellow. Dr. Scott earned his B.S. from Miami University and his Ph.D. from the Institute of Paper Chemistry.

Dedication

This book is dedicated to all who have contributed to the advancement of the practice of papermaking chemistry, especially those who have contributed so much to my understanding of the discipline: Joe Marton, Ed Strazdins, John Swanson, John Penniman, Tom Lindstrom, Bob Stratton, Jerry Gess, Allan Springer, John Unbehend, Ken Britt, Kasy King, Bob Gill, Oscar May and many others who have relentlessly championed the "cause" during the past four decades.

Special recognition and gratitude go to Barry Crouse, who has provided consistent encouragement to the author over the years and who reviewed the entire draft version of this manuscript and made many helpful suggestions.

The excellent support of the TAPPI staff in this and many other projects is warmly acknowledged.

– July 1995

Table of Contents

Chapter 1

Introduction to Wet End Chemistry

A recent review estimated that the North American paper industry spent approximately $725 million on papermaking additives in 1993.[1] This impressive figure indicates the importance of the nonfibrous components of a papermaking furnish. Broadly speaking, wet end chemistry additives and practices are focused on two objectives:

- Achievement of paper properties
- Enhancement of paper machine runnability

The first objective is aimed at meeting product specifications set by customers. Of primary importance in this regard is the efficient application of functional additives. The second category is most important to paper machine productivity — a key to profitability. Chemical processing aids assist in this endeavor.

1. Wet end chemistry and paper properties

Every paper grade has a set of paper property specifications that are agreed to by customer and manufacturer. Paper has a number of inherent property limitations. Wet end chemistry enables the papermaker to overcome many of these deficiencies in order to meet a customer's needs more closely.

a. Categories of paper properties influenced by wet end chemistry

Most paper properties that are affected by wet end chemistry can be included under one of the following five categories:[2]

- Structural properties (basis weig™ht, caliper, formation, directionality, two-sidedness, porosity, roughness, dimensional stability)
- Mechanical properties (tensile strength, tearing resistance, bursting strength, stiffness, folding endurance, internal bonding strength, surface strength)
- Appearance properties (color, brightness, opacity, gloss)
- Barrier and resistance properties (sizing)
- Permanence properties (durability, color reversion, chemical stability)

The following discussion provides examples of how wet end chemistry facilitates the achievement of the above properties.

b. Structural properties

Basis weight is controlled primarily by a paper machine's stock metering system. However, the uniformity of the machine direction basis weight profile is strongly affected by first-pass retention. Modern closed loop retention control systems that can reduce machine direction basis weight variability are being increasingly utilized, particularly in fine paper manufacture.

Formation (the property) is a function of the three dimensional distribution of materials throughout the sheet. A well-formed sheet is uniformly opaque when viewed by transmitted light. Formation can be adversely affected by over flocculating the fiber components of a furnish. This produces a blotchy appearance and can degrade sheet strength and its interaction with printing inks. Thus, the papermaker must avoid excessive use of strongly flocculating retention aids. In very long fibered stocks, such as those used in the production of nonwovens, polymeric formation aids are sometimes used.

Two-sidedness is another structural property that is affected by wet end chemistry. One of the factors that effects two-sidedness is a nonuniform z-direction (the 'thin' direction) distribution of materials in the sheet. Specifically, two-sided sheets have a fines distribution that is higher on the felt side of the sheet than on the wire side, with a related fines gradient in the z-direction.

Two-sidedness occurs primarily when web drainage occurs from one side only during sheet formation. Initially, only long fibers and the fines attached to them are retained by the relatively coarse forming fabric. Later, the formed web becomes the filtering medium and smaller particles are sieved out and retained. Thus, web plies near the wire side contain less fines material than the plies formed later which are near to the felt side. Increasing the attachment of fines to fibers by flocculation will, of necessity, even out the fines distribution since more fines will be retained during the initial sheet formation stages. The term 'two-sidedness' is also used to refer to the fact that dyes tend to concentrate more on the felt side and impart a deeper shade to that side of the sheet in many instances. Again, this phenomenon is caused by the different fines concentration in the wire and felt sides.

Mineral filler particles fill in small pores in the sheet and even out surface nonuniformities. Hence, increased filler levels usually reduce paper roughness and porosity. It follows that these two properties are also affected by fines and filler retention.

c. Mechanical properties

Mechanical properties (strength properties) are functions of bond strength, fiber strength, and sheet formation. If formation is allowed to degrade beyond a certain extent, then strength properties are adversely affected. This is a second order effect, however, compared to the influence that wet end chemistry has on bond strength. Specifically, fiber-fiber hydrogen bonding can be greatly enhanced by the use of dry strength additives such as starch. The use of such additives has been common for decades and will become more important in the future as the recycled fiber content of furnishes increases. Consequently, the papermaker must understand the factors that affect starch retention and set up system conditions accordingly. Other chemical additives, such as fillers and sizing agents, interfere with fiber-fiber bonding and reduce paper strength.

It is also commonly experienced that neutral and alkaline papermaking conditions promote improved fiber-fiber bonding relative to acid papermaking. Many companies have converted from acid to alkaline papermaking during the past decade.

Many paper grades have to retain a high level of strength when saturated with water. In other words, they must have "wet strength." Paper has no inherent wet strength due to the nature of fiber-fiber bonds. Hence, wet strength must be achieved through the use of additives. Such additives, called wet strength resins, are widely used in the paper industry. They are cationic, polymeric materials that are added to the furnish prior to sheet formation, and must be retained and cured in the paper web.

d. Appearance properties

Dyes and mineral pigments (fillers, optical brightners, colored pigments) are a principal means of influencing the appearance of paper, as described by the properties color, opacity and brightness. Chemical additives have less effect on gloss. These materials either scatter light more effectively than fibers or selectively absorb specific visible light wavelengths. Dyes interact specifically with visible light and this reaction is affected by a wide range of chemical and physical factors. Hence, the papermaker must closely control the chemical environment and physical conditions of a furnish in order to avoid color changes. It is also important that conditions be established that will promote the ion exchange retention mechanism of dyes.

In order to attain high opacity and brightness, it is necessary to retain high levels of mineral filler pigments in the formed web. Mineral pigment light scattering efficiency is also a strong, inverse function of pigment aggregation. These two factors pose a dilemma to the papermaker since the flocculation chemistry that promotes filler retention also promotes filler aggregation. Good understanding of the interactions that occur among fillers, fiber and retention aids is imperative in order to achieve the required balance.

e. Barrier and resistance properties

One of the major weaknesses of paper is that wood fibers have absolutely no natural resistance to water penetration. Hence, when this property is required (as it is for most paper grades), it must be achieved either by adding a sizing agent prior to sheet formation or treating the formed paper with a water-resistant coating or saturant. The former practice falls into the realm of wet end chemistry.

Internal sizing (wet end sizing) entails retention of either an alum-rosin sizing agent or a synthetic sizing agent in the paper web when it is formed. The sizing agent is mixed with the fibers and other ingredients prior to the paper machine headbox. This agent, which imparts hydrophobic character to the fiber surfaces, must be retained rapidly and to a high degree in order to be effective and avoid machine cleanliness and runnability problems. The effectiveness of sizing agents is influenced by a number of furnish and machine factors.

f. Permanence

Traditionally, permanence has been of interest primarily to librarians and archivists. While still important to these groups, others have become aware of the problem of paper aging and the matter is receiving increased attention.

Chemistry plays a large role in paper permanence. Indeed, it is well-established that alkaline paper is far more permanent than acid paper, under practically all circumstances. It is also recognized that the chemical and physical condition of fibers, impurities remaining in poorly washed pulps, heavy metal ions, and calcium carbonate fillers also play roles in determining paper permanence. These all fall into the category of wet end chemistry. Thus, much of the work being done to increase permanence has focused on improvements in wet end chemistry and wet end chemistry control.

g. Closing comments

The above discussion, while abbreviated, nevertheless illustrates the central role played by wet end chemistry in the attainment of the paper property values required by grade specifications. This connection will be referred to often in the chapters that follow.

2. Wet end chemistry and machine efficiency (runnability)

Wet end chemistry plays both positive and negative roles in paper machine runnability. On the one hand, wet end chemistry can be used to enhance drainage, minimize entrained air and foam, keep machines clean, and keep the solids levels down in recycled white water streams. Conversely, when out of control, the same wet end chemistry phenomena produce machine deposits, scale and foam spots, and reduce drainage, machine cleanliness, and pressing efficiencies. Some typical examples of these situations are the following.

a. Drainage

Drainage from the forming web is strongly influenced by the nature of the fiber-fines and fines-fines flocs present. If the flocs are large and porous, then they will retain water tenaciously and retard drainage. If the flocs are small and tight, then they tend to give up water readily and enhance drainage. Since the retention system determines floc structure, it is a major factor in the drainage performance of a machine.

b. Deposits and scale

Deposits and scale usually arise from out-of-control wet end chemistry. Typical examples include chemical additive overdosing, charge imbalances, chemical incompatibility and the shifting of chemical equilibria. All of these phenomena can lead to the formation of precipitates or colloidal aggregates that produce deposits and scale. While there are numerous approaches to treating the symptoms of deposits, the best approach is to determine what is out of control and then to fix it.

c. Foam and entrained air

Wood fibers contain substances that will stabilize air that gets entrained in papermaking stock slurries. Several common chemical additives will do likewise. Whenever this happens, a number of adverse effects (reduced drainage, increased slime, increased foam) can appear. As with deposits, the best approach to such problems is to find the source and eliminate it. When this is impossible, there are both mechanical and chemical solutions that can be adopted. In either instance, knowledge of the wet end chemistry factors involved will be invaluable.

3. Current trends

A major wet end chemistry trend for the past decade has been the conversion of fine papermaking from acid to alkaline conditions. This process has proceeded until it is thought that nearly 85% of the current coated and free sheet fine paper manufactured is made alkaline. This activity has entailed a complete review of the chemical additives employed in papermaking. One major aspect of this is the replacement of clay and titanium dioxide fillers by calcium carbonate — with a parallel increase in loading levels.

Another trend which has strong bearing on wet end chemistry practices is the increased use of recycled fibers. These fibers are often contaminated with deinking chemicals or other substances. Deposit problems often increase, as does the consumption of additives by contaminants. Since the recycled fibers are often relatively weak, the need for dry strength resins increases.

Another trend is increased chemical additive use in grades other than fine papers. For example, fillers are being used in a wide range of groundwood grades, as are retention aids. Linerboard is also seeing increased additive use — especially in instances where high recycled fiber content exists

A related area where the future promises advances is wet end chemistry process control. New on-line sensors are being developed and papermakers are experimenting with new process control strategies and incorporating wet end chemistry process control with other paper machine control schemes.

4. References

[1] Young, J. "Papermaking additive expenditures hold steady, await market return." Pulp and Paper, 46-48 (March, 1994).

[2] Abbott, J., Scott, W. E., and Trossett, S. *Properties of Paper: An Introduction.* Second Edition, TAPPI PRESS, Atlanta, 1995.

Chapter 2

Water: The Primary Papermaking Medium

Even though papermakers tend to concentrate on the important properties of wood fibers and chemical additives when thinking about wet end chemistry, it is clear that water is an equally important ingredient. Practically all papermaking employs water as the suspending medium. The swelling effects of water on wood fibers leads directly to the development of strong fiber-fiber bonds. Water dissolves or hydrates most important papermaking furnish components. It is therefore not surprising to discover that much of wet end chemistry involves application of the classical water chemistry taught in introductory chemistry courses. Consequently, we will begin our study of wet end chemistry by reviewing some basic water chemistry concepts important to papermaking. Included will be a brief review of the structure of water and its hydrogen bonding ability. The chemical equilibrium concepts of pH, acidity and alkalinity will also be treated. Finally, two properties of water solutions, specific conductance and hardness, will be examined. More extensive treatments of water chemistry can be found in any general college chemistry textbook. References [1] and [2] in this chapter also contain more thorough treatments of the subject.

1. The geometry of water molecules

Water has the very well-known molecular formula, H_20, where the two hydrogen atoms are each bonded to the oxygen. The molecular geometry is shown in **Fig. 2-1**.

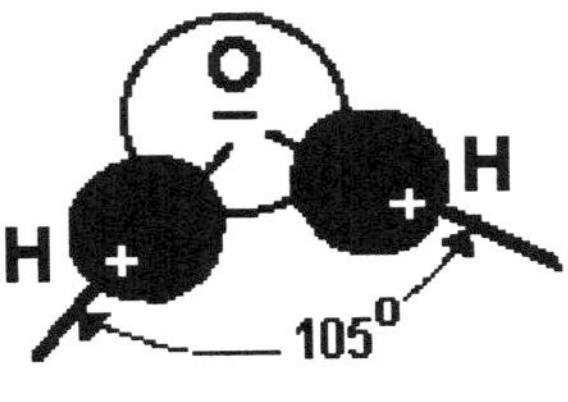

Figure 2-1. The geometry of the liquid water molecule, H_2O. Each hydrogen atom is bonded to the oxygen atom. The O-H bond length is 0.096 nm and the H-O-H bond angle is 105°.

The electrons in an O-H bond are not shared equally between the hydrogen and oxygen, but tend to reside closer to the oxygen most of the time. This leads to polarization of the O-H bond and, in turn, the entire molecule.

The polarity of molecules can be determined by measuring their dipole moments. **Table 2-1** lists the dipole moments of some common molecules.

Table 2-1. Dipole moments of some molecules

Compound	Dipole Moment (debye units, D)
Sodium chloride	9.0
Water	1.85
Methanol	1.70
Carbon tetrachloride	0

2. The hydrogen bonding ability of water molecules

The polarity of the water molecule leads to their association through hydrogen bonding interactions (**Fig. 2-2**).

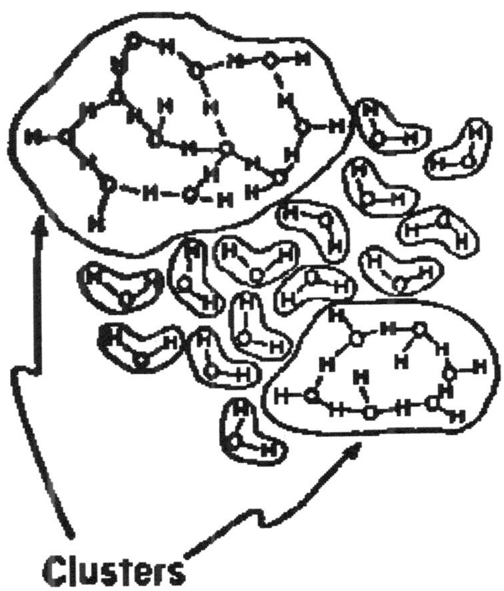

Figure 2-2. Strong hydrogen bonding between water molecules leads to the formation of many clusters in liquid water and produces many of the unique properties of water.

This association, in turn, produces many of the unique properties of water listed in **Table 2-2**.

Table 2-2. Some unique properties of water

Substance	Specific Heat	Freezing Point, °C	Boiling Point	Latent Heat of Evaporation, cal/gram
Water	1.00	0	100	540
Methanol	0.57	-98	65	263
Ethanol	0.54	-117	79	204
Benzene	0.39	6	80	94

Water also has a high surface tension — a feature which is important in the development of fiber-fiber bonding in paper and in the wetting of paper surfaces. Surface tension will be discussed in more detail in Chapter 5.

Water can also hydrogen bond with many other substances — especially those containing hydroxyl (OH) functional groups. This is particularly relevant to papermaking since papermaking fibers contain such groups. We will refer to the hydrogen bonding ability of water when talking about dry strength development and wet strength. Water's abilty to interact with other substances also contributes to its solvent properties. Indeed, water is often referred to as the "Universal Solvent."

3. Chemical equilibria in a water environment

Many molecules dissociate into two or more ions when dissolved or suspended in water. Indeed, the water molecule itself is dissociated to a small extent. Some molecules, such as NaCl, dissociate essentially one hundred percent and only Na^+ and Cl^- ions exist in solution. Other molecules dissociate to a lesser degree and the equilibrium solution contains both ions and undissociated molecules. Water is one such molecule. **Fig. 2-3** shows the water equilibrium.

Chemical Equilibrium in a Water Environment

pH

$$H_20 \rightleftharpoons H^+ + OH^-$$

ca. 55 M　　　10^{-7} M　10^{-7} M

$$[H^+][OH^-] = 1 \times 10^{-14}$$

Figure 2-3. The dissociation of water and its equilibrium ion concentrations. The product of the H^+ and OH^- concentrations always equals 10^{-14}.

a. A dynamic equilibrium

It is important to remember that the water equilibrium is a dynamic phenomenon. In other words, H_20 molecules are constantly dissociating into H^+ and OH^- ions, while H^+ and OH^- ions are recombining into H_20 molecules. It is the product of the ion concentrations that remains constant.

Addition of an acid to water contributes H^+ ions to the mix. **Fig. 2-4** illustrates the response of the system when sufficient HCl is added to make the resulting H^+ concentration 10^{-3} M. In brief, the momentary ion product $[H^+][OH^-] = (10^{-3})(10^{-7}) = 10^{-10}$, which is considerably greater than 10^{-14}. In order to alleviate this non-equilibrium situation, OH^- and H^+ combine to form H_2O molecules, effectively "shifting" the dynamic equilibrium to the left in **Fig. 2-3**. This occurs until the final ion product equals 10^{-14}. The final equilibrium ion concentrations are then $[H^+] = 10^{-3}$ M and $[OH^-] = 10^{-11}$ M.

Effect of Acid on Equilibrium

$$H_20 \rightleftharpoons H^+ + OH^-$$

ca. 55 M　　　10^{-3} M　10^{-11} M

Figure 2-4. Effect of acid addition on the ionic equilibrium in water.

b. Definition of pH

It is somewhat cumbersome to work with scientific notation when referring to hydrogen and hydroxide ion concentration. Therefore, many years ago the notion of pH was developed in order to simplify things. The pH was defined as: $pH = -\log_{10}[H^+]$. **Table 2-3** demonstrates the common pH range.

Table 2-3. The pH values encountered in natural systems.

Hydrogen Ion Concentration	Hydroxide Ion Concentration	pH	Acid, Alkaline, or Neutral
10^{-1}	10^{-13}	1	Acid
10^{-2}	10^{-12}	2	...
....	......	......	...
10^{-7}	10^{-7}	7	Neutral
....	......	......	...
10^{-13}	10^{-1}	13	Alkaline
10^{-14}	10^{0}	14	...

Papermakers talk about *acid papermaking*, *neutral papermaking* and *alkaline papermaking*. These terms refer to the pH of the headbox furnish.

4. Acidity and alkalinity

The quantities *acidity* and *alkalinity* are often confused with the pH-related terms *acid* and *alkaline.* They are different. The pH is a measure of the instantaneous $[H^+]$ in a system. *Total Acidity,* on the other hand, is a measure of the total amount of H^+ that can be produced through equilibrium shifting ionization reactions. For example: organic acids, mineral acids, salts which hydrolyze to form acids and carbon dioxide (both free and combined) are capable of undergoing reactions that contribute H^+ to water. (Total Acidity is determining by measuring the amount of NaOH required to titrate a sample to pH 8.0.)

Total Alkalinity is the total capacity of a sample to contribute OH^- to water. Carbonates, bicarbonates, hydroxides, borates, silicates and phosphates all contribute to the total alkalinity of a system. (Total Alkalinity is defined as the amount of acid required to reduce a sample to pH 4.0. Note that if the titration is only carried out to pH 8.0, all hydroxides and one-half the carbonates are reacted.) Alkalinity is thought to be important in alum-rosin sizing as a mediator of alum ionization and in AKD sizing as a reaction catalyst.

Figs. 2-5 and 2-6 contain examples of hydrolysis reactions often encountered in papermaking that produce acidity and alkalinity.

Acidity Example

$$H_20 \rightleftharpoons H^+ + OH^-$$

Alum Ionization

$$Al_2(SO_4)_3 \rightleftharpoons 2Al^{+3} + SO_4^{-2}$$

Alum Hydrolysis

$$Al^{+3} + OH^- \rightleftharpoons Al(OH)^{+2}$$
$$Al(OH)^{+2} + OH^- \rightleftharpoons Al(OH)_2^{+1}$$
$$Al(OH)_2^{+1} + OH^- \rightleftharpoons Al(OH)_3$$

$$[Al^{+3}][OH^-] = 9 \times 10^{-33}$$

Figure 2-5. An example of a reaction that produces acidity — the hydrolysis of alum.

Alkalinity Example

$$H_20 \rightleftharpoons H^+ + OH^-$$

$CaCO_3$ Ionization

$$CaCO_3 \rightleftharpoons Ca^{+2} + CO_3^{-2}$$
$$[Ca^{+2}][CO_3^{-2}] = 9 \times 10^{-9}$$

Reaction Between Carbonate Ion and Hydrogen Ion

$$CO_3^{-2} + H^+ \rightleftharpoons HCO_3^{-1}$$
$$HCO_3^{-2} + H^+ \rightleftharpoons H_2CO_3^{-2}$$
$$H_2CO_3^{-2} \rightleftharpoons H_2O + CO_{2\,(gas)}$$

Figure 2-6. An example of a reaction that produces alkalinity — the hydrolysis of calcium carbonate.

5. Ionization of carboxyl groups in water

The carboxyl group [-COOH] represents an ionizable functional group that occurs in wood fibers and is particularly important to wet end chemistry through its influence on the surface charge of fibers.

The carboxyl group ionic equilibrium is influenced by the pH of the system (**Fig. 2-7**). Increasing the pH (i.e. increasing the $[OH^-]$ in the system) shifts the carboxyl ionization equilibrium to the right, resulting in an increase in $[COO^-]$ and a decrease in [COOH]. The ratio of the concentrations of the various species always stays the same, however, and is expressed by the equilibrium constant K_{eq}.

Decreasing the pH has the opposite effect — a reduction in $[H^+]$ and $[RCOO^-]$. In fact, at pH's below approximately 2.8, the equilibrium shifts almost entirely to the left and there is practically no $RCOO^-$ left. This has important implications to fiber surface charge — as will be discussed in Chapter 5.

Effect of pH on Carboxyl Groups

$$RCOOH \rightleftharpoons H^+ + COO^-$$

$$\frac{[H^+][RCOO^-]}{[RCOOH]} = 2 \times 10^{-5}$$

Low pH

$$RCOOH \leftleftharpoons H^+ + RCOO^-$$

High pH

$$RCOOH \rightleftharpoons H^+ + RCOO^-$$

Figure 2- 7. Effect of pH on carboxyl ionization. Increased pH leads to increased ionization. Decreased pH produces the opposite effect. Carboxyl ionization affects fiber surface charge.

6. Specific Conductance

Papermaking systems contain a variety of dissolved ions and are capable of conducting electricity. *Specific Conductance*, or conductivity, is defined as the conductance of a material between opposite sides of a cube, 1 cm in each direction. The traditional units for specific conductance have been $ohm^{-1}cm^{-1}$ or mho cm^{-1}. The current preferred units are milli-Siemens, where 1 mS = 10 μmhos.

Specific conductance determinations can be used to measure the total dissolved ion content of a sample — both inorganic and organic. Often, a linear relationship exists between dissolved salt concentration and specific conductance. **Table 2-4** contains some typical data.

Table 2-4. Typical conductance values for NaCl solutions

Concentration, Wt. %	Specific conductance, mS/cm
0.10	1.7
0.50	8.2
1.00	16.0
2.00	30.2
4.00	57.3
8.00	105
16.0	179

Paper mill conductivities may vary from less than 100 μmhos in very clean mills to as high as 10,000 μmhos in highly closed recycled mills. Conductivity can be correlated with a number of effects in the paper mill. **Table 2-5** lists several.

Table 2-5. Factors related to conductivity of papermaking process streams.

- Corrosivity of process streams
- Carryover from pulp mill and bleach plant
- Build-up of dissolved substances during white water close-up

7. Hardness

Hardness refers to a special subset of dissolved ions and their effects. The traditional definition of hardness referred to "the soap-consuming capacity of water." More recently, hardness has been defined as the "concentration of calcium and magnesium cations in water, independent of the anions present, expressed as ppm $CaCO_3$."

Hardness is often determined by titration with disodium ethylenediamine tetraacetate (EDTA) plus Calgamite indicator. The endpoint is a red-to-blue transition.

Hardness is an indication of the scaling potential of process streams and also interferes with alum-rosin sizing and wet strength development at very high levels.

8. Summary

Water is a critical ingredient in papermaking. It is the suspending medium from which paper is formed and it is the medium within which all of the furnish components are dissolved or dispersed prior to sheet formation. It promotes the formation of fiber-fiber bonding. The wet end chemist must be very familiar with the classical water chemistry concepts of pH, acidity, alkalinity, solubility, conductance, and hardness and understand how they affect the performance of papermaking materials.

9. References

[1] Water. Franks, F. Royal Society of Chemistry Paperbacks. London: The Royal Society of Chemistry, 1983.

[2] The Nalco Water Handbook. Ed. by Kemmer, F. McGraw-Hill Book Company, New York.

Chapter 3

Basic Properties of Papermaking Fibers and Fiber Fines Important to Wet End Chemistry

A papermaking furnish contains fibers, fiber fines, functional additives, and chemical processing aids. The papermaker purchases these materials and blends them into the furnish in the ratios required for a given grade of paper. The process water in which the various ingredients are suspended is another important component. Water plays a direct role in many papermaking steps — e.g., dissolving additives and hydrating and swelling fibers during refining. It also provides the environment for all papermaking chemistry interactions.

Finally, the furnish also can contain a variety of substances, referred to as "interfering substances", that interfere with the action of chemical additives. These substances are introduced to the papermaking furnish in pulp mill and bleach plant carryover, are brought into the system with various types of broke and secondary fibers, and introduced with process water.

In this chapter, we will review the properties of fibers and fiber fines that are important to papermaking chemistry. These materials participate in most of the coagulation and flocculation phenomena that occur during papermaking. They also significantly affect the performance of the functional additives to be discussed later in the book. We will emphasize those aspects of fiber morphology and fiber chemistry that are important to papermaking chemistry. The characteristics of fiber fines will be discussed in detail because of the central role they play in papermaking chemistry.

1. Papermaking fibers

Many different types of fibers have been used to make paper, including mineral, animal, synthetic, and vegetable fibers. Vegetable fibers are by far the most important and, of these, wood fibers find the greatest use. In the US, the split between softwood and hardwood use is about two-to-one in favor of softwoods.

a. Fiber morphology

Softwoods. The most important morphological characteristics of softwood fibers are fiber length (3-7mm), which contributes to their tendency to flocculate when suspended in water, fiber width or diameter (20-50μm), and fiber wall thickness (3-7μm). The fiber diameter is approximately in the size range of colloidal particles and contributes toward the relatively high specific surface area of fibers. Softwood is comprised mainly of fiber elements.

Hardwoods. Hardwoods are shorter than softwoods (1-2 mm), but still have a strong tendency to flocculate in water suspension. Like softwoods, their diameters lie in the range of colloidal materials (10-40μm). Unlike softwoods, hardwoods also contain a large percentage of nonfibrous elements. These elements include vessel segments, ray cells, and parenchyma cells. While they have many other influences in papermaking and on paper properties, these elements exert their greatest effects on papermaking chemistry through their behavior as primary fines, a topic to be discussed later in this chapter.

Fiber ultrastructure. **Fig. 3-1** contains a schematic drawing of the cross-section of a wood fiber. It can be seen that the wood fiber wall consists of different layers: the primary wall and the S1, S2, and S3 secondary walls. The middle lamella is located between the fibers. The lumen is the hole in the center of the fiber.

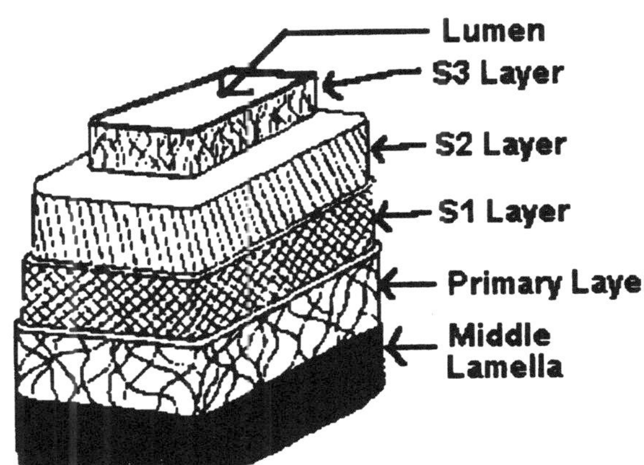

Figure 3-1. Cross-section of a wood fiber showing the layered structure of the fiber wall. M = middle lamella; P = primary layer; S1 = secondary 1 layer; S2 = secondary 2 layer; S3 = secondary 3 layer. The hole in the center of the fiber is called the lumen. The composite thickness of the fiber wall (less the lumen) is 3-7μm.

Fiber wall thickness is very important to the development of paper properties. Trees that grow in regions having long growing seasons tend to produce fibers that are relatively thick-walled (coarse). On the other hand, shorter growing seasons produce thinner-walled (slender) fibers. In general, thick-walled fibers are less conformable than thin-walled fibers and produce higher tearing resistance and lower tensile strength. They refine more slowly and scatter light less well.

b. Fiber chemistry

Chemical composition. Many papermaking chemistry properties of fibers and fines derive from their chemical composition. Wood pulp fibers contain cellulose, hemicellulose, and lignin, arranged in a complex mixture in the cell wall. In addition, the wood from which the fibers are derived usually contains organic solvent-soluble resins and fatty materials (designated extractives) that are not completely removed during pulping and can have profound effects on the papermaking process. The relative amounts of each material in wood are 40-50% cellulose, 20-30% hemicellulose, 20-30% lignin, and 1-10% extractives (**Table 3-1**).

Table 3-1. Chemical composition of three wood species[a]

Main Components	Spruce	Pine	Birch
Cellulose	41	40	40
Glucomannan	18	17	3
Xylan	8	7	30
Other carbohydrates	4	5	4
Total Carbohydrates	69	67	74
Lignin	27	27	20
Extractive Substances	2	4	3
Total	100	100	100

[a]*Eklund, D. and Lindstrom, T. Paper Chemistry, An Introduction, DT Paper Science Publications, Helsinki, p.21(1991).*

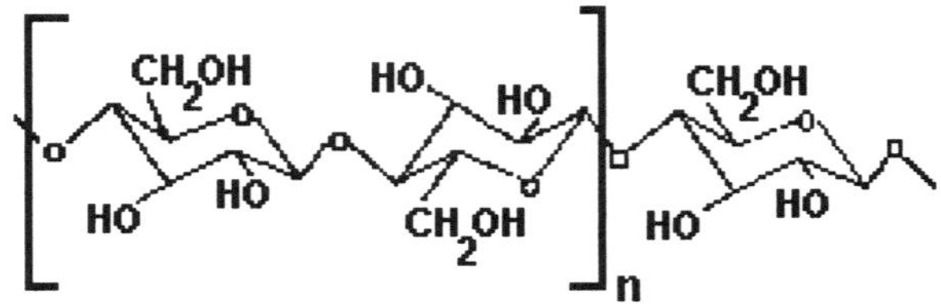

Figure 3-2. The cellulose molecule. Note that the glucose units are linked together in such a manner that a flat, ribbon-like shape arises. This shape is very good for packing together to form a compact and strong wood fiber wall.

Cellulose. Cellulose is the principal structural component of a wood fiber. Chemically, it is a polysaccharide composed of glucose sugar units bonded end-to-end through carbons 1 and 4 by an acetal linkage called a *glycosidic linkage*. The configuration of the linkage is β. A drawing of the cellulose molecule is presented in **Fig. 3-2**. Cellulose molecules are very long, typically containing 10,000 glucose units.

Cellulose chains existing at fiber surfaces contain hydroxyl groups that participate in fiber-to-fiber bonding in paper. Cellulose hydroxyl groups on the fiber surface and within the fiber wall also interact strongly with water and are important to the swelling phenomena associated with refining.

Hemicelluloses. Hemicelluloses are also polysaccharides, but differ from cellulose in that they contain a variety of sugars (xylose, galactose, mannose, arabinose) and linkages and linkage positions. They are often branched. They have lower molecular weights than cellulose, usually containing fewer than 400 sugar units. **Fig. 3-3** contains a drawing of a typical hardwood xylan molecule.

Hemicelluloses interact strongly with water and contribute greatly to the swelling of wood fibers during beating. They also promote the development of fiber-fiber bonding in paper. The carboxyl groups on xylan glucuronic acid

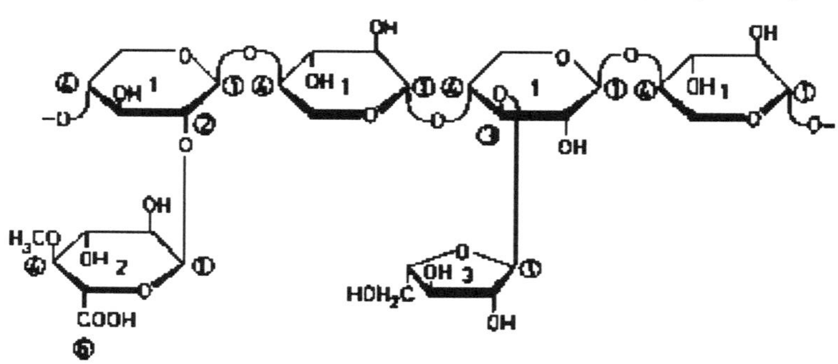

Figure 3-3. Molecular structure of a softwood xylan hemicellulose. The structure is branched and contains xylose, glucuronic acid, and arabinose sugars.

groups are the principal source of negative fiber surface charge.

Lignin. Lignin is the third component of the wood fiber wall. It plays a major role in holding fibers together in wood. Its chemical constitution is greatly different from cellulose and hemicellulose, being comprised of phenylpropane units for the most part. It contains many inter-unit ether and carbon-carbon bonds. Lignin has a network structure and a very high molecular weight in wood (**Fig. 3-4**).

Residual lignin in papermaking fibers adversely influences their ability to participate in interfiber bonds; provides an inhibiting influence on the swelling of fibers in water; and contributes to the formation of a bulky, rapid-draining mat on the paper machine. The free phenolic groups in the structure contribute to fiber negative surface charge when they ionize.

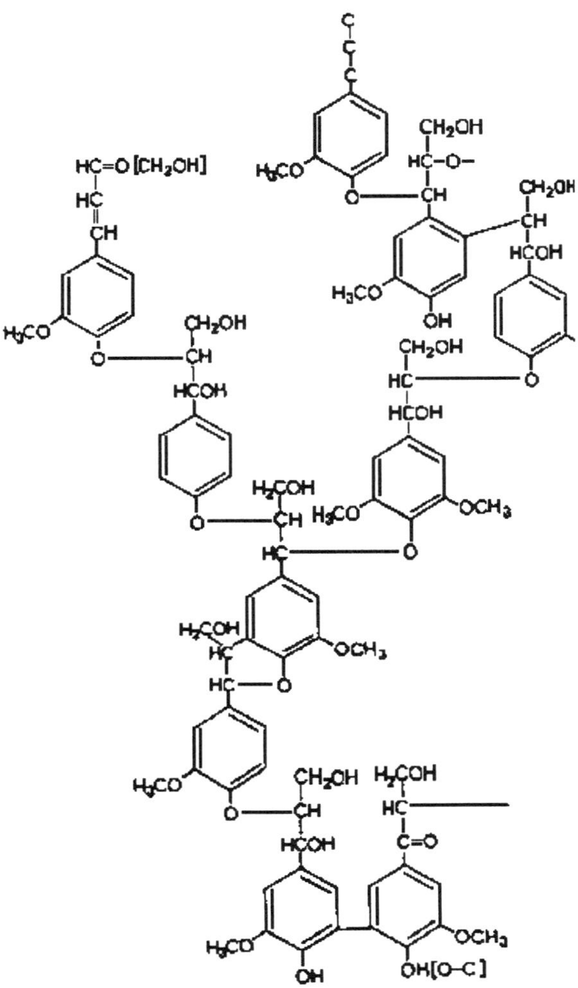

Figure 3-4. A segment of the molecular structure of lignin. The molecule is a three-dimensional network comprised of phenylpropane units linked through ether bonds and carbon-carbon bonds.

Location of components within the fiber wall. The average chemical composition of each layer of the fiber wall is presented schematically in **Fig. 3-5**. It is evident that the chemical composition varies greatly as you go from the outside of the fiber (middle lamella) to the S3 layer.

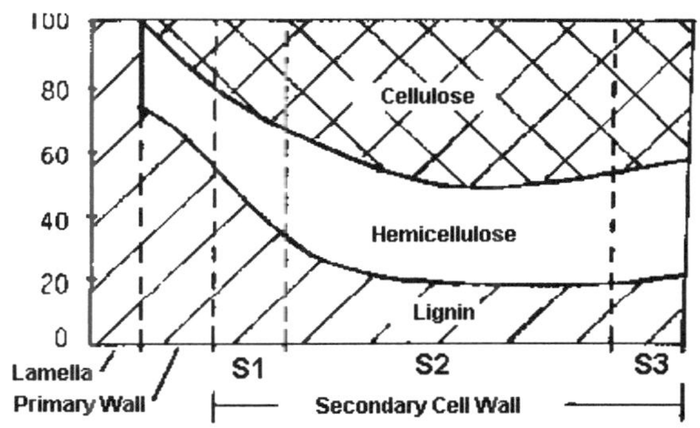

Figure 3-5. Distribution of chemical components across the fiber wall

The cellulose and hemicellulose contents are very low in the outer fiber wall near the middle lamella and increase continuously as you move in toward the lumen. On the other hand, lignin concentration is greatest in the middle lamella and primary layer. It decreases as you move into the fiber wall.

The chemical characteristics of any particular fiber surface depends greatly on the mechanical and/or chemical processing treatments that the fiber has undergone. Severe processing may disrupt the outer wall structure and the chemical properties will be predominantly those of the S2 layer. Conversely, mild chemical or mechanical processing will leave substantial portions of the primary layer intact and it will control the surface properties of the fiber.

Extractives. In addition to the chemical constituents of the wood fiber wall and middle lamella, wood contains smaller amounts of other materials, known as *extractives.* (The name derives from the fact that they can be removed from wood by extraction with neutral solvents.) A wide range of different substances are included under the extractive heading: fatty acids, fats, resin acids, and volatile terpenes. These materials play important roles in the life of a tree, such as food and energy storage and protection from microbial attack. **Fig. 3-6** shows the molecular structures of some typical extractives.

Extractives provide valuable by-products in the kraft pulping process. However, their ef-

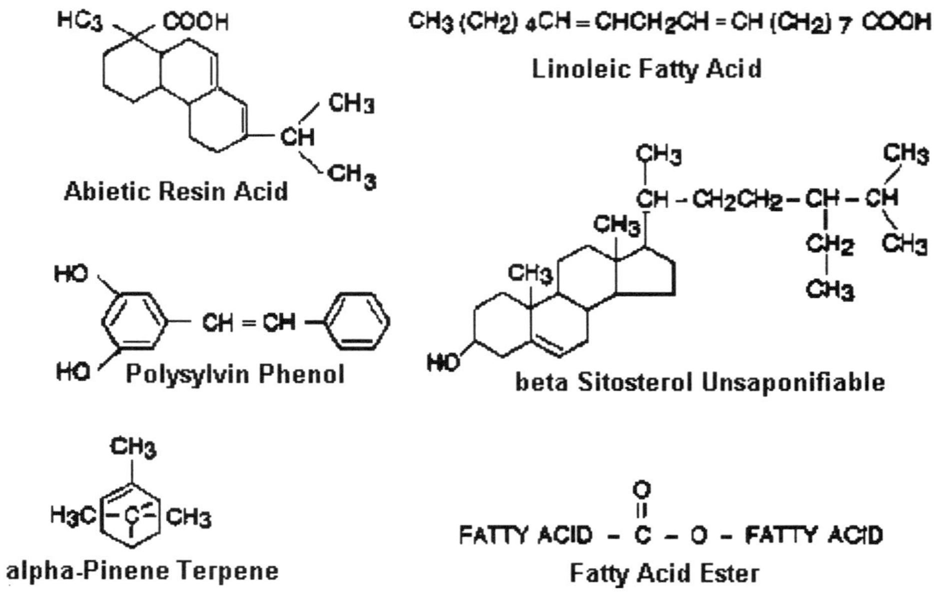

Figure 3-6. The molecular structure of some typical extractive found in wood

fects on papermaking are all adverse. They contribute to foaming, deposit problems, desizing, strength losses, and decreased paper quality.

c. Influence of pulping processes on fibers

Pulping processes modify the chemical components of wood fibers in a variety of ways, depending on the process employed and yield achieved. Mechanical pulping processes produce few chemical changes in fibers. Consequently, fiber surface composition and chemistry depend mainly on the degree of structural disruption that occurs during pulping. The inner cell wall region of mechanical pulp fibers is inaccessible and, consequently, they do not swell much during beating and refining.

Chemical pulping processes produce both chemical and structural changes in fibers. Structurally, the dissolution of lignin and hemicelluloses during pulping results in more of the internal fiber wall surface becoming accessible to water and other materials in the papermaking process.

Chemical pulping changes the chemical composition of fibers (**Table 3-2**) and also results in a number of chemical modifications to cell wall constituents. For example, cellulose may be degraded by hydrolysis and oxidation reactions, resulting in polymer chains of reduced molecular weight and containing carbonyl and carboxyl groups. Hemicelluloses are dissolved, deacetylated, and hydrolyzed during chemical pulping. Extractives are mostly removed by chemical pulping processes and recovered for their economic value. As mentioned above, sufficient quantities of extractives are often carried over to the papermaking process to cause problems. Much of the lignin is dissolved. The carbonyl and carboxyl content of the remaining lignin depends on the pulping and bleaching processes employed. For example, pulps that have undergone an oxygen stage during pulping or bleaching have greater carboxyl contents than pulp produced by more traditional processes.

Table 3-2. Typical chemical composition of kraft pulps[a]

Component	Pine	Birch
Cellulose	35	34
Glucomannan	4	1
Xylan	5	16
Other carbohydrates	—	—
Total Carbohydrates	44	51
Lignin	3	2
Extractive Substances	0.5	0.5
Total	47	53

[a]Eklund, D. and Lindstrom, T. Paper Chemistry, An Introduction, DT Paper Science Publications, Helsinki, p. 21(1991).

2. Fiber fines

Most papermaking furnishes contain fine fiber particles as well as intact fibers. In the past, papermakers have defined fines in many different ways. For example, Jaycock[1] defined all soluble or colloidal material in a furnish as fines.

Others[2] have taken all material that passes through a 60-mesh screen as fines. Most recent work[3,4] defines fines to be that part of the furnish passing through a 200-mesh (76μm) screen —the definition to be followed in this book.

Pulp fines vary widely in morphology, but often are rod-shaped, semicrystalline, fibrillar particles having widths of 0.1-0.5μm and lengths reaching several tens of micrometers. Fines also occur naturally, particularly in hardwoods, where the small rectangular-shaped parenchyma and ray cells represent a substantial percentage of the total weight of a fiber sample.

It is suggested that the following terminology be adopted when talking about fiber fines.

Primary fines: Fines that occur naturally in a wood pulp sample. These include parenchyma and ray cells and some hardwood vessel elements.

Secondary fines: Fines generated by refining. They usually have the geometrical characteristics discussed above. They can also include "veil-like" portions of cell wall layers stripped off by the refining process.

Tertiary fines: Fines that are generated during the flow of stock around the paper machine white water system. This material arises as a result of the action of pump impellers and tank agitators. Tertiary fines also arise from the "fractionation" that occurs on the wire during sheet formation.

It is believed that, in a pure fiber/water system, secondary and tertiary fines have the same inherent chemical composition as their parent fibers. This is not the case in a typical papermaking system, as will be discussed shortly.

The fiber fines concentration in any particular furnish depends on factors like the amount and type of recycled fibers employed, the extent of refining applied, the amount of broke utilized, the first-pass retention performance of the paper machine, and the amount of saveall cloudy white water used throughout the system.

3. Fiber and fines characteristics important to papermaking chemistry

The papermaking chemistry phenomena related to fibers and fines can be directly related to their surface and colloid chemistry. The characteristics of importance are sorption capacity, swelling behavior, and ion exchange properties. These are affected in turn by the specific surface area, surface charge, and surface chemical composition of fibers and fines.

a. Specific surface area

Because of their small particle size, fines have a large surface area per unit weight. Marton[5] reported that the specific surface area of fines is 5-8 times that of fibers. **Table 3-3** contains sample data to illustrate this point. Fines play a major role in papermaking chemistry because many additives act through adsorption processses that are strongly affected by surface area.

Fines also have a major influence on paper machine drainage performance due to their high water sorption capacity (2-3 times that of fibers).[4]

Fines retention is important because fines contribute to paper strength and other properties. Fines retention is also crucial to efficient materials utilization since fines may represent as much as 40-60% of the weight of a furnish in some instances. You are referred to the excellent series of papers by Marton[5–8] for a review of the importance of fines in papermaking chemistry.

Table 3-3. Surface area measurements of pulp fractions.[5]

Pulp Fraction (BET/N_2-After solvent exchange)	Specific Surface Area (m^2/g)
Fiber Fraction	9.9
Fines Fraction (Hydrodynamic)	46.8
Pulp Blend, 410 CSF	2.1
Fiber Fraction	1.2
Fines Fraction (16.5%)	8.0

b. "Preferential" additive adsorption by fines

Since much (or most) of the surface area in a papermaking furnish can be associated with fines, it is not surprising to find that chemical additives tend to adsorb on them preferentially. **Table 3-4** illustrates this point.[9]

Table 3-4. Adsorption of wet end additives by fines

	Relative Adsorption Intensity		
Additive	Fibers	Filler Clay	Fines
Cationic Starch	1	4	5
Rosin Soap Size	1	4	16
Dispersed Rosin Size	1	16	20
Alum	1	2	3

c. Surface charge

Types of charged groups. The surface charge on fibers and fines is a complex function of the chemical composition of their surfaces, its state of ionization, and the nature and amounts of additional substances adsorbed. For the simplest case, we will consider a suspension of well washed fibers and fines in very pure deionized water. Under these conditions, the fibers and fines always take on a negative charge due to the ionization of surface groups associated with hemicelluloses, oxidized cellulose, and oxidized lignin. **Table 3-5** lists the principal types of ionizable groups that can be present in wood pulp fibers and their respective ionization constants.

Table 3-5. Types of ionizable groups in wood

Acid Group	pK_a (25°C)
Hemicellulose & Lignin Carboxyl Groups	4-5
Lignin Phenolic OH Groups	9.5-10.5
Sugar Alcohol Groups	13.5-15.0
Hemiacetal OH Groups	12.0-12.5
Lignosulphonate Groups	100% ionized in water

Table 3-6 contains typical carboxyl group levels in virgin and recycled pulps.

Table 3-6. Typical carboxyl group contents of wood pulps and recycled fibers

Virgin Pulps	Carboxyl Number (meq/100g)
Bleached Softwood Kraft	2-4
Bleached Hardwood Sulfite	3-8
CTMP	20-40

Recycled Fibers	Carboxyl Number (meq/100g)
Deinked Bl. N. Kraft	7.4
Deinked Bl. S. Kraft	7.3
Deinked Tissue	8.6
Merchant Recycled Pulp	8.9
Waste Newsblank	19.9

The effect of pH on fiber ionization. The pH has a profound effect on the state of ionization of these groups. If we lower the pH below 2.75, most of the carboxyl groups take on a proton and the charge on the fibers and fines becomes essentially zero. Conversely, if we increase the pH to the 7-8 range, many of the carboxyls are neutralized and the surface charge becomes more negative. This process can be readily followed by measuring the zeta potential or electrophoretic mobility of the slurry. **Fig. 3-7** illustrates the effect of pH on the dissociation of fiber charge-producing groups.

Note that lignin phenolic groups become ionized (charged) at pH levels above 9. Therefore, in certain papermaking systems (e.g. linerboard) these groups might contribute significantly to fiber charge. It is also important to notice that sugar hydroxyl groups do not contribute any

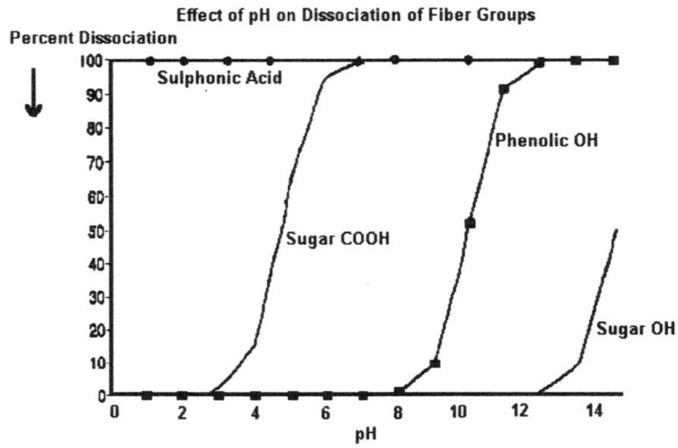

Figure 3-7. The effect of pH on the dissociation of fiber charge-producing groups

negative charge to fibers at normal papermaking pH levels.

The "clean" environment described above does not exist under papermaking conditions. Pulps are rarely washed free of extraneous pulp mill or bleach plant chemicals and reaction products. Even when pulp has been well washed, it will adsorb dissolved substances from the water in which it is suspended. These sorption phenomena involve both ionic and nonionic (Van der Waals) interactions. In most instances, such materials are anionic in nature, causing pulp to take on a greater negative charge. In the case of dissolved salts, the process is called ion exchange (or base exchange).

d. Ion exchange behavior of wood fibers and fines

The ion exchange capacity of cellulose fibers and fines is important to many papermaking phenomena, among which are electrical and electrokinetic properties; adsorption of dyes, aluminum and polyelectrolytes; rosin sizing, and fiber and fines flocculation. The strength of attraction of wood fibers for various ions occurs in the following order:

$$N(CH3)_4^{+1} < Li^{+1} < Na^{+1} < K^{+1} < Ag^{+1} < Ca^{+2} = Mg^{+2} = Ba^{+2} < A1^{+3}$$

Least tightly bound   Most tightly bound

Thus, $A1^{+3}$ binds more strongly to wood fibers than Ca^{+2}, which binds more strongly than Na^{+1}. The charge-neutralizing power of these cations occurs in similar order. The adsorption of $A1^{+3}$ ions shifts the fiber charge further in the positive direction than does the adsorption of Na^{+1} ions. Indeed, it is possible to neutralize the fiber charge with $A1^{+3}$.

Rosen[11] studied the ion exchange behavior of pulps over a wide pH range and concluded that carboxyl groups were responsible for ion adsorption in the pH range of 3 to 7. In the 9.0 to 11.8 range, phenolic hydroxyls also were important. Other studies[10] have shown that the ion-exchange capacity of wood fibers results mainly from the presence of naturally acidic hemicelluloses and that sulfonate groups also contribute to ion exchange reactions in certain pulps. (Note that this is essentially the same conclusion that was derived from **Fig. 3-7.**) Finally, studies have shown that soda-oxygen pulps have a greater affinity for cations than kraft pulps, as a result of the higher carboxyl contents of soda-oxygen pulps.[12]

One then concludes that under normal papermaking conditions only the carboxyl and sulfonate groups, and, occasionally phenolic groups, are involved. However, under the higher pH conditions of bleaching and pulp washing, lignin phenolic hydroxyls can also contribute to ion exchange phenomena.

4. Summary

Pulp fibers and fines participate in papermaking chemistry phenomena through chemical, colloidal, and surface interactions. The most important factors to consider are the chemical composition of the fiber and fines surfaces, and the size and surface area of the particles. Since they have an exceedingly high surface area per unit weight compared to fibers, fines tend to dominate many papermaking chemistry phenomena through their high adsorption capacity. A parallel result is that, even though fibers and fines initially have similar chemical makeup, the high adsorption of substances by the fines fraction results in a final difference in characteristics between the two fractions. Fines also have a greater specific swelling tendency in water and can exert a significant influence on the drainage performance of a furnish.

Fibers and fines have a naturally negative surface charge that can be modified by changing the degree of ionization of surface carboxyls through a pH change or by the ion exchange and adsorption of charged species.

5. References

[1] Jaycock, M. J. Papers presented at 1st and 2nd Intl. Seminars on Papermaking Chem. Amsterdam (1977) and New York (1978).

[2] Hinton, A. J. and Quinn, M. Fiber surface characteristics as they affect two-sided dyeing of paper, Paper Techn. 5(1):60(1964).

[3] Britt, K.W. and Unbehend, J. E. New methods for monitoring retention, Tappi J. 59(2):67 (1976).

[4] Htun, A. J. and Deruvo, A. The implication of the fines fraction for the properties of bleached kraft sheets, Svensk. Papperstidn. 16:507(1978).

[5] Marton, J. and Marton, T. Wet end starch: adsorption of starch on cellulosic fibers, Tappi J. 59(12):121(1976).

[6] Marton, J. The role of surface chemistry in fines-cationic starch interactions, Tappi J. 63(4):87(1980).

[7] Marton, J. and Marton, T. Effect of fines on rosin sizing of paper, Tappi J. 66(12):68(1983).

[8] Marton, J. and Marton, T. Some new principles to optimize rosin sizing, Tappi J. 65(11):105(1982).

[9] Scott, W. E. Fines management and control in wet end chemistry. Tappi Journal, 69:11, 31-34(1986).

[10] Ekwall, P. and Bruun, H.H. Surface chemical studies of the conditions of rosin sizing, Tappi J. 37(7):303(1954).

[11] Rosen, A. Adsorption of sodium ions on kraft pulp fibers during washing, Tappi J. 58(9): 156(1975).

[12] Xuan, N.N. Venkatesh, V. Gratzl, J.S. and McKean, W. T. The washing of soda-oxygen pulps: sorption phenomena of cations, Tappi J. 6(18): 53(1978).

Chapter 4

Interfering Substances

A papermaking furnish consists of water, fibers and fiber fines, functional additives, chemical processing aids, and interfering substances. Interfering substances, which are sometimes referred to as "anionic trash" or "cationic demand," consist of colloidal and dissolved materials that adversely affect the papermaking process in a variety of ways through deposit formation or interference with chemical additives. They can also cause spots and other defects in paper and reduce paper strength.

The level of interfering substances is usually high in newsprint, linerboard, and corrugating medium mills due to their high concentrations of dissolved lignin and hemicellulose. Recycled fibers also can contain large amounts of contaminants as a result of incomplete removal of impurities or carryover of deinking chemicals.

We will just introduce this important subject at this point. A more detailed discussion of the effects of interfering substances in papermaking will be presented in subsequent chapters. References [1-4] also provide more complete treatments of the subject.

1. Sources and types of interfering substances

Interfering substances can be classified into the following three groups according to their source:

- Substances dissolved or dispersed during pulping and bleaching of wood
- Substances dissolved or dispersed during the refining and storage of pulp
- Other substances originating in fresh water, additives, broke, and recycled fibers

a. Substances released during pulping and bleaching

Kraft pulping and bleaching decompose lignin and polysaccharides into low molecular weight fragments that dissolve in the pulping liquor and can be present in poorly washed pulps. Fatty acids and resin acids also are present. These materials contribute strongly to deposit problems. They do not interfere greatly with additives.

Higher molecular weight anionic wood components are also dissolved during kraft and groundwood pulping and bleaching. These materials form complexes with cationic additives and also adsorb onto fiber surfaces and interfere with additives. This situation is particularly acute in the case of mechanical pulp manufacture, where hemicelluloses, lignins, proteins, pectins, and other materials exist in process white water. These substances cause such systems to have a very high demand for cationic additives.

b. Substances released during the refining and storage of pulps

More wood substances dissolve from unbleached pulps than bleached pulps during storage and refining. Nevertheless, quantities of dissolved material levels on the order of 1-3% (by weight) can be found in bleached kraft pulp suspensions.

c. Substances in fresh process water, additives, broke, and recycled fiber

These substances are similar to the materials described above. In addition, hydrophobic resins, surface sizing starch, coating adhesives, and dispersants can come from broke. Recycled fibers have their own particular set of substances (e.g. hot melt adhesives, glues, plastic particles, tapes, printing ink binders) which give rise to deposits called "stickies." Dispersants can be introduced with fresh filler slurries.

2. Treatment of interfering substances

When the level of interfering substances is particularly high, the usual practice is to neutralize them with a charge-neutralizing agent like alum, polyaluminum chloride, or a low molecular weight, high charge density cationic polymer prior to the addition of other cationic chemical additives.

3. Summary

As mills close up white water systems and use higher levels of recycled fibers, they find that the level of dissolved and suspended interfering substances increases. This often leads to decreased additive efficiency, lower paper quality, and poorer paper machine runnability. Interfering substances can arise from pulping and bleaching carryover, from pulp storage and refining, and from outside sources such as broke and secondary fiber. It is important that the papermaker be cognizant of the nature and level of such materials in a given mill and be able to solve problems attributed to them.

4. References

[1] Auhorn, W., Linhart, D., Degen, H., and Lorz, R., Anionic Trash: Controlling Detrimental Substances, Tappi J. 70(10):79 (1987).

[2] Springer, A., Noe, J., and Wegner, T., Fundamental Strategy for Control of Retention and Drainage on a Modern Paper Machine, Proc. 1987 TAPPI Papermakers Conf., Atlanta, p. 137, TAPPI PRESS, Atlanta.

[3] Proc. 1987 TAPPI Papermakers Conf., Atlanta, p. 141, TAPPI PRESS, Atlanta.

[4] Lindstrom, T. and Eklund, D. Paper Chemistry: An Introduction. 52-62, DT Paper Science Publications, Grankulla, Finland, 1991.

Chapter 5

Surface and Colloid Science Concepts Important to Papermaking Chemistry

The objective of this chapter is to introduce the basic surface and colloid science concepts that operate in papermaking chemistry in order to provide a foundation for future study. The discussion first presents a case for categorizing most papermaking chemistry phenomena as hydrophobic or hydrophilic colloid systems. Next, various mechanisms proposed for the aggregation behavior of hydrophobic colloids are defined and their relevance to papermaking introduced. This is followed by a brief discussion of the adsorption of polymers onto solid surfaces. The chapter next deals with the surface chemistry concepts related to sizing. The final section discusses surface activity and other phenomena related to association compounds. In all instances, the discussion will be introductory in nature, with the emphasis on general concepts and important terminology. The information presented in this chapter will be applied throughout the remainder of the book.

1. Definition of papermaking chemistry

Papermaking chemistry can be defined as the "surface and colloid chemistry of papermaking furnish components." To demonstrate the basis for this definition, we will consider some characteristics of a typical furnish.

Table 5-1. "Resident species" of a papermaking furnish

1. Dissolved electrolytes
2. Suspended fibers
3. Suspended fiber fines
4. Suspended filler particles
5. Water
6. Surface-active molecules (either alone or as aggregates). (Examples: detergents, dispersants, wood extractives, defoamers)
7. Dissolved polyelectrolytes (Examples: polyacrylamides, cationic starch, wet-strength agents, dry strength agents)
8. Aggregated sizing molecules.

2. Resident species of a typical papermaking furnish

Table 5-1 lists many of the "resident species" of a papermaking furnish. These species are present at various concentrations and combine to make up a headbox consistency of 0.3% to 4.0%, depending upon the grade of paper being manufactured. The environment can be relatively static (e.g. a high density stock chest) or quite dynamic (e.g. a high-turbulence headbox).

3. Important interactions among furnish components

Furnish components interact with one another in a variety of ways (**Table 5-2**). Interactions may be desirable or undesirable. For example, the deposition of an alum/rosin size precipitate onto fibers is desirable, but the formation of an alum scale precipitate is undesirable. In addition, the papermaker may or may not be able to control the interactions listed in the table.

4. Wet end intermolecular forces

Interactions between two particles involve both attractive and repulsive forces. Chemists usually refer to two categories of attractive interactions at the atomic and molecular level: namely *primary* and *secondary* bonding interactions. Primary bonding forces unite atoms into molecules or chemical compounds. In organic compounds these bonds have energies of 420-840 kJ/mol of bonds formed, operating at interatomic distances of 0.1-0.2 nm.

Secondary bonding forces involve energies of 840 kJ/mol operating over distances of 0.2-0.3 nm. Hydrogen bonding is a special case of secondary bonding. A hydrogen bond has an energy range of 12-21 kJ/mol and a bond distance of 0.28 nm. It is important in the water-fiber interactions and fiber-fiber interactions that

Table 5-2. Examples of important interactions among furnish components

1. Aggregation of fibers, fillers, and fines
2. Adsorption of dissolved polymers onto fibers, fines and fillers
3. Aggregation of pitch and sizing molecules
4. Adsorption of pitch and sizing molecules onto fibers, fines, and fillers
5. Neutralization of negative charges on suspended and dissolved anionic materials
6. Establishment of equilibria between dissolved inorganic salts and insoluble ionic products
7. Development of micelles composed of surface-active molecules
8. Sorption of water by fibers, fines, and starches

lead to bonding in paper. Secondary bonding forces are referred to collectively as Van der *Waals Forces.*

In **Table 5-2**, the equilibrium between dissolved ions and their insoluble precipitates involve primary, ionic bonds. Similarly, charged polyelectrolytes may also participate in ionic interactions, but to a lesser degree than salts. The remainder of the interactions in the table involve secondary bonding.

5. Dimensions of furnish components

All the furnish components listed in **Table 5-1** have at least one very small dimension (**Table 5-3**). Such particles have high specific surface areas (surface area per unit weight). For example, cubes of density 1.0 gm/cm^3 having edge lengths in the 10.0 μm to 0.1 μm range, which are typical dimensions for papermaking components, will have specific surface areas ranging from 0.6 m^2/g to 600 m^2/g. Further discussion of the relative dimensions of papermaking components can be found in Linhart.[1]

Table 5-3. Smallest dimension of furnish components

Component	Size	Dimension
Fiber	10-20 μm	Width
Fibril	<1 μm	Width
Nonfibril fiber fine	<1-2 μm	Diameter
Filler particle	0.1-10 μm	Diameter
Dispersed rosin size aggregate	<1 μm	Diameter
Dissolved poly-acrylamide molecule	<1 μm	Length

Solvent-exchanged, water swollen fibers and fines have specific surface areas ranging from 10 m^2/g to 50 m^2/g. Other particulate papermaking materials, such as fillers, also have high specific surface areas. These will be described in greater detail in Chapter 10.

6. The "colloidal" state

When the surface area-to-mass ratio of a dispersed particle becomes large, then an appreciable fraction of its molecules lie in the interface between the particle and the dispersion medium. These "surface" molecules participate in secondary bonding interactions with other molecules identical to themselves located in the bulk of the particle and with the molecules of the surrounding medium. Whenever the two kinds of interactions are unbalanced, then excess energy exists at the particle surface. This excess energy, in turn, strongly influences the behavior of the entire system of dispersed particles.

When the surface chemical properties of a dispersed system are important to the overall chemical behavior of the system, the study of the system involves a branch of physical chemistry known as *colloid chemistry.* Thus, papermaking chemistry is really colloid chemistry because of the large specific surface areas involved and because most papermaking chemistry interactions occur at particle surfaces.

In summary, many papermaking furnish particles have colloidal dimensions and high specific surface areas. In addition, many of the interactions that occur among these particles involve secondary bonding forces. These are two primary traits of colloid systems and it is appropriate to define wet end chemistry in terms of surface and colloid science.

7. Classification systems in colloid chemistry

Having established that papermaking chemistry can be considered to be colloid chemistry, it is convenient to consider papermaking chemistry in terms of common colloid systems. For example, most colloid systems can be classified as either *lyophobic* or *lyophilic.* **Table 5-4** lists the principal characteristics of the two systems.

Table 5-4. Characteristics of lyophobic and lyophilic colloid systems

Lyophobic Systems

1. Suspensions of particles — not solutions
2. Little attraction or affinity between the solvent and particle
3. Unstable with respect to the formation of aggregates
4. Interface exists between the particle and suspending medium that strongly affects system behavior

Lyophilic Systems

1. True solutions of very large molecules or aggregates of small molecules
2. Strong attraction between solvent and particle
3. No true interface between particle and medium

The prefix "lyo-" means "solvent" and since papermaking is carried out in water, the relevant colloid systems are designated *hydrophobic* and *hydrophilic.* **Table 5-5** lists some typical examples of hydrophobic and hydrophilic papermaking systems.

Table 5-5. Examples of hydrophobic and hydrophilic papermaking systems

Hydrophobic

1. All pigments dispersed in water
2. Fines dispersed in water
3. Rosin size in water
4. Dispersed rosin size emulsion

Hydrophilic

1. Starch dissolved in water
2. Gums dissolved in water
3. Hemicelluloses dissolved in water
4. Surfactants, dispersants, wetting agents dissolved in water
5. Retention, drainage, and formation aids dissolved in water

Two types of hydrophilic colloids are important in papermaking. The first type includes materials like retention and drainage aids that affect the behavior of hydrophobic substances. Dry strength and wet strength additives also fall into this category of hydrophilic materials. The second type of hydrophilic colloids are the *association compounds*, such as wetting agents, surfactants, defoamers, and dispersants. These will be discussed in greater detail later in the chapter.

a. Further comments about the use of the term "hydrophobic"

Further comments are in order about the application of the term "hydrophobic" to the common papermaking systems listed in **Table 5-5**. Pigments and fiber fines are known to be wettable and to interact with water. Indeed, the surface of a fiber fine is covered with a gel formed by solvated hemicellulose and cellulose molecules. Even rosin size dispersion must interact with water to a certain degree or it would be impossible to form a stable suspension. Nevertheless, dispersions of these materials do exhibit the characteristics of lyophobic colloids listed in **Table 5-4** and do not have any of the properties listed for lyophilic systems in **Table 5-4**.

Another way to look at the matter is to realize that individual particles may have wettable, water-attracting (hydrophilic) surfaces, yet produce dispersed systems having hydrophobic characteristics.

b. Stability of hydrophobic dispersions

The study of hydrophobic dispersions is concerned principally with the *rate* at which they revert to their equilibrium, aggregated state. Indeed, all of the hydrophobic systems listed in **Table 5-5** become aggregated during papermaking. In practice, the papermaker wishes to control the aggregation process in order to achieve a uniform distribution of materials throughout the sheet, while at the same time retaining the maximum percentage of incoming materials in the sheet. Uncontrolled aggregation leads to poor paper quality and deposit problems.

The stability of a hydrophobic colloid dispersion is indicated by its ability to remain dispersed and is measured by the time it takes for the particles to settle out of the liquid phase. This phase separation may be due to gravity sedimentation of the primary dispersed particles or the aggregation of primary particles into larger aggregates that sediment. The main types of aggregation mechanisms are listed in **Table 5-6**.

Table 5-6. Hydrophobic colloid aggregation mechanisms

1. Coagulation: The destabilization of a colloidal suspension with salts or very low molecular weight, high charge density polyelectrolytes.
2. Flocculation: The destabilization of a colloidal suspension by bonding the particles together with a long chain polymer.

The terms coagulation and flocculation are often used interchangeably. However, since they represent two distinct aggregation mechanisms, it seems logical to apply different terms to them. That practice will be followed in this book.

8. Surface charge on hydrophobic particles

An electrostatic charge develops at the solid/water interface when solids are dispersed in water. In papermaking systems, this charge may arise from dissociated surface carboxyls and sulfonic acid groups and the adsorption of substances, such as hemicelluloses, dissolved lignin, retention aids, and cationic starches onto surfaces. Mineral fillers, such as clay and titanium dioxide, also develop surface charge by ionization of the particle/water interface and by adsorption of other charged substances.

The description of charges at solid/liquid interfaces was first outlined by Helmholtz[2] in 1879 and was elucidated in subsequent years by Gouy[3], Chapman[4], and Stern[5]. **Fig. 5-1** depicts a currently accepted representation of the phenomenon.[6–8] The surface potential is determined by the type and surface density of charged groups. Papermaking particles are generally anionic in nature. Therefore, **Fig. 5-1** is drawn with a negative charge at the surface.

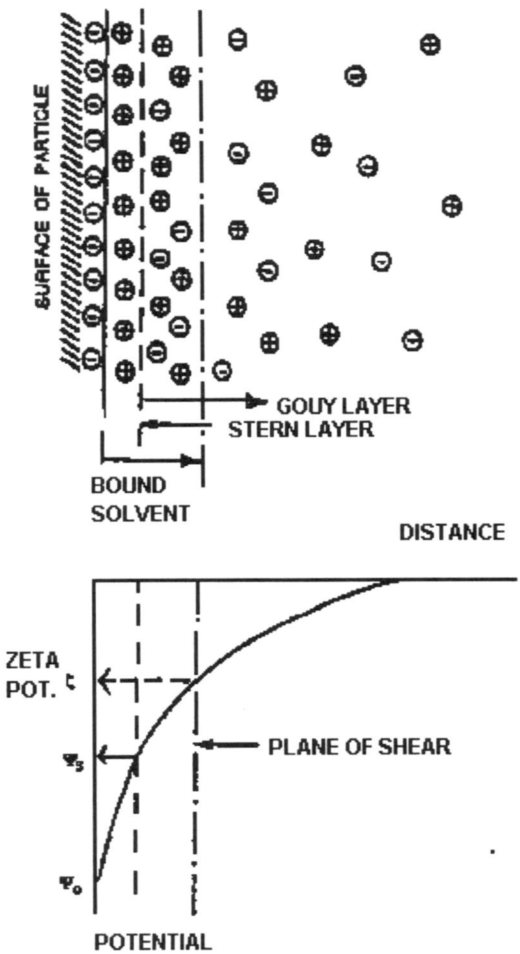

Figure 5-1. Representation of the electrical double layer existing at the surface of a hydrophobic particle suspended in water. The zeta potential,ξ,is taken to be the potential at the slip plane, within which counterions bound to the particle travel with the particle and outside of which counterions are free to move independently of the particle.

Ions of opposite charge to that of the particle surface (counterions) are held closely adjacent to the surface by electrostatic and Van der Waals forces. The potential energy drops off rapidly in this region. A hydrodynamic slip plane exists between the tightly bound counterions and solvent molecules adjacent to the surface (Stern Layer) and the rest of the solution. The less ordered, diffuse portion of the system is called the Gouy-Chapman Region. The potential falls off at a slower rate in this region than in the Stern Layer until zero potential is realized in the bulk of the solution.

The potential at the interface plane between the Stern Layer and Gouy-Chapman region is called the *zeta potential*, ξ. The common methods of measuring surface charge actually give values for zeta potential rather than surface charge. (Charge measurement is discussed in Chapter six.) The common name applied to the entire system is the *electrical double layer*.

There is no overall net charge in the suspension due to the requirement of electro-neutrality. However, local areas of positive and negative potential exist close to the surface of the particles being observed. The strength of these potentials and the distances involved determine the resistance of hydrophobic suspensions to aggregation.

9. Importance of the electrical double layer

The stability of a hydrophobic suspension is measured by its ability to remain dispersed over a period of time. This stability depends upon the relative magnitudes of the repulsive and attractive forces that exist between the suspended particles and how likely the particles are to collide with one another.

Attractive and repulsive forces arise from a number of sources. Van der Waals forces are always attractive for particles of the same chemical composition. Electrostatic forces arising from the interaction of particle electrical double layers are always repulsive if the particles have the same charge. They are attractive for particles of opposite charge. Stabilizing and destabilizing forces can also be developed by adsorbed polyelectrolytes.

The Van der Waals attractive forces and double layer repulsive forces were combined by Derjaguin, Landau, Verwey, and Overbeek into what is now referred to as the *DLVO Theory*.[8] This theory considers the attractive and repulsive forces that exist between two particles and describes the potential energy as a function of distance between them.[8] **Fig. 5-2** illustrates this treatment.

In the figure, the electrostatic repulsion energy, V_R, is represented by the upper dashed curve. The Van der Waals attraction energy, V_A, is the lower dashed curve. And the Total Potential Energy, V_T, is the solid curve. The depth of the minimum in the Total Potential Energy Curve is an indication of the stability of a particle-particle aggregate. A shallow minimum suggests that the aggregate can be disrupted more readily than when a deeper minimum exists.

10. Aggregation and stabilization of hydrophobic dispersions

We will now discuss how aggregation can be caused to occur. We will also describe how a system can be stabilized if a dispersed condition is desired.

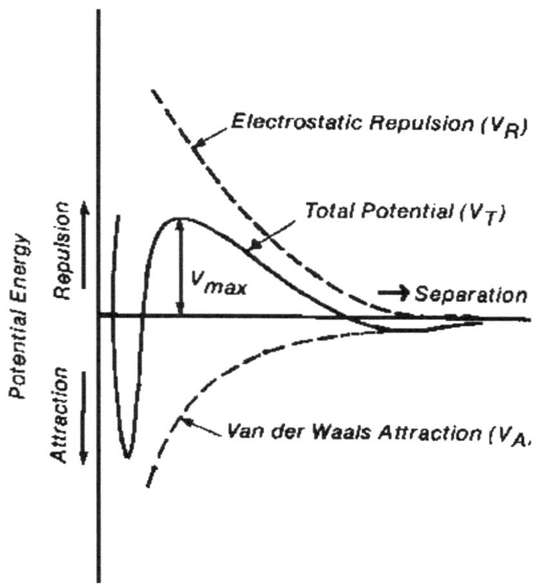

Figure 5-2. Representation of interparticle interaction energy as a function of the distance of separation between two particles. The figure illustrates the combining of attractive and repulsive potential energies. The electrostatic repulsion energy (V_R) is represented by the upper dashed curve. The Van der Waals attraction energy (V_A) is the lower dashed curve. The Total Potential Energy (V_T) is the solid curve. The depth of the minimum in the Total Potential Energy Curve is an indication of the stability of the particle-particle aggregate. A shallow minimum suggests that the aggregate can be disrupted more readily than when a deeper minimum exists.

a. Coagulation by charge neutralization

As discussed above, a layer of strongly adsorbed counterions and a diffuse region of counter- and co-ions make up the electrical double layer of particles suspended in water. This layer has a certain thickness for a given set of conditions. The thickness of the double layer decreases when electrolytes are added to the suspension due to an increase in the availability and number of counterions. Increased counterion concentration reduces the zeta potential and, thereby, the repulsion component of the total potential. Particles can then approach one another more easily and aggregation becomes more likely. Aggregations by this mechanism is referred to as *charge neutralization* because the addition of excess counterions moves the net potential towards a state of zero charge. Charge neutralization is a coagulation phenomenon. (See **Table 5-6**.)

There are two further points to be made regarding this mechanism. First, there exists a critical concentration of salt where the double layer is sufficiently suppressed for coagulation to proceed. This is normally referred to as the *critical coagulation concentration*, or C.C.C.

The second point deals with the relative effectiveness of different cations in reducing the double layer thickness of anionic surfaces. Shultz and Hardy found that the valence magnitude of the ion charged oppositely to the surface had the principal effect on colloid stability.[9] They found that the higher the valence, the more effective was the ion in reducing surface charge — with the ratios of ionic valence +1, +2, and +3 required to achieve the C.C.C. being 4000:100:15.[9] Thus, Al^{+3} salts (e.g. alum $Al_2(SO_4)_3$) are more effective than Ca^{+2} salts (e.g. $CaCl_2$), which, in turn, are more effective than Na^{+1} salts (e.g. NaCl).

Of importance to the papermaker is that coagulation produces tight aggregates where the particles come into close contact. Such aggregates tend to drain well and produce a well-formed sheet.

b. Patch model

When low molecular weight (100,000-1 million) cationic polyelectrolytes with high charge density (>4 meq charge/g) are mixed with anionic particles, the polymer molecules are believed to adsorb completely onto the particle surfaces and form a "patch" of positive charge (**Fig. 5-3**).[10] This patch effectively reverses the anionic charge at that site on the particle surface while the rest of the particle remains anionic. (The overall potential may also remain anionic.)

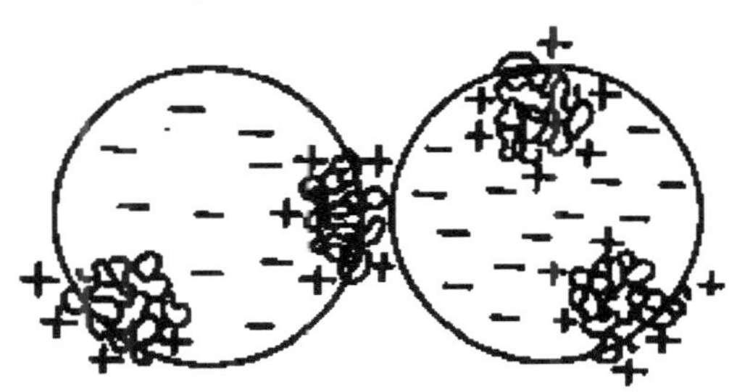

Figure 5-3. Schematic representation of the model for patch flocculation behavior. The positively charged patches on one particle are attracted to negative regions on the surface of another particle and electrostatic attraction draws them together so that Van der Waals forces can cause them to aggregate.

After the polyelectrolyte is adsorbed, collision of the positively charged patch with an uncovered negative area on another particle leads to coagulation. The aggregation rate is greater than that produced by coagulation with a simple electrolyte.

Patching, being a charge neutralization phenomenon, produces closely packed aggregates that promote good formation and good drainage.

c. Flocculation by bridging polymers

The flocculation of colloidal materials by polyelectrolytes of high molecular weight (>1 million) is presumed to take place by a bridging mechanism.[11,12] In this mechanism, the polymer adsorbs onto the particle surface with a series of loops and tails extending into the liquid phase (**Fig. 5-4.a**). These loops and tails extend well beyond the double layer. Flocculation then occurs by adsorption of the extended loops and tails onto the negative surface of a second particle (**Fig. 5-4.b**). Bridging flocculation is dependent on the collision frequency of the two particles and repulsion between their double layers does not come into play. Flocculation is the primary mechanism for high molecular weight retention aids. It is a very important phenomenon in paper making.

Figure 5-4.a

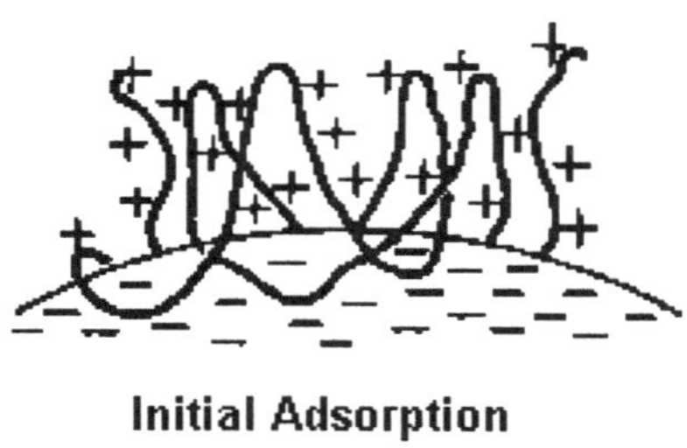

Figure 5-4.b

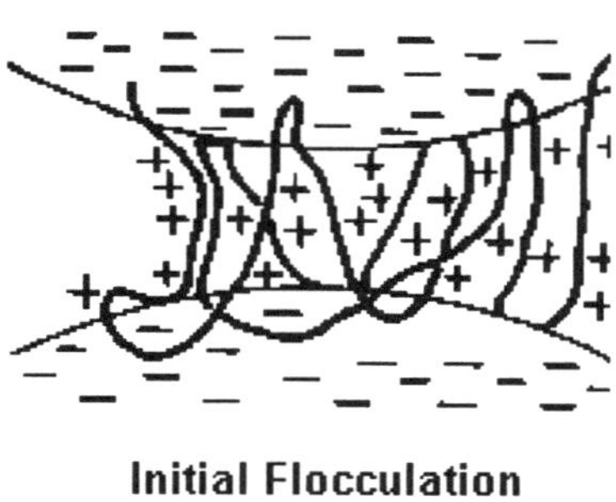

Figure 5-4. Schematic representation of the model for bridging flocculation behavior. (a) The positively charged high molecular weight polyelectrolyte partially adsorbs onto the surface of the negative particle. Loops and trains of the polymer extend out into the surrounding medium. (b) The extended loops and trains of the adsorbed polyelectrolyte adsorb onto a second particle forming a "bridge" between the two.

The charge density and molecular weight of a bridging polymer are very important. Higher molecular weights promote increased bridging because longer polymer chains extend farther into the surrounding water and away from the particle surface. Polymer charge density influences the attraction of the polymer to the particle surface and the strength with which the macromolecule bonds to the surface.

Combining a low molecular weight, high charge density cationic polymer with a high molecular weight, low charge density anionic polymer is often done to produce the strongest aggregates. Such systems are referred to as *dual polymer* systems. In this instance, the cationic polymer forms cationic patches that serve as anchor points for the bridging anionic polymer.

d. Steric stabilization

In addition to stabilizing a hydrophobic colloid suspension by the development of a high surface charge, a system may also be stabilized when long chain polymers are adsorbed onto particle surfaces in a manner that the solvated polymer loops and tails extending away from one particle interfere with the solvated loops and tails from another particle, preventing the particles from approaching closely enough to aggregate(**Fig. 5-5**).[13] Such polymers are normally nonionic in nature and are of high molecular weight. This phenomenon is often called *protective colloid* action.

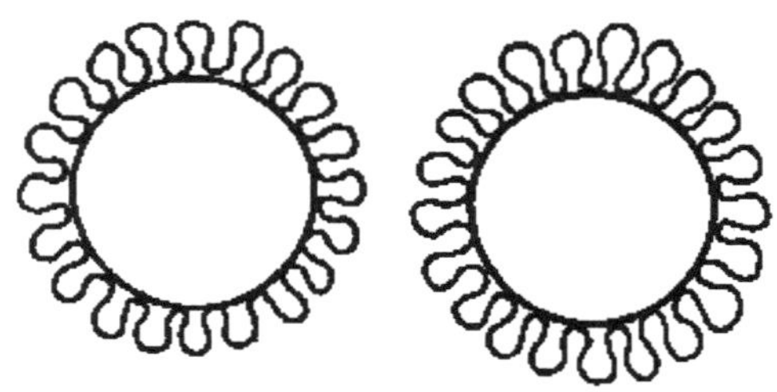

Figure 5-5. Schematic representation of the model for steric stabilization. In this instance the surfaces of particles are covered by adsorbed high molecular weight polymer material that extends out past the electrical double layer. When two particles approach one another, the extended polymer and its associated solvation water provides a steric barrier that holds the particles apart far enough that Van der Waals attractive forces cannot operate.

11. Adsorption of polymers from solution by solid surfaces

In the preceding section we discussed the patch and bridging aggregation mechanisms. Both require dissolved polymers to attach themselves to the surface of the particles that are to be aggregated. This process is called *polymer adsorption.*

Most adsorbing polymers of interest to papermakers are positively charged (cationic). Examples include cationic retention aids, cationic starch, and cationic wet strength resins. These materials all have a natural attraction for negatively charged fibers and fines. For the most part, nonionic and anionic polymers are not attracted to these particle surfaces and any adsorption that occurs must be promoted by the use of a cationic substance such as alum or a low molecular weight, high charge density cationic polymer. Consequently, our treatment of this subject will be focused primarily on cationic polymers.

a. General mechanism

In a system containing dispersed papermaking particles (fibers, fines, fillers, size particles) and dissolved cationic polymers (cationic starch, retention aids, wet strength resins) there is always movement of the particles and molecules caused by mixing forces and diffusion. These motions lead to particle-particle and particle-polymer collisions. Since cationic polymers and anionic fibers are oppositely charged, an attractive force develops that promotes polymer diffusion to fiber surfaces. Once the polymer contacts the surface, other forces come into play that bond the molecule to the surface (hydrogen bonding, Van der Waals forces).

Some important factors that affect polymer adsorption are listed in **Table 5-7**. Notice that the list includes not only polymer attributes, but also variables associated with water and the papermaking system. Subsequent chapters will delve more deeply into these subjects with respect to specific additives, retention systems, and wet end chemistry process control.

Table 5-7. Factors that affect polyelectrolyte adsorption

- **Polymer Characteristics**
  (Backbone conformation and flexibility, charge density, molecular weight, chain length)
- **Fiber Characteristics**
  (Number of available adsorption sites, surface charge, surface area, fines content)
- **Chemical Environment**
  (pH, conductivity, interfering substances, anionic polymers)
- **Physical Environment**
  (Shear and turbulence, furnish consistency, order of addition)

b. Polymer adsorption isotherms

Polymer adsorption occurs very rapidly when fibers and cationic polymers are mixed together — on the order of a few seconds or less for the major portion of the reaction. A plot of adsorption versus time produces the type of curve illustrated in **Fig. 5-6**. Such a curve is called a Langmuir adsorption isotherm. The specific isotherm obtained for a given system depends upon the factors listed in **Table 5-7**.

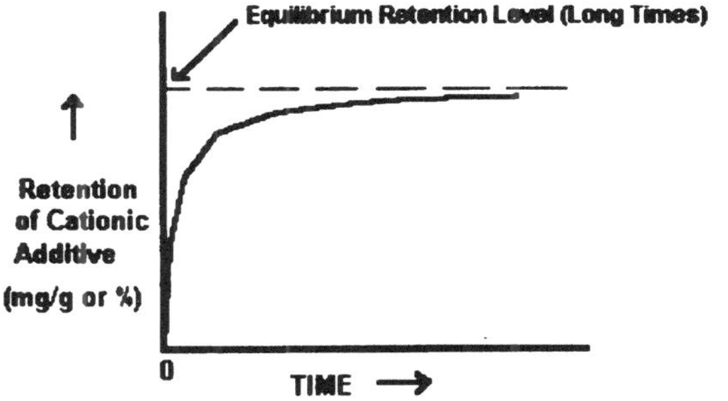

Figure 5-6. Typical cationic polymer adsorption isotherm for cationic polymeric additives on papermaking fibers. Such a curve is called called a Langmuir adsorption isotherm.

The "equilibrium retention level" in the above figure varies greatly from system to system, depending upon the conditions listed in **Table 5-7**. Experience has shown that for most cationic additives, very high equilibrium retention levels can be achieved at addition levels below 1%. For example, Marton and Marton found the behavior illustrated in **Fig. 5-7** in a lab experiment with a single cationic starch.[14]

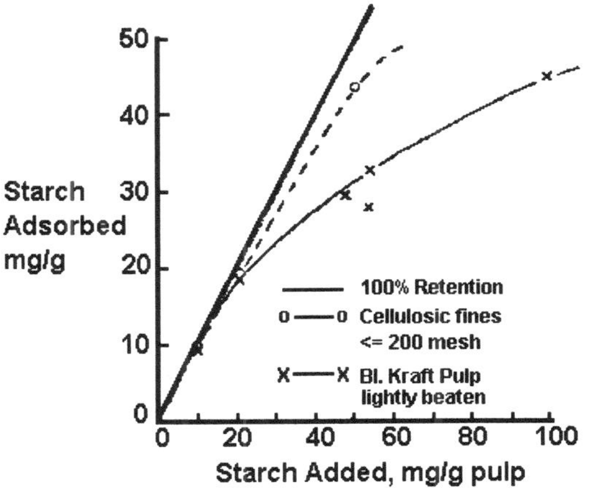

Figure 5-7. Effect of starch addition level on adsorption. (One minute contact time.)[14]

c. The effect of fines level on cationic starch adsorption by pulp [15]

Cationic starch and other cationic polymers tend to adsorb more on the fines in a furnish due to their high specific surface area. **Table 5-8** illustrates this point.

Table 5-8. Effect of fiber and fines surface area on polymer adsorption[15]

Ratio, fiber:fines	Total surface area, m^2/100 g	Relative cationic starch adsorption (fiber:fines)
75:25	290	38:62
65:35	358	27:73

An excellent discussion of all of the factors that influence polymer adsorption on papermaking particles can be found in Reference [16].

12. Surface science and papermaking

We now turn to another aspect of the surface and colloid science of papermaking — namely, surface chemistry. In Chapter 2, we mentioned that water has a high surface tension. We now return to that subject.

a. Source of surface tension

Liquid surfaces are in a state of tension which causes them to contract to their minimum surface area. For example, liquid droplets adopt spherically shaped forms. This phenomenon expresses itself in other ways as well. For example:

- Water can be added to a glass until it appears to "pile up" on top without spilling over.
- Water bugs "walk" on the surface of a pond without breaking through.
- A water surface will support a steel pin, even though the density of steel is much greater than the density of water.

What is the origin of this behavior? **Fig. 5-8** schematically depicts the situation that exists in a container of water. Bulk water molecules are, on average, surrounded in all directions by a uniform molecular environment. This means that they undergo balanced attraction and repulsion interactions with surrounding water molecules. On the other hand, surface molecules are not uniformly surrounded by water molecules because the concentration of water in the vapor phase above the surface is much lower than in the bulk phase. Therefore, surface molecules do not experience balanced intermolecular interactions on all sides. In particular, there are far fewer interactions above the surface than below it.

Conceptually, one can imagine that the greater degree of interaction with bulk molecules leads to the exertion of a downward force on surface molecules, with a resultant surface "contraction". This contraction produces an apparent "surface tension".

b. Papermaking phenomena and surface tension

Table 5-9 lists several important papermaking phenomena that are influenced by surface tension. The list contains phenomena related both to paper quality and process efficiency.

Table 5-9. Papermaking phenomena affected by surface tension

- Sizing
- Foaming
- Air entrainment
- Detergency
- Fiber-fiber bonding
- Wetting and absorbency
- Wet web strength
- Corrugating medium gluability

c. Surface energy

The following scenario represents another way to consider surface tension. The unbalanced forces on surface molecules cause them to migrate from the surface to the bulk. A reverse diffusion of molecules from the bulk to the surface also occurs. Equilibrium exists when the two rates are equal. In fact, there is a net migra-

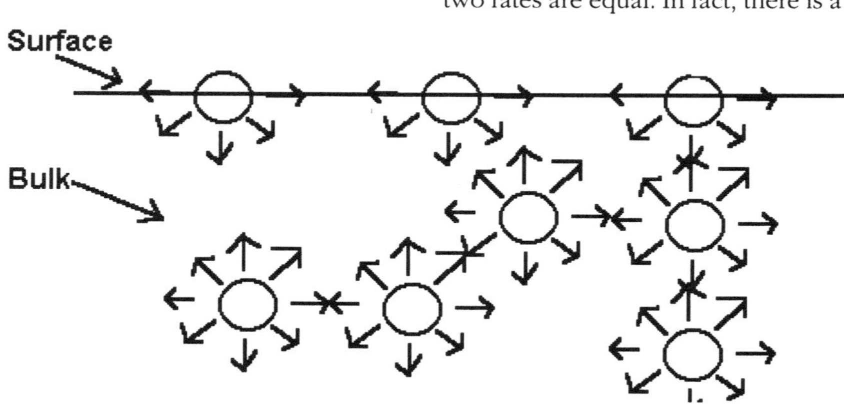

Figure 5-8. Forces acting on surface and bulk molecules

tion of molecules from the surface. Thus, the surface density is lower than the bulk density and the surface molecules are farther apart than the bulk molecules. This gives rise to an excess interaction energy in the surface - a surface energy arises. This surface energy means that it takes work to create new liquid surface. The following relationship exists between the work energy, W, required to produce a new surface and the created surface area, A.

$$W = \gamma A$$

The proportionality constant, γ, is referred to as the surface tension of the liquid. It has units of mN/m when referred to as a force. It can also be referred to as surface free energy, which has units of mJ/m^2. Surface tension values for several common liquids are listed in **Table 5-10**.

Table 5-10. Surface tension of common liquids

Liquid	Surface Tension, γ (20°C, mN/m)
Perfluoropentane	9.9
Diethyl ether	17.0
Ethanol	22.3
Methanol	22.5
Acetone, Methyl ethyl ketone	24.0
Carbon tetrachloride	26.4
Benzene	28.9
Ethylene glycol	46.5
Glycerol	63.3
Water	72.9
Mercury	484

c. Interactions between liquids and solid surfaces

Many papermakers are interested in how water and other fluids interact with paper surfaces. Internal sizing, surface sizing, and coating all are related to this topic. If a liquid drop is placed on a solid surface, the equilibrium condition is expressed by a thermodynamic relationship called the Young Equation (**Fig. 5-9**).

The Young Equation defines the contact angle at the liquid-solid-vapor point in terms of three surface tensions. In papermaking, we believe that sizing influences the surface energy of the solid paper surface and, hence, affects γ_{SV} and γ_{SL}. If the liquid (usually water) surface tension is taken to be constant, then it follows that the contact angle, θ, gives an indication of sizing degree. In fact, a contact angle test is used to measure sizing for certain papers. We will return to this important topic in Chapter 12 when we discuss internal sizing. For more thorough treatments of the subject, you are referred to the general colloid chemistry reference books listed at the end of this chapter.

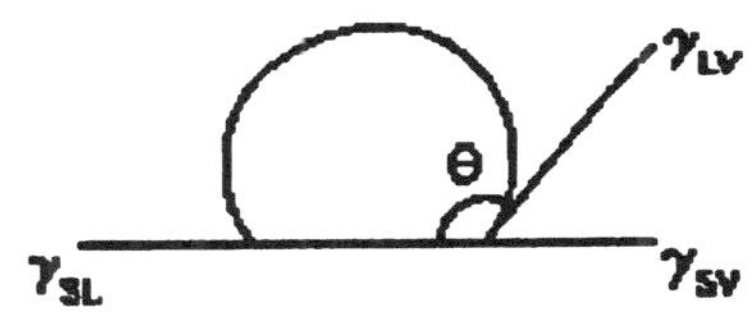

Figure 5-9. The equilibrium condition for a stationary drop on a solid surface — the Young Equation

$$\gamma_{SV} - \gamma_{SL} = \gamma_{LV} \cos \theta$$

where:

γ_{SV} = surface tension at the solid—liquid vapor interface

γ_{SL} = surface tension at the solid—liquid interface

γ_{LV} = surface tension at the liquid—vapor interface

13. Association colloids

One way to produce a colloidal system is to have a large number of small solute molecules associate together to form an aggregate which is very much larger than the individual solvent molecules and falls into the colloidal size range. If the solute molecules associate by Van der Waals forces, then the aggregate has special properties and is referred to as an *association colloid*. Solutions formed by soaps and detergents fall into this category. Most surface active substances used as wetting agents, detergents, and emulsifying agents also behave like association colloids. Since true molecular solutions are involved, association colloids fall into the hydrophilic class of colloids. **Table 5-11** lists some typical applications of these materials in the paper industry.

Table 5-11. Applications of association colloids in the paper industry.

1. Pulp washing improvement
2. Pulp fluffing improvement
3. Paper softening
4. Defoaming and antifoaming
5. Absorbency improvement in papers
6. Deposits control
7. Felt washing and conditioning
8. Corrosion inhibition
9. Yankee dryer adhesion control.

The molecules of association colloids are *amphipathic*. In other words, they have a "dual" character and exhibit both hydrophilic and hydrophobic traits. The hydrophilic portion of the molecules normally dissolves in water to form molecularly dispersed solutions. The hydrophobic part is chemically similar to materials normally insoluble in water. The compounds in **Table 5-12** are good examples of this. In all cases the hydrocarbon portion of the molecule resembles liquid hydrocarbons or benzene derivatives which are sparingly soluble in water. The salt-like parts resemble sodium sulfate, sodium formate, or ammonium bromides, all of which are water soluble.

Many association colloid materials utilized by the papermaker are called *chemical processing aids*. These materials are utilized primarily to enhance the cleanliness of a paper machine system and to improve machine runnability. These substances are extremely important and their application programs are often complex. A good review of chemical processing aids can be found in the notes from the 1986 TAPPI Chemical Processing Aids Seminar and a recent TAPPI publication "Chemical Processing Aids: A Practical Guide".[17]

Table 5-12. Association colloids compounds.

Anionic
- Sodium alkyl aryl sulfonates
- Sodium cetyl sulfate
- Sodium oleate
- Sodium dodecyl sulfate

Nonionic
- Nonylphenoxypoly(ethyleneoxy)
- Ethanol

Cationic
- Cetyltrimethylammonium bromide

Amphoteric
- Dodecyl-beta-alanine

a. Surface activity

Amphipathic association molecules often cause a dramatic, spontaneous lowering of the surface or interfacial tension, even when present in very small concentrations. This occurs because the system tends to decrease its surface energy whenever possible, which can occur if the amphipathic molecule orients itself on the surface with its polar group oriented into the water and its hydrocarbon group extending into the air or oil interface. Materials that exhibit this behavior are called *surfactants*.

b. Micelle formation

As was mentioned above, association compounds form micelles of colloidal dimensions. It is thought that micelles have structures similar to that shown in **Fig. 5-10**. Micelle formation involves a delicate balance between the hydrophobic and hydrophilic portions of the molecule and depends as well on the solvent properties and concentration of the substance. The concentration at which micelle formation occurs spontaneously is called the "critical micelle concentration".

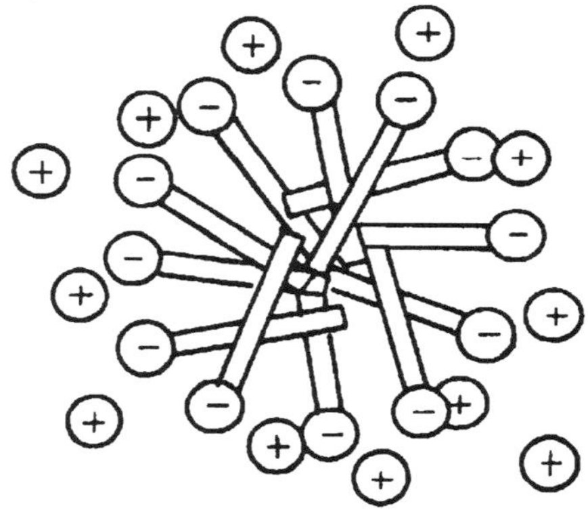

Figure 5-10. Schematic representation of micelle formation from surfactant molecules in water.

c. Detergency

Detergents are used for the removal of dirt from solid surfaces by surface chemical means. Good examples of detergents in the paper industry are the many different felt and fabric cleaning systems available today. A satisfactory detergent must possess the following properties:[18]

- Good wetting characteristics in order that the detergent intimately contacts the surface to be cleaned.
- Ability to remove or to help remove dirt into the bulk of the liquid.
- Ability to solubilize or disperse removed dirt and prevent it from being redeposited onto the cleaned surface or from forming a scum.

The removal of dirt or deposits from solid surfaces involves a combination of mechanical action and surface chemistry. In the absence of a detergent, water is often unable to remove a deposit due to its high surface tension and poor wetting action. When a detergent is added, the hydrophobic part of detergent molecules adsorb onto the deposit and onto the solid surface and reduce the adhesion of the deposit to the surface. It is then possible to dislodge the deposit by mechanical action. The dislodged deposit particles remain in suspension because detergent molecules form adsorbed layers on both the particles and cleaned surface.[19]

The most successful detergents form micelles, and this originally led to the opinion that micelles were directly involved in detergent action, their role probably being that of solubilizing oily material. However, detergent action is dependent upon the concentration of unassociated surfactant and practically unaffected by the presence of micelles (other than as a reservoir for replenishing the unassociated surfactant adsorbed from solution). It appears, therefore, that the molecular properties of surfactants associated with detergency also lead to micelle formation as a competitive, rather than as a contributing, process.[20]

d. Emulsions

An emulsion is a suspension of droplets of one liquid in another. If a vessel containing two immiscible liquids is shaken vigorously, both liquids are broken up into droplets whose size depends upon the viscosity of the liquids, the interfacial tensions, and the vigor of the shaking. The coalescence of the droplets is retarded in the presence of small amounts of additional components, called emulsifiers, and a stable emulsion is formed. Depending upon which liquid is affected most by the emulsifier, either an oil-in-water (O/W) or a water-in-oil (W/O) emulsion is produced. Many retention aids are delivered as water-in-oil emulsions. Synthetic and rosin sizes are O/W emulsions.

The most important physical property of an emulsion is its stability. Emulsion stability refers to three phenomena: creaming, coagulation, and emulsion breaking by droplet coalescence. In all three instances the emulsion separates into two phases. It is the function of the emulsifying agent to facilitate emulsification and promote emulsion stability. The emulsifying agent forms an adsorbed film around the dispersed droplets which helps to prevent coagulation and coalescence. The stabilizing mechanisms are usually complex and may vary from system to system. Factors which favor emulsion stability are low interfacial tension between the dispersed droplets and dispersing medium, a mechanically strong and elastic interfacial film between the droplets and medium, electrical double layer repulsion between dispersed droplets, narrow droplet size distribution, and a high viscosity in the dispersing medium. The detailed mechanisms involved in emulsification are quite complex. Refer to References [22 and 23] for further treatment of the subject.

e. Foams

Air gets mixed with paper stock at many points in a paper mill and can be a source of significant problems. Among the many troubles related to air are decreased washer efficiency, decreased pumping capacity, excessive foaming, decreasing drainage rate, sizing reduction, increased pitch and deposit problems, and pinholes in the sheet.

An excellent discussion of current theoretical ideas about the action of defoamers can be found in Reference [17]. For a surfactant-type defoamer to work, it must spread rapidly and thoroughly on the film surface — thereby displacing the liquid in the film and thinning the film to the point of mechanical instability. It is predicted that, as the defoamer becomes saturated with foam stabilizer, the driving force for spreading decreases and the defoamer becomes less effective. This is indeed observed in cases involving tightly closed white water systems. There has not been widespread application of the theory to date. Instead, the formulation of defoamers and antifoams relies heavily on the considerable body of existing empirical knowledge.

14. Summary

This concludes our brief review of relevant colloid and surface science concepts. We began by defining papermaking chemistry as the surface and colloid chemistry of the components of a papermaking furnish. We then explained that this definition applies because the dimensions of papermaking furnish components fall into the range of colloidal particles and the types of chemical interactions occurring among particles in the wet end involve secondary bonding forces characteristic of colloidal systems.

Next, papermaking furnish systems were classified as being either hydrophobic or hydrophilic in nature, with the single most important characteristic of hydrophobic systems being their tendency to aggregate. It was pointed out that the papermaker wishes to control such aggregation during furnish preparation and sheet formation. The basic mechanisms of aggregation were described.

Adsorption of cationic polymers is a critical step in the function of many additives. Consequently, the fundamentals of this process were next reviewed. This was followed by a discussion of association colloids and surface science. These subjects were then combined in the subject of surface active materials.

15. References

[1] Linhart, F., Dimensional Analysis — A Novel Approach to Understanding Retention Aids and Their Mode of Action, 1989 TAPPI Papermakers Conf., Washington, DC, p.277.

[2] Helmholtz, H., Wied Am., 7:337 (1879).

[3] Gouy, G., Am. Phys. 7:129 (1917).

[4] Chapman, D.D., Phil. Mag. 25(6):475 (1913).

[5] Stern, O., Elektrochem. 30:508 (1924).

[6] Williams, D.G. and Swanson, J.W., TAPPI CA Rept. 57, p. 47 (1975).

[7] Voyutrsky, S., Colloid Chem., Mir. Rub., Moscow (1978).

[8] Overbeek, J.Th.G., J. Coll. Interface Sci. 58(2):408 (1977).

[9] Hiemenz, P.C., Principles of Colloid and Surface Chemistry, 2nd ed., Marcel Dekker, New York (1986).

[10] Gregory, J., J. Coll. Interface Sci. 42(2):448 (1973).

[11] LaMer, V.K. and Healy, T.W., Rev. Pure Appl. Chem. 13:112 (1963).

[12] Fleer, G.J. and Lyklema, J., J. Coll. Interface Sci. 46(1):1 (1974).

[13] Napper, D.H., J. Coll. Interscience Sci. 58(2):390 (1977).

[14] Marton, J. and Marton, T. "Wet end starch: adsorption of starch on cellulosic fibers". Tappi Journal, 59:12, 121-124(1976).

[15] Scott, W. E. "Fines management and control in wet end chemistry". Tappi Journal, 69(11):30-34(1986).

[16] Eklund, D. and Lindstrom, T. "Retention Mechanisms", pp. 151-179, In Paper Chemistry: An Introduction. DT Paper Science Publications, Grankulla, Finland, 1991. 305 pages.

[17] 1986 TAPPI Papermaking Chemical Processing Aids Seminar, New Orleans, TAPPI PRESS, Atlanta; "Chemical Processing Aids in Papermaking: A Practical Guide". Edited by K. Hipolit. TAPPI PRESS, Atlanta, GA, 1992.

[18] Shaw, D.J., "Introduction to Colloid and Surface Chemistry," 3rd Ed., Butterworths, London, p. 139142 (1980).

[19] Kushner, L.M., and Hoffman, J.I., Synthetic Detergents, Scientific Am. 26-30 (Oct. 1951).

[20] Adamson, A.W., "Physical Chemistry of Surfaces," 3rd Ed., Wiley-Interscience (1976).

[21] Becher, R., "Emulsions: Theory and Practice," 2nd Ed. A.C.S. Monograph 162, Reinhold (1965).

[22] Sherman, P., Ed., "Emulsion Science," Academic Press (1968).

Chapter 6

Charge Measurement Practices in the Paper Industry

1. Introduction

Papermakers have long been interested in the measurement of charge and its application to wet end chemistry control. For example, ten years ago a survey by the Retention and Drainage Subcommittee of the TAPPI Papermaking Additives Committee concluded that furnish charge was the most needed on-line measurement.[1] In the period since that time, several advances have been made in the area of off-line and on-line charge measurement and the papermaker now has several instruments available for this purpose.

Why are papermakers so interested in charge measurement? A partial answer to this question was provided in the preceding chapter where it was pointed out that the charge on colloidal particles plays a major role in the stability of their dispersions. Papermaking phenomena affected include:

- fines retention
- filler retention
- size retention
- pigment slurry preparation
- size emulsification

The adsorption and retention of polyelectrolyte chemical additives by papermaking particles is also strongly influenced by electrostatic charge. Examples of such additives include:

- dry strength additives
- wet strength resins
- polyelectrolyte retention aids
- synthetic organic dyes

Electrostatic charge also plays a key role when dissolved anionic contaminants react with cationic additives and interfere with their function. Note that no particles are involved in these interactions. Instead, polymer-polymer reactions occur.

In summary, the electrostatic charge on papermaking materials play key roles in many important wet end chemistry phenomena and the measurement and control of these charges is a key element in wet end chemistry process control.

The following topics will be covered in this chapter:

- The sources of electrostatic charge on papermaking components
- Surface charge and dissolved charge definitions
- Basic charge measuring technologies and procedures
- Applications of charge measurement in papermaking

Additional information about this subject can be found in Reference [2].

2. The source and nature of electrostatic charge on papermaking materials

Many papermaking materials become electrostatically charged when dissolved or suspended in water. Included in the list are fibers, fiber fines, fillers, retention aids, dry strength resins, wet strength resins, wood lignin, wood hemicellulose, dispersants, and alum. Depending upon whether discrete particles or dissolved molecules are involved, one is concerned either with *surface charge* or *dissolved charge*. Both play important and different roles in wet end chemistry process control.

a. Surface charge on papermaking fibers and fiber fines

In kraft pulp fibers, carboxyl groups located on hemicelluloses represent the largest source of charge. A typical range for carboxyl content in wood pulps is 5-10 meq/100 g wood. Unbleached pulps also contain lignin phenolic groups. Mechanical pulps, such as CTMP, contain sulfonic acid groups that originate from sodium sulfite treatment during pulping.

With the exception of sulfonic acid groups, which are completely ionized at all pH values, the charged groups on fibers and fillers are mostly weak acids and their extent of ionization depends upon the pH of the system. **Fig. 6-1** illustrates this effect.

For comparison purposes, the ionization curve for a typical alcohol hydroxyl group is included in order to illustrate that under normal papermaking conditions, sugar hydroxyl groups do not contribute to fiber charge. It is also worth noting that below pH 2.7, papermaking fibers and fines have essentially zero surface charge due to the complete protonation of all carboxyl and phenolic groups.

b. Surface charge on mineral pigments and fillers

For some mineral pigments, such as clay, isomorphous substitution of ions in the crystal lat-

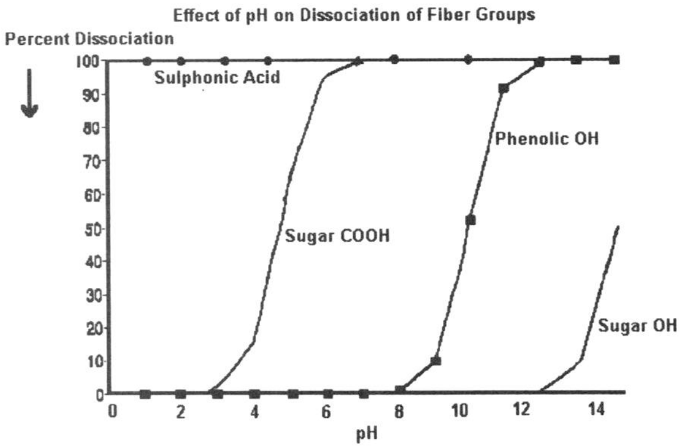

Figure 6-1. The effect of pH on the degree of ionization of carboxyl, phenolic, and sulfonic acid groups. For comparison purposes, hydroxyl ionization is included to illustrate that under papermaking conditions, these groups do not contribute to fiber charge.

tice leads to charge imbalance and a net particle charge. Clay exhibits both anionic and cationic characteristics. The ratio of the two types of sites depends on the pH and the chemical environment. Under acidic conditions, the edges are positively charged, while the faces are negative. For some other pigments, ion adsorption produces a surface charge. **Table 6-1** lists the isoelectric pH (pH at which the particle is electrically neutral) for several different minerals.

Table 6-1. The isoelectric point of some mineral surfaces [2]

Mineral	Isoelectric pH
TiO_2	
anatase	6.0±0.3
rutile	6.7±0.1
Kaolinite	2.0
$CaCO_3$ (calcite)	8.3±0.1

Another approach to effecting filler surface charge is to treat the particles chemically with a substance that will impart a strong surface charge to them. This is often done to enhance their dispersability. Recently, some calcium carbonate and clay filler pigments have been given cationic surface charges to improve their retention in the forming web.

c. The zeta potential of dispersed particles

The electrical double layer was introduced in Chapter 5 as part of the discussion about the stability of colloidal dispersions (**Fig. 5-1**). The zeta potential was defined as the potential at a slip plane, within which counterions bound to the particle move with the particle and outside of which counterions are free to move independently of the particle. The notion of a slip plane implies relative motion within the electrical double layer. The term *electrokinetics* refers to such situations. Most particle charge measurement technologies employ electrokinetic principles.

d. Dissolved charge on polyelectrolytes

The electrostatic charge associated with dissolved polyelectrolyte additives, such as retention aids, wet strength resins and dry strength resins, is either derived from carboxyl groups (anionic) or from a variety of quaternary or tertiary nitrogen functional groups (cationic). These may be fully or partially ionized, depending upon the pH and the chemical nature of the group. The molecular structures of such functional groups will be presented later in the book in the chapters dealing with dry strength resins, wet strength resins, and retention aids.

Dissolved wood hemicelluloses contain glucuronic acid carboxyl groups and dissolved wood lignin contains acidic phenolic groups. The ionization behavior of both types of groups is affected by pH according to the patterns shown in **Fig. 6-1.**

Dissolved polyelectrolyte molecules have no discernible surface. Hence, there is no interface between the molecule and the surrounding medium and no electrical double layer. Thus, the concept of zeta potential has no meaning as far as dissolved polyelectrolytes are concerned. Rather, dissolved charge measurements deal with the quantity of charged groups in a given polyelectrolyte solution.

3. Measuring the zeta potential of suspended papermaking particles (electrokinetic measurements)

Three electrokinetic methods are employed by papermakers to measure the zeta potential of papermaking particles:

- microelectrophoresis
- streaming potential
- AC streaming current

Of the three, microelectrophoresis was the first method developed and has been the most commonly used. More recently, AC streaming current has been widely adopted for end-point detection in dissolved charge determination. Also, an on-line streaming potential sensor is now available and its application in wet end chemistry monitoring and control is being explored. We will discuss all three approaches.

The following discussion of charge measurement practice is generic in nature and no spe-

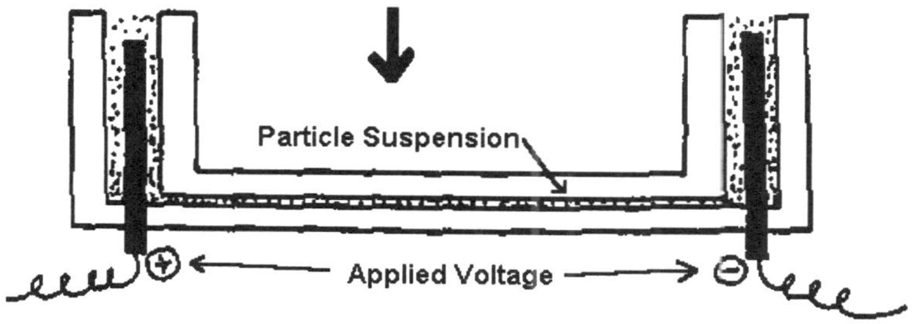

Figure 6-2. A schematic drawing of an electrophoresis cell. The particle velocity tracking arrangement can be manual and involve a microscope, human observer, and timer. Or, it may be automated and involve lasers and sophisticated electronics.

cific instruments are mentioned. You are referred to Reference [4], which contains a thorough compilation of available instruments.

4. Microelectrophoresis

In microelectrophoresis, a particle dispersion is placed in a specially designed cell; an electric field is placed across the cell; and, the migration velocity of the dispersed, charged particles is measured. The quantity determined by the measurement is called the *electrophoretic mobility* (μm/s/V/cm). A schematic drawing of a microelectrophoresis cell is presented in **Fig. 6-2**.

a. Calculation of zeta potential

For most purposes, electrophoretic mobility data is sufficient for wet end chemistry control and it is not necessary to compute the zeta potential. However, if it is desirable to convert electrophoretic mobilities into zeta potentials, the equation can be used.

$\xi = f\mu\eta/\varepsilon$ (A form of the Helmholtz-Smoulochowski Equation)

where:

ξ = zeta potential (mV)

μ = electrophoretic mobility

ε = dielectric constant of the medium

η = viscosity of the medium

f = constant whose value varies from 1.0 to ca. 0.33, depending on the double layer thickness-to-particle radius ratio, the magnitude of the zeta potential, the particle shape and particle size. For most papermaking cases, f is assumed to be 1.0.

For more complete discussions of this subject, see References [2] and [3].

Typical values of electrophoretic mobility in papermaking systems range from -3 to +3, while zeta potentials typically range from -40 mV to +40 mV.

b. Instrumentation for microelectrophoresis measurements

Since the first widely accepted microelectrophoresis instrument was introduced in the 1960's, an array of new devices have been introduced, ranging from manual, single particle instruments to automated instruments that provide multi-particle information about the distribution of zeta potentials existing in a heterogeneous suspension. Both portable and nonportable products are available. An excellent review of suppliers and products associated with zeta potential measurements has been provided by Sanders.[4]

c. Practical considerations in microelectrophoresis measurements

The preferred microelectrophoresis cell has a rectangular cross-section. Also, both the particles and the suspending liquid are influenced by the electric field. The particles migrate toward the electrode of charge opposite to their (particle) surface. The liquid next to the cell surface moves also because the cell surface is charged (usually negative) and the dissolved counterions in the electrical double layer at the cell interface move toward the oppositely charged electrode under the influence of the imposed electric field.

The flow along the cell wall creates a backflow in the center of the cell. There are zero-flow regions at the interface between the upper and lower cell surface flows and the backflow. These regions are called the *stationary layers*. It is critical that microelectrophoresis measurements be made on particles that lie within stationary layers. Otherwise, the liquid flow velocity will affect the migration velocity of the particles in the electric field. Each instrument has instructions for locating the stationary layers in a cell.

Microelectrophoresis measurements require that the particles lie completely in the stationary layer and that they not settle in the cell.

These requirements mean that measurements cannot be made on papermaking fibers, but only on the fines fraction of a sample.

There is disagreement among investigators regarding the assumption that fibers and fines have the same zeta potential in a papermaking system. It seems reasonable to expect that if the fines are derived from the secondary fiber wall and that they are suspended in very clean, distilled water, that they should have the same chemical composition and, hence, zeta potential as the fiber wall itself.

On the other hand, if the fines are primary fines, or if they have very different chemical composition from the fiber, or if they have been exposed to adsorbing chemical additives, then it would not be surprising to find that they have a different zeta potential from fibers. Indeed, measurements with modern, multi-particle instruments have revealed that there is a distribution of zeta potentials in a given papermaking system.

Temperature gradients must be absolutely avoided in the cell or convection currents in the sample slurry will invalidate the measurements. High conductivities and long measurement times give rise to this problem. Consequently, measurements should be made as rapidly as possible and the polarity of the electrical field alternated often. Modern, automated instruments have these features.

In general, samples should be sufficiently dilute that individual particles do not interfere with one another when they migrate in the cell. Also, since the zeta potential is a function of ionic environment, dilutions must always be made with filtrate collected from the sample under investigation. Dilution with distilled or deionized water will alter the ionic conditions and change the zeta potential.

5. Streaming potential (Fig. 6-3)

In the a streaming potential measurement, liquid is forced, under a pressure gradient, through a plug formed from fibers, fines, and other furnish components. Electrodes are placed on either side of the plug to measure the streaming potential established across the plug by the flowing liquid. The zeta potential can then be calculated from the following equation,

$$\xi = (SP)(C)\eta/[\varepsilon\Delta P]$$

where:

SP = streaming potential

η = viscosity of the flowing liquid

ε = dielectric constant of the liquid

C = conductivity of the liquid

ΔP = liquid pressure drop across the pad

Both on-line and laboratory streaming potential instruments are available.[4]

A thorough evaluation of two on-line streaming potential instruments is presented in Reference [5]. It was found that a good correlation existed between off-line microelectrophoresis and on-line streaming potential zeta potential values for four different pulps under a variety of furnish conditions. The on-line instruments facilitated the investigation of both steady-state and time-dependent surface charge phenomena.

a. Practical considerations regarding streaming potential measurements

Since streaming potential measurements are made on a formed pad, all the components — fines included — are contained in the measurement. This has advantages for on-line instrumentation because no stock filtration is required prior to making a measurement. Also, the issue of whether or not fines and fibers are the same is avoided.

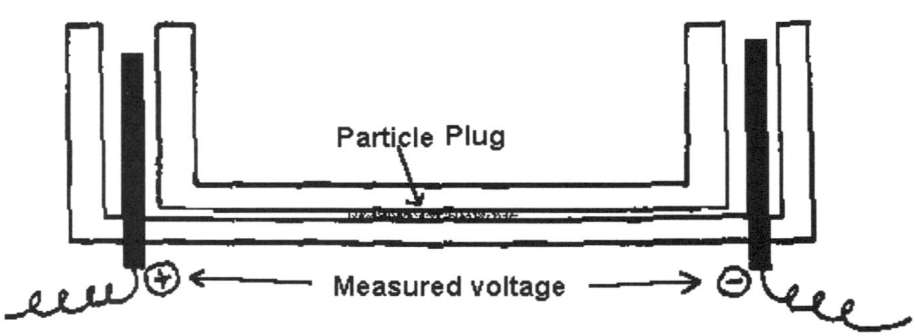

Figure 6-3. Schematic drawing of a streaming potential measuring cell. Liquid is forced under pressure through a pad formed from a papermaking stock sample and held between two electrodes. The streaming potential set up by the flowing liquid is measured.

Anything that affects the pressure drop across the formed pad will affect the instrument reading and it is very important to prevent such an occurrence when setting up a streaming potential instrument. The exact precautions to be taken depends on the configuration of the instrument and the method used to create the DP across the pad.

Other variables that can interfere with streaming potential measurements include foam and entrained air in the stock sample. Air bubbles in the formed pad tend to impede the flow of liquid during the measurement and leads to erratic results. Stock temperature is also important due to its effects on the conductivity and viscosity of the liquid forced through the plug. It must be measured and corrected.

6. AC streaming current measurement

The streaming potential can be measured over a thin slit rather than a packed plug of fibers. One popular version of this approach utilizes a piston to force liquid back and forth through a small clearance between the piston and enclosing cylinder. The pressure created by the piston motion produces an AC current that can be detected by electrodes imbedded in the sides of the cylinder. **Fig. 6-4** contains a schematic drawing of such a device.

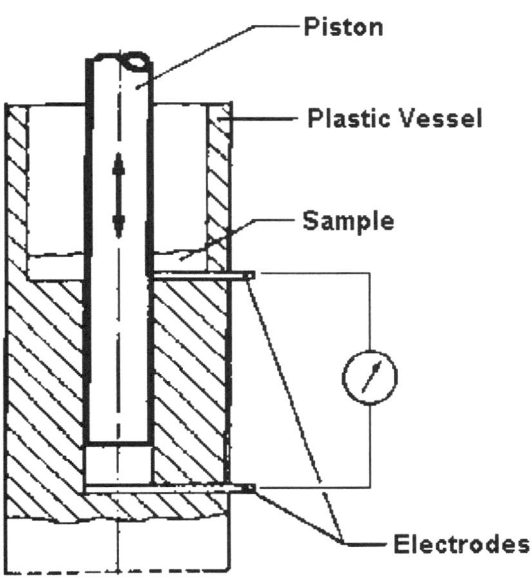

Figure 6-4. A schematic drawing of an AC streaming current detector. Particles and dissolved counterions located in the annulus near the cylinder wall and piston walls move back and forth in an oscillating fashion and cause an AC current to be produced. Electrodes imbedded in the cylinder walls measure this current.

One major consideration when using a streaming current instrument is the absolute requirement for cleanliness. Failure to recognize this fact results in spurious and unreliable results. It is common to coat the piston and cylinder walls with Teflon in order to avoid contamination of these surfaces.

7. Streaming current versus microelectrophoresis and streaming potential

It has been found that the streaming current measurement does not necessarily reflect the exact zeta potential of suspended particles, but rather indicates whether a system is cationic or anionic. It has been shown, however, that the point of charge reversal (i.e. the isoelectric point) of a suspension does correlate with the charge reversal point on the piston and cylinder surfaces when polyelectrolytes are used to change the charge in the system. Consequently, streaming current measurements have been widely adopted as end-point indicators in cationic demand determinations. This topic will be covered more completely next in our discussion of dissolved charge measurement.

8. Dissolved (Soluble) Charge Measurement

a. Background

The terms *anionic trash* and *cationic demand* appeared in the 1970's when environmental legislation required papermakers to close white water systems and increase water reuse in the paper mill. This practice lead to increased concentrations of dissolved and suspended solids in mill water systems. Since many of the suspended and dissolved substances were colloidal and carried anionic charges, they reacted with cationic additives and reduced their effectiveness. **Fig. 6-5** illustrates this phenomenon schematically.

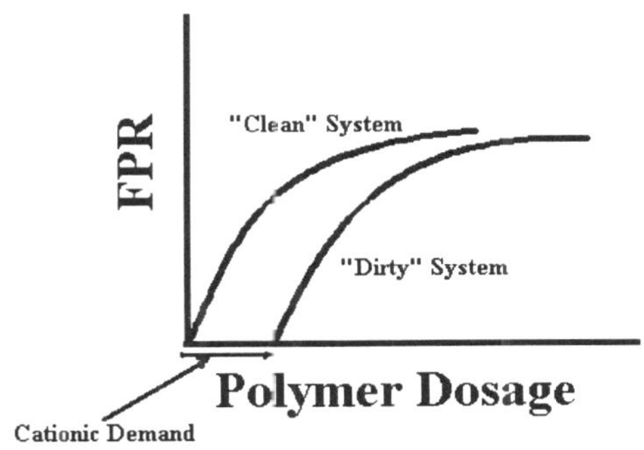

Figure 6-5. A typical effect of interfering substances (cationic demand) on the effectiveness of a retention aid polymer.

Later it was found that recycled cationic materials could also reduce the efficiency of chemical additives and the term *interfering substances* was coined to cover all types of such materials.

It soon became clear that a method was needed to measure the amount of interfering substances in a papermaking system and predict their effects on cationic additives. Several different approaches were developed and put into practice.

As is often the case when different approaches are taken to solve a given problem, the terminology that arises around these methods can lead to confusion. Such is the case in this instance. Therefore, one of the principal objectives of this chapter is to distinguish among the various measurement techniques and terminology being employed in the paper industry today.

b. Colloid titration — the underlying procedure

Even though several different measurement approaches have been adopted for determining the level of interfering substances in papermaking systems, they all are based on a procedure called *colloid titration*. This procedure is based on the fact that in many instances, oppositely charged, dissolved polyelectrolytes will form one-to-one charge complexes (**Fig. 6-6**). This reaction has been widely applied in polyelectrolyte characterization by polymer scientists. Reference 6 contains an excellent literature survey on this subject.

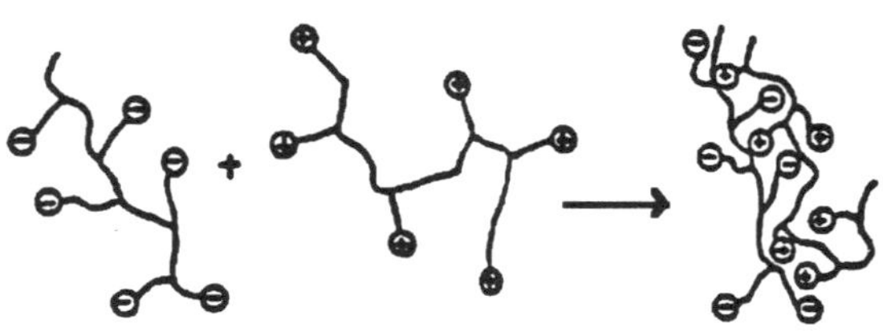

Figure 6-6. Schematic illustration of one-to-one charge complex formation between dissolved anionic and cationic polyelectrolytes

Experimentally, colloid titration usually involves titrating a sample with a standard cationic or anionic polymer that will react with its oppositely charged counterpart in the sample to form a one-to-one charge complex. An endpoint detection method is employed that indicates the point at which the first bit of excess titrant is added. The amount of titrant charge added up to that point is equal to the amount of oppositely charged, dissolved charge in the sample.

c. Application of colloid titration to papermaking systems

The search for a method of determining the level of interfering substances in papermaking systems lead naturally to the adoption of techniques based on colloid titration. Strazdins[7] and Halibisky[8] were the leaders in this regard. The former coupled colloid titration with microelectrophoresis endpoint detection and referred to the determination as a *cationic demand* procedure. The latter researcher combined two colloid titration steps — one for anionic substances and one for cationic substances — into a parameter called the *colloid titration ratio*. Other workers have developed their own variations on these two themes during the past two decades and, as would be expected in such situations, a certain amount of confusion has arisen around the terminology and jargon being applied in this area today. Therefore, it will be helpful at this point to distinguish among the various terms and practices before delving into experimental details.

9. Definitions of various dissolved charge measurement terms in use today

Table 6-2 lists the most important terms to be defined.

Table 6-2. Terminology associated with dissolved charge measurement

- Dissolved charge
- Soluble charge
- Total charge
- Cationic demand
- Anionic demand
- Colloid titration ratio (CTR)

a. Dissolved charge; soluble charge

Earlier in the chapter, *dissolved charge* was introduced in terms of the charged groups associated with dissolved polyelectrolytes in papermaking systems (e.g. lignins, hemicelluloses, dispersants, recycled size press starches, dry strength and wet strength agents, retention aids). It was pointed out there that these dissolved polyelectrolytes contain cationic and anionic functional groups whose extent of ionization may be pH dependent. These hydrophilic substances contain no true "surface" and there is no well-defined interface between the dissolved molecule and the water medium. Consequently, no electrical double layer (or zeta potential) will develop. The charged groups on dissolved polyelectrolytes are capable of undergoing charge-charge interactions with other polyelectrolytes, which is the primary mechanism by which interfering substances interfere with additives.

The same charge-charge interaction that causes interference with additives can be employed to determine the amount of interfering substances dissolved in a sample of filtered stock taken from a papermaking system.

b. Total charge

Many of the same anionic functional groups that exist on dissolved polyelectrolytes also occur on fibers, fines and fillers, and they react with cationic additives to an extent dependent upon their quantity and degree of ionization. When the quantity of charged functional groups associated with the particles in a slurry sample is added to the quantity of dissolved polyelectrolyte charges in the sample, the sum is referred to as *total charge.*

c. Cationic demand; anionic demand

These terms refer to measurement of the specific capacity of a test specimen to neutralize cationic or anionic polymer titrants, respectively. One may refer to the dissolved cationic and anionic demand, the total cationic and anionic demand, or the cationic and anionic demand of just the solids in a sample. The specific parameter measured depends upon how the sample is treated prior to carrying out a colloid titration determination.

d. Colloid titration ratio

The term colloid titration ratio (CTR) refers to the practice of calculating the ratio of the anionic demand to the cationic demand of a sample. The value of the ratio indicate whether a system is "net anionic" or "net cationic."

Practitioners who have extensive experience with this parameter find it useful for analyzing the charge state of a papermaking system.[9] The CTR can refer to the dissolved charge, the solids charge, or the total charge — depending upon how a sample is treated prior to carrying out the colloid titration determination.

10. Cationic and anionic demand are extensive quantities

There are two types of variables that describe the state of a system: extensive variables and intensive variables. *Extensive variables*, such as mass and volume, depend upon the size of the system being investigated. Extensive quantities are additive. Dissolved charge, cationic demand, and anionic demand are extensive quantities.

Intensive variables do not depend upon the size of a system. Temperature, pressure, density, zeta potential, and pH are all intensive variables. Intensive quantities are not additive.

These considerations speak to how charge data is applied to process control in papermaking. For example, it is valid to calculate a dissolved charge balance around a papermaking system. On the other hand, one should not do a zeta potential balance.

11. Determination of dissolved cationic demand and anionic demand

a. Dissolved cationic demand procedure

Table 6-3 presents a common procedure for determining the dissolved cationic demand of a sample. This procedure employs a *back titration* technique where a known, excess quantity of standard cationic polymer is added to the sample and reacts with the anionic material present. The remaining, unreacted cationic polymer is then determined by titration with a standard anionic polymer. The endpoint is the point where an excess of standard anionic polymer is first added to the solution. A color change (blue-to-pink) can be used to identify this point. **Fig. 6-8** illustrates the calculation of dissolved cationic demand from experimental values.

Table 6-3. A common dissolved cationic demand procedure

- Filter the sample to remove all suspended solids.
- Dilute the filtrate with distilled or deionized water.
- Add an excess amount of standard cationic polymer to an aliquot of sample filtrate.
- Titrate the excess cationic polymer with a standard anionic polymer. Determine the complexation endpoint with an indicator dye. (The color change is blue-to-pink.)
- Determine the amount of standard anionic polymer consumed by the indicator dye color change alone.

Fig. 6-7 illustrates schematically the polymer-polymer reactions involved in the colloid titration method.

The dissolved cationic demand (meq/l) is computed from the following relationships.

b. Determination of dissolved anionic demand

The procedure followed to determine the dissolved anionic demand of a sample is analogous to the cationic demand procedure. **Table 6-4** contains the specific steps.

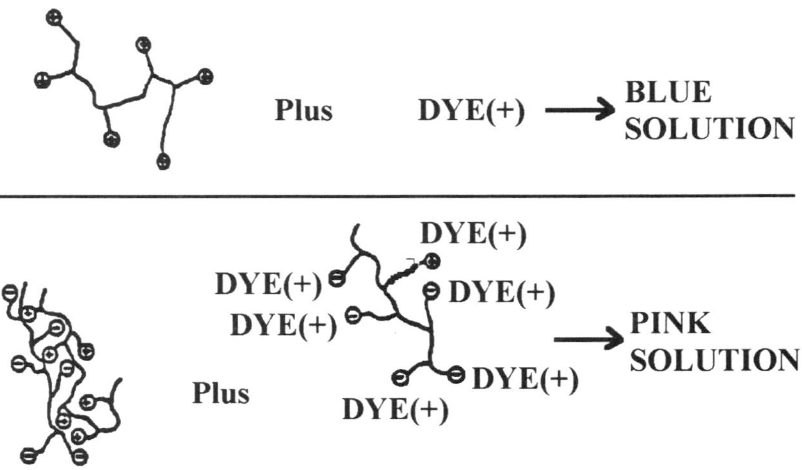

Figure 6-7. Illustration of the dye-polyelectrolyte reactions in the most common colloid titration method. The top portion of Figure 6 depicts the situation prior to the addition of anionic titrant to a solution of cationic polymer and blue indicator dye. The dye does not react with dissolved cationic polyelectrolytes and so the mixture of the two substances is blue. The bottom illustration shows the situation when a cationic polymer solution has been titrated with an anionic polymer to the point where an excess of anionic titrant occurs, which reacts with the dye to produce a pink complex. The blue-to-pink transition occurs as soon as all of the cationic substance has been complexed with anionic titrant and an excess of anionic polymer occurs in the system.

$$\boxed{\begin{array}{l}\text{Cationic}\\\text{Charge}\\\text{Demand}\\\text{(meq/l)}\end{array}} = \frac{\left[\begin{array}{l}\text{Vol. Cat.}\\\text{Polymer}\\\text{Added}\\\text{(ml)}\end{array} \times \begin{array}{l}\text{Cat. Polymer}\\\text{Concentration}\\\text{(meq/l)}\end{array} - \left(\begin{array}{l}\text{Vol. An.}\\\text{Polymer}\\\text{Used}\\\text{(ml)}\end{array} - \begin{array}{l}\text{Vol. An.}\\\text{Polymer to}\\\text{Titrate Dye}\\\text{(ml)}\end{array}\right) \times \begin{array}{l}\text{An. Polymer}\\\text{Concerntration}\\\text{(meq/l)}\end{array}\right]}{\begin{array}{l}\text{Sample}\\\text{Volume}\\\text{(ml)}\end{array}}$$

Figure 6-8. Calculation of dissolved cationic demand from experimental data. The experimental values needed are the volume of excess cationic polymer initially added, the concentration of the cationic polymer solution, the volume of anionic polymer required to back-titrate the unreacted cationic polymer, the anionic polymer concentration, the sample volume, and the volume of anionic polymer required to produce a color change in the dye.

12. Calculation of Colloid Titration Ratio (CTR)

The colloid titration method was first applied in papermaking by Halibisky.[8] Reference [9] contains a very good description of the method. The *Colloid Titration Ratio (CTR)* is equal to the ratio of anionic demand to cationic demand. According to Wallace[7], typical values for alkaline paper machines are 0.20-0.30, and typical values for acid paper machines are 0.3-0.5. The difference between the two ranges can be explained in terms of the increased ionization of anionic groups on fibers and fines under alkaline pH conditions compared to acid papermaking pH conditions. This results in increased cationic demand under alkaline conditions. Since the anionic demand does not change appreciably, the anionic:cationic ratio decreases.

Table 6-4. A common dissolved anionic demand procedure

- Filter the sample to remove all suspended solids.
- Dilute the filtrate with distilled or deionized water.
- Add an excess amount of standard anionic polymer to an aliquot of the diluted filtrate. (Part of the anionic polymer complexes with dissolved cationics and part remains free in solution.)
- Titrate the excess anionic polymer with a standard cationic polymer. Determine complexation endpoint with an indicator dye. (The color change is pink-to-blue.)

13. Material requirements

Table 6-5 lists the materials required to carry out the colloid titration procedure with a dye indicator endpoint.[6]

Table 6-5. Material requirements of the colloid titration method with a dye indicator endpoint

Material	Requirements
Anionic and Cationic Polymers	Known charge density Molecular Wt.: 1,000 to 1 million Flexible backbone.
Indicator Dye	Must bind to a standard polymer. Must bind with less strength than polymer-polymer binding. Must undergo color change when bound.

Empirically, it has been found that polymer molecular weight and charge density affect the results. Therefore it is important to select a single standard polymer for all work. **Fig. 6-9** contains the chemical structures of some typical standard cationic and anionic polymers and the most common indicator dye.

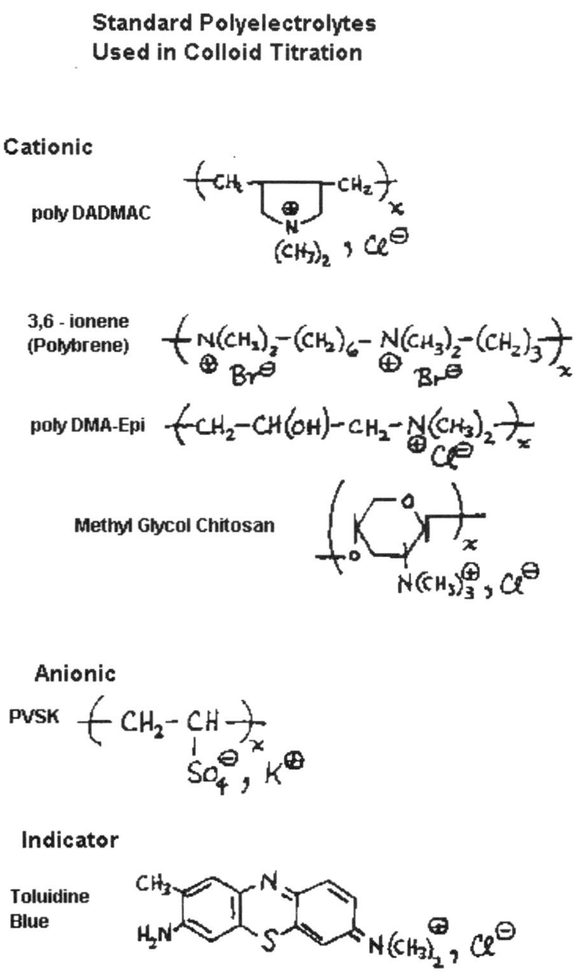

Figure 6-9. Some common polymers and the indicator dye used in the colloid titration method.

14. Disadvantages of colloid titration with indicator dye endpoint

The colloid titration method employing the dye indicator endpoint has not been widely adopted by the paper industry. **Table 6-6** lists some reasons why this is so.

Table 6-6. Some disadvantages of the dye indicator endpoint method in colloid titration

- "Vague" endpoint (pink/purple)-to-blue)
- Paper mill dyes may interfere
- Manual test
- Tedious, time-consuming titrations required

a. Use of microelectrophoresis endpoint detection

These disadvantages lead to a search for endpoint detection methods other than a dye color change. Two approaches have been employed most frequently. The first approach was to combine colloid titration with a microelectrophoresis zeta potential endpoint determination.[7] **Table 6-7** contains a procedure for such a method.

Table 6-7. Use of microelectrophoresis for end point detection in the colloid titration of dissolved charge (cationic demand)

- Filter the sample to remove all suspended solids.
- Dilute an aliquot of the filtrate with distilled or deionized water.
- Add a small amount of a "marker" colloid e.g. (TiO_2) to the diluted filtrate.
- Titrate with a standard cationic polymer to zero zeta potential (or zero electrophoretic mobility).
- Report the results as the quantity of standard cationic polymer to bring the sample to zero zeta potential per unit volume of sample.

This method was adopted by several chemical suppliers and used to analyze many papermaking systems during the 1970's and 1980's. It is still preferred by many wet end chemists.

b. Use of AC streaming current endpoint detection

During the past few years, it has become common practice to employ an AC streaming current detector zero point reading as the endpoint in colloid titration measurements. When this technique is combined with an automatic titrater, a completely automated colloid titration procedure can be carried out. The apparent ease of carrying out this determination has lead to its wide adoption by many chemical suppliers and papermakers.

A key requirement when this approach is taken is that the streaming current detector be cleaned scrupulously after each run in order to avoid cross-contamination of samples and spurious results. This is the greatest source of experimental error in the method. Unfortunately, this fact is often overlooked when the method is applied on a routine basis in the mill. All of the general precautions that apply to the colloid titration method also apply to the automated approach, as well. Since many people make measurements on samples containing solids, it is important that the cautions described in the next section be understood, also. Reference [10] contains a good review of this subject.

15. Measurement of total charge (dissolved charge plus particle charge)

The following procedure can be used to determine the total cationic demand of a slurry sample "as is," i.e. without filtering out the solids. Interchange the roles of the cationic and anionic polymers in this procedure to determine the total anionic demand.

There are some serious drawbacks to applying the colloid titration method to solid particles. These are discussed below.

Table 6-8. Measurement of the total cationic demand

- Measure an aliquot of slurry sample.
- Add excess standard cationic polymer to the aliquot.
- Filter the sample to remove all solids.
- Titrate the filtrate with standard anionic polymer to determine the amount of excess cationic polymer.
- Determine the amount of anionic polymer required to change dye from blue to pink in a blank sample. (If dye indicator endpoint employed.)
- Calculate the total cationic demand of the sample.

16. Measurement of particle charge by colloid titration

Particle charge can be determined by dividing a sample into two parts and then measuring the total charge of one portion and the soluble charge of the other. The particle charge is calculated by subtracting the soluble charge from the total charge.

This measurement is theoretically valid, but there are a number of cautions that apply. It is very important that these be understood in order to avoid experimental errors and the collection of spurious data.

a. Cautions when doing colloid titration on particles

While particle charge measurements can be carried out by the colloid titration method, there are reasons to be cautious. For example,

A one-to-one charge reaction may not occur between a standard polymer and the particle surface.

Papermaking fibers and fines are porous and polymers diffuse into them. This phenomenon is strongly influenced by the molecular weight of the polymer — lower molecular weights penetrating faster and to a greater extent. This effect is time-dependent.

A fading endpoint occurs at times when a dye indicator is used. This is possibly due to the removal of standard cationic polymer from the particle surface by the standard anionic polymer.

17. Applications of charge measurement in papermaking

Recent years have seen greatly increased application of charge measurement by papermakers and chemical suppliers. Evidence of this heightened interest is that 28 papers were presented on various aspects of the subject at the TAPPI Papermakers Conference during the three year period 1992-1994. Both off-line and on-line instruments were discussed. It is an exciting area, and many possible process control applications of charge measurements are being developed. Some examples follow.

a. Applications of off-line charge measurements

Chemical suppliers and papermakers are using colloid titration and microelectrophoresis data to guide them in setting product addition rates and for trouble-shooting wet end chemistry process and product problems (retention, sizing, coloring, drainage, opacity, brightness).[11–14] One particularly widespread use of colloid titration is in the determination of interfering substance concentrations in process streams.[15–17]

b. Applications of on-line charge measurements

Papermakers have long been interested in on-line charge measurement[1] and recent streaming potential instrumentation advances[18,19] are being applied increasingly. One application of on-line monitoring that has been cited by several workers involves the apparent inverse correlation between variability in headbox zeta potential and paper machine runnability — the greater the variability, the poorer the runnability. (References [18-20,22]) On-line zeta potential measurements have also been used in coated broke treatment closed loop control.[21]

On-line charge measurements on process feed streams can also be used to reduce the electrokinetic variability of the headbox furnish. King suggests that the following parameters be monitored:[22]

- Zeta potential before and after chemical addition points to determine their interaction with fibers and fines
- Cationic demand and zeta potential variability of incoming broke and pulp streams — both before and after treatment.
- Cationic demand and zeta potential of incoming filler streams to determine the uniformity of dispersant levels and particle surface charge.

King also strongly emphasizes the importance of maintaining a constant headbox zeta potential on the slightly negative side. A slightly negative charge is preferred because this situation indicates that one has not overdosed cationic additives at points leading up to the headbox. If one tries to control to a neutral zeta potential, then the possibility of overshooting on the cationic side exists. Many systems become unstable when cationic because subsequent addition of cationic additives leads to redispersion of fines and loss of properties. The adverse effects on paper machine retention and runnability of allowing headbox zeta potential to swing back-and-forth between cationic and anionic are also well known.

18. Summary

This concludes our discussion of charge measurements in the paper industry. Charge measurements are but one element of the larger subject of wet end chemistry process control, which will be addressed later. However, it is critical to note that charge measurement technology is in rapid evolution at the moment and you are urged to keep up with current technical literature and maintain close contacts with instrument suppliers in order to become aware of advances as they occur. While this is an admonition that applies to all subject areas in this book, it is particularly relevant here.

19. References

[1] Scott, W. E. Tappi J. 67(11): 72(1984).

[2] Lindstrom, T. "Electrokinetics of Papermaking." Chapter 3 in Paper Chemistry, J. C. Roberts, editor. Chapman and Hall Publishers, New York, 25-43, 1991.

[3] Hunter, R. J. Introduction to Modern Colloid Science, Oxford Science Publications, Oxford University Press, 1993, 238-240.

[4] Sanders, N. Course Notes from TAPPI Introduction to Wet End Chemistry Short Course, November 8-10, 1993, Portland, OR.

[5] Hand, V., Koethe, J., Kuchibhotla, S., and Scott, W. "An evaluation of two pad-forming paper stock charge analyzers." Proceedings of the 1993 TAPPI Papermakers Conference, Atlanta, GA, 591-598 (April 18-21, 1993).

[6] St. John, M. R. and Gallagher, T. M. "Evaluation of the charge state of paper machine systems using the charge titration method." Proceedings of the 1992 TAPPI Papermakers Conference, Nashville TN, April 5-8, 1992, pp. 479-503.

[7] Strazdins, E. "Microelectrophoresis theory and praxis." Proceedings of the 1992 TAPPI Papermakers Conference, Nashville TN, April 5-8, 1992, pp. 503-505.

[8] Halibisky, D. D. "Surface charge measurement of the headbox furnish by titration." Notes from 1976 TAPPI Retention and Drainage Short Course, Minneapolis, October, 1976.

[9] Wallace, W. G. "Practical applications of charge analysis in papermaking." Proceedings of the 1992 TAPPI Papermakers Conference, Nashville TN, April 5-8, 1992, pp. 445-456.

[10] Ojala, T. "Charge measurement of different furnishes using polyelectrolyte titration with a streaming current detector." Proceedings of the 1993 TAPPI Papermakers Conference, Atlanta, GA, April 18-21, 1993, 613-626.

[11] Farley, C. A. "Microelectrophoresis as a wet end teaching tool." Proceedings of the 1992 TAPPI Papermakers Conference, Nashville TN, April 5-8, 1992, pp. 509-512.

[12] Patten, P. and Hipple, B. "Understanding and optimizing wet end chemistry through particle charge characterization." Proceedings of the 1992 TAPPI Papermakers Conference, Nashville TN, April 5-8, 1992, pp. 5-12.

(continued)

[13] Schneider, J. and Huang, F. "The use of anionic modified potato starch for ionic charge control." Proceedings of the 1992 TAPPI Papermakers Conference, Nashville TN, April 5-8, 1992, pp. 599-612.

[14] Patten, P. and Lee, D. T. "Charge analyses: powerful tools in wet end optimization." Proceedings of the 1993 TAPPI Papermakers Conference, Atlanta GA, April 18-21, 1993, pp. 555-568.

[15] Medvecz, P. J. "Review of the chemical composition, process implications, measurement techniques, and control strategies for interfering substances in the paper industry." Proceedings of the 1993 TAPPI Papermakers Conference, Atlanta GA, April 18-21, 1993, pp. 451-459.

[16] Glittenberg, D., Hemmes, J., and Bergh, N. "Use of cationic starch in systems containing high levels of anionic trash." Proceedings of the 1993 TAPPI Papermakers Conference, Atlanta GA, April 18-21, 1993, pp. 639-654.

[17] Smith, D. C. "Chemical additives for improved compression strength of unbleached board." Proceedings of the 1993 TAPPI Papermakers Conference, Atlanta GA, April 18-21, 1993, pp. 393-404.

[18] Hand, V., Koethe, J., Kuchibhotla, S., and Scott, W. "An evaluation of two pad-forming paper stock charge analyzers." Proceedings of the 1993 TAPPI Papermakers Conference, Atlanta GA, April 18-21, 1993, pp. 591-598.

[19] Penniman, J. G. "On-line monitoring of microparticulate papermaking process chemistry." Proceedings of the 1994 TAPPI Papermakers Conference, San Francisco CA, April 24-27, 1994, pp. 229-236.

[20] Miyanishi, T. "On-line zeta potential analyses of a fine paper machine and a newsprint machine." Proceedings of the 1994 TAPPI Papermakers Conference, San Francisco CA, April 24-27, 1994, pp. 313-324.

[21] Gossage, R. "On-line treatment control of coated broke slurries." Proceedings of the 1993 TAPPI Papermakers Conference, Atlanta GA, April 18-21, 1993, pp. 551-554.

[22] King, C. A. "Charge and paper machine application." Proceedings of the 1992 TAPPI Papermakers Conference, Nashville TN, April 5-8, 1992, pp. 507-512.

Chapter 7

Coloring Materials

Colorants are used in the manufacture of paper for a variety of reasons, including eye appeal, color coding, and brand identification. The selection of suitable colorants depends upon a number of factors such as end-use requirements, physical and chemical properties, and handling characteristics. An overview of color as a paper property can be found in Reference [1].

Coloring is done to accomplish the following two general goals:

- produce paper having a given color or shade
- produce white paper having a desired tint

Accomplishment of one or both of these two goals is not a simple task. Many process variables can interfere with dyes and it is important to be knowledgeable about the properties of specific dyes used in a given furnish. It is also important to be aware that dyes can adversely interact with other additives — primarily through charge interactions — and interfere with them.

The selection of dyes as the first additive to be covered in the book was guided by their relatively simple chemical structures and the fact that they interact with the components of a papermaking furnish primarily by straightforward ion exchange mechanisms. We will move to increasingly complicated additives in subsequent chapters.

Papermakers utilize a number of different coloring materials to meet their dyeing needs. They fall into the following five categories:

1. Colored pigments
2. Acid dyes
3. Basic dyes
4. Direct dyes
5. Fluorescent dyes

These categories differ according to chemical and physical properties, fastness, affinity for papermaking fibers, effective pH range, and solubility. Each colorant has papermaking advantages and disadvantages. However, as a class, water-soluble direct dyes are the most commonly used dyes in the paper industry. Pigments are used sparingly for internal dying, but find applications in the preparation of colored coatings. We will briefly discuss the most important characteristics of each of the five categories listed above and then turn to the role of dyes and colored pigments in papermaking chemistry.

The history of dyes and colored pigments dates back to the earliest days of civilization when natural coloring substances that could be readily extracted from plants were applied to fabrics. The synthetic colorant industry is usually dated to 1856 when William Henry Perkin accidentally discovered the dyestuff mauve while attempting to simplify the procedure for synthesizing quinine. General reviews of dyeing can be found in References [2] and [3]. Also, anyone wishing to study the subject of color in depth is encouraged to read Reference [4], which is considered by many to be the "definitive" publication. References [5] and [6] review the applications of dyes in papermaking.

In this discussion we will concentrate on the properties and papermaking chemistry of the colorants most widely used in the paper industry.

1. Colored pigments

Colored pigments fall into three classifications: natural inorganic pigments, synthetic inorganic pigments, and synthetic organic pigments. As a class, the third category is by far the most important to the paper industry. For reasons that we will discuss later, colored pigments do not find extensive use for the wet-end coloring of paper in the US (they are more commonly used for this purpose in Europe) but, are applied more often in coating formulations where they are superior to dyes because they do not have a tendency to follow the water into the sheet when coating is applied.

Table 7-1 lists some of the more important properties of colored pigments. The low tinctorial value of colored pigments means that high loading levels are required to achieve strong colors, which leads to a weakening of the sheet. The lack of affinity produces poor retention unless retention aids are used. Alum has often been used for this task, although ordinary filler pigment retention aids will also work. When alum is used, it is necessary to keep the pH of the system below 5.0 in order to keep the alum in an active form for retention.

Table 7-1. Properties of colored pigments

1. Excellent lightfastness
2. Low tinctorial value
3. No direct affinity for fibers
4. Contribute to two-sidedness

2. Synthetic dyes

a. Acid dyes

This is the name given to a large group of dyes for nylon and wool and one or two for paper. Acid dyes are all water-soluble salts (usually sodium or potassium salt form) of colored aromatic organic acids which dissociate in water to form colored anions. Most acid dyes are azo dyes. A typical structure for an acid dye is presented in **Fig. 7-1**.

Acid dyes are similar to direct dyes (see Sec. 1c) and the two groups merge together with no clear boundary. Acid dyes generally contain more acid groups which gives rise to their greater solubility in water, compared to direct dyes. **Table 7-2** lists some characteristics of acid dyes.

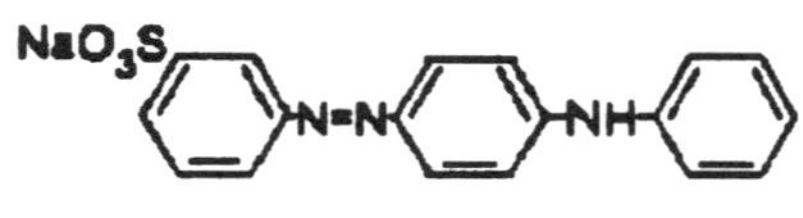

C.I. ACID YELLOW 36
C.I. 13065

Figure 7-1. Typical acid dye utilized for paper coloring

Table 7-2. Properties of acid dyes

1. High water solubility
2. Neutral or slightly alkaline water solution
3. Fair-to-good lightfastness
4. Low affinity for wood fibers
   - dye retention aids must be used
   - white water will usually be colored
5. Contribute to two-sidedness
6. Sensitive to heat (migrate on dryers)
7. Excellent for calender dyeing

b. Basic dyes

Basic dyes are the salts (chlorides, hydrochlorides, sulfates, and oxalates) of color bases. **Fig. 7-2** contains the chemical structure of a typical basic dye. They are soluble in aqueous acid, and this is often the reason for the use of acetic acid to produce concentrated solutions. Basic dyes constitute the only class of dyes generally applied to paper whose members are soluble to any appreciable extent in methyl-, ethyl-, or isopropyl alcohols, as well as compounds having similar solvent characteristics, among which are oils and waxes. **Table 7-3** lists some important characteristics of basic dyes.

Table 7-3. Properties of basic dyes

1. Cationic in nature
2. No affinity for cellulose
3. High affinity for acidic materials such as the lignin present in groundwood and unbleached pulps
   - Retention aids required for retention on bleached wood fibers
   - Can produce "granting"
4. Highest tinctorial values
5. Brightest of all dyes
6. Poor-to-fair light stability
7. Inferior resistance to acid, alkali, and chlorine

c. Direct dyes

Direct dyes are sodium salts of dye acids. Chemically, they are similar to acid dyes, and are known as direct dyes because of their affinity for cellulose, which distinguishes them from acid dyes which have no affinity for cellulose. **Fig. 7-3** contains an example of a direct dye.

The affinity of direct dyes for cellulose fibers can be attributed in part to the low solubility of the materials. In fact, aqueous solutions of the dyes are often highly colloidal rather than true solutions, and some direct dyes may even gel when a solution is left standing. The classification of acid dyes and direct dyes is often arbitrary and dyes are frequently encountered that have more affinity for cellulose than the average acid dye and less than the average direct dye. These are usually classified as acid dyes.

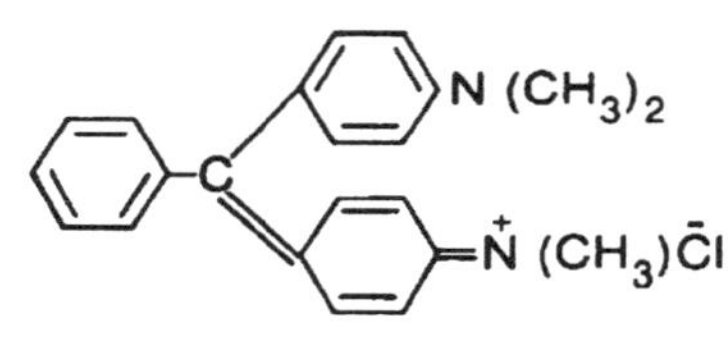

C.I. BASIC GREEN 4
C.I. 42000

Figure 7-2. Typical basic dye utilized for paper coloring

Cationic direct dyes can also be utilized.[7] These dyes retain the planar molecular structure of acid direct dyes, but cationic groups have

Direct Blue 218

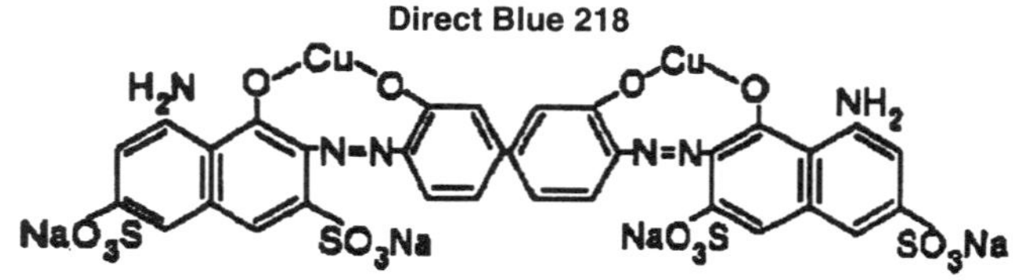

Figure 7-3. Typical direct dye utilized for paper coloring

been incorporated in the structure to increase their affinity for cellulose fibers. An example of one such material is illustrated in **Fig. 7-4. Table 7-4** lists some characteristics of direct dyes.

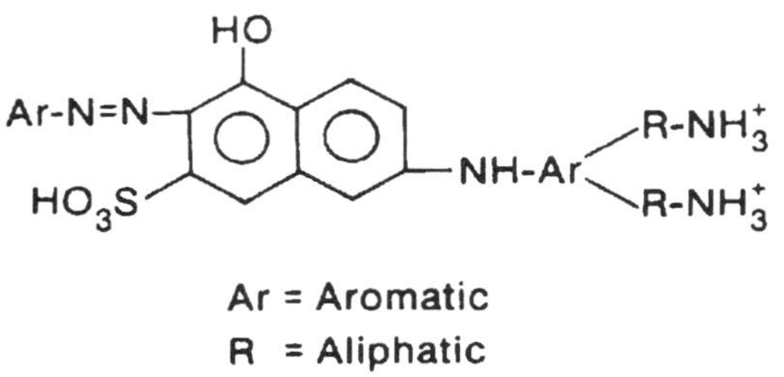

Figure 7-4. Typical cationic direct dye utilized for paper coloring

Table 7-4. Characteristics of direct dyes

1. Strong affinity for cellulose. Retention can be improved by:
   - the use of alum or other retention aids
   - heating the stock after adding the dye
   - adding a cationic group to the acid dye molecule
2. Superior lightfastness
3. Intermediate tinctorial values
4. Tendency to granite if not handled correctly
5. Amenable to a wide range of pH

d. Fluorescent dyes

These dyes are referred to as "paper whites" or "optical brighteners" and are used by papermakers to brighten white papers. They are very effective when used with highly bleached pulps, but are ineffective when applied to unbleached pulps and groundwood. (See below for an explanation of why this is so.)

Fluorescent dyes absorb light in the ultraviolet region of the spectrum (below 370 nm) and re-emit the light in the visible blue range (peaking at 435 nm), giving a fluorescent effect that produces a bright white in daylight, masking the inherent yellowness of the bleached pulp.

Any material that absorbs ultraviolet light will lower the efficiency of fluorescent dyes. For example, lignin absorbs ultraviolet light and the higher the lignin content of the pulp, the less the effect from the brightener. Hence, groundwood pulps and unbleached pulps interfere with the effectiveness of fluorescent dyes.

Titanium dioxide also absorbs ultraviolet light and will reduce the fluorescence of these dyes. Most clays tend to counteract the fluorescence and reduce the effect of the white dye. Fillers such as calcium carbonate and aluminum trihydrate reflect ultraviolet light, thereby enhancing the effect of the dyes. A high pH (above 5.5) also aids in achieving maximum whiteness.

2. Dyes and papermaking chemistry

The relationship between dyes and papermaking chemistry can be considered from two viewpoints. First, there are those factors that affect the actual incorporation of the dye in the sheet of paper, i.e. the retention of the dye. Second, other components and conditions of a furnish and paper machine system can influence the development of color by a dye. We will consider the factors related to dye retention first.

a. Dye retention

Dye retention is closely related to the retention of fines and fillers.[8] This follows from our previous discussion of the relative surface areas of fibers, fines, and fillers, and the fact that additives tend to adsorb onto components according to their respective surface areas. As was discussed above, direct dyes have a natural affinity for cellulose fibers due to their (the dye) low solubility. Whenever alum is added to a slurry containing these dyes, a positive charge is imparted to the dye through interaction with the alum. This increases the attraction of the dye for the anionic fibers.

Other mechanisms assist in the bonding of direct dye molecules to cellulose fibers. You will recall that cellulose fibers have a multiplicity of surface groups that are capable of participating in hydrogen bonds. Direct dyes are all linear and coplanar in structure, i.e. long, flat molecules. They contain groups such as -OH and $-NH_2$ which are capable of hydrogen bonding to cellulose hydroxyl groups. In addition, Van der Waal's forces play a significant role in bonding dye molecules to a fiber surface.

Cationic direct dyes have a strong cationic charge that can interact directly with negative charge sites on fibers. In addition, dye suppliers have also advocated the combination of cationic and anionic direct dyes to produce "synergistic" mixtures that are reported to be more efficient than the individual components.[7]

Acid dyes are similar to direct dyes but lack molecular coplanarity. They also have higher numbers of solubilizing groups that prevent good bonding between the dye molecule and fiber surfaces. The negative charge on the dye is the same as the charge on fibers and there is no electrostatic attraction between them. That is why cationic retention additives must be used with acid dyes.

Basic dyes carry a positive charge which means that they are electrostatically attracted to fibers. The cationic groups react with any negatively charged acid group present in unbleached pulps and mechanical pulps. Bleached pulps do not have a very high concentration of such groups and they are less attractive to the dyes.

In most instances, the papermaker desires to retain as much of an added dye as possible. When a dye is added to a furnish, an adsorption equilibrium is set up between the dispersed fibers and fines and the suspending water. With light shades, the equilibrium is such that almost all of the dye (direct dye) is adsorbed by the fibers. With deep shades it is often necessary to add something to shift the equilibrium toward the fibers. Alum and/or size will do this. Other actions that promote dye retention include heating the stock, adding salt to the dyed pulp, adding alum to the furnish after dye addition, and adding a dye retention aid (a "promoter").

Numerous dye retention aids are on the market as proprietary products. They are usually low molecular weight, high charge density, cationic polymers. Wet-strength resins and cationic starches also impart a positive charge to a dye and improve its retention.

b. Interactions that affect color development

The second type of interaction between papermaking chemistry and dyes includes those interactions that affect the color development obtained from the dye. An example of such an interaction might involve interaction with residual chlorine present in a paper machine system as a microbiological control agent. Unless the residual level of these materials is controlled carefully, they can react with dyes and render them ineffective. **Table 7-5** lists other furnish variables that can affect shade development by dyes.

A discussion of each of these variables is beyond the scope of this chapter. The principal objective here has been to introduce the notion that the performance of dyes on the paper machine is very much affected by the chemistry of the wet-end system.

Table 7-5. Furnish variables affecting dye performance

1. Fiber composition
2. Type and amount of fillers
3. Degree and amount of refining
4. Contact time before the sheet is formed
5. Type and amount of sizing agent
6. Alum and aluminum ion concentration
7. pH, total acidity
8. Type and amount of resins and starches

c. The effect of dyes on other additives

As the above discussion indicates, dyes are charged substances — either cationic or anionic — that can react with other charged additives. Indeed, we saw earlier that a polymer-dye reaction can be used to detect the endpoint of a colloid titration determination. In most instances, dyes bring anionic charges to a papermaking system. Hence, it is critical not to add them close to the addition point of a cationic additive. Thus, the same general precautions must be applied to dyes that are employed with other charged additives in order to avoid undesirable cationic-anionic reactions.

3. Summary

Dyes are important additives in many grades of paper. Since dyes are sensitive to many furnish and process factors, it is critical that the papermaker understand the conditions required to achieve satisfactory performance from the particular dyes being utilized. Equally important is the fact that dyes can interfere with other additives, primarily through ionic charge-type reactions and the papermaker must take care that the order of addition and points of addition of various additives and dyes are compatible with one another.

4. References

[1] Abbott, J. and Scott, W. E. "Appearance Properties." Chapter 4 in Paper Properties: An Introduction, Edition Two, TAPPI PRESS, Atlanta, GA. 1995.

[2] Lips, H.A., Dyeing, in Pulp and Paper: Chemistry and Chemical Techn., 3rd Edn., Vol. III, (J.P. Casey, Ed.), John Wiley, p. 1627-1666 (1981).

[3] Eklund, D. and Lindstrom, T., Paper Chemistry: An Introduction Chapter 10, "Dyes and Optical Whiteners", 265-280, DT Paper Science Publications Grankulla, Finland, 1991.

[4] Judd, D. B. and Wyszecki, G., Color in Business, Science and Industry, Wiley-Interscience, New York (1975).

[5] Colling, F. "The use and mis-use of dyestuffs and fluorescent whitening agents." Chapter 8 in Applications of wet end chemistry, edited by C. O. Au and I. Thorn, Blackie Academic and Professional, pp. 120-138 (1995).

[6] Murray, S. G. "Dyes and fluorescent whitening agents for paper," Chapter 9 in Paper chemistry edited by J. C. Roberts, Blackie Academic and Professional, pp. 132-161 (1991).

[7] Brady, B.P., "Dye Synergism: A Combined Effort," Proc. 1987 TAPPI Papermakers Conf., Atlanta, GA, p. 17, TAPPI PRESS, Atlanta.

[8] Britt, K.W., Some New Approaches to Improving Retention in Paper Sheet Formation, Pulp and Paper 51(4) (1977).

Chapter 8

Dry Strength Additives

In Chapter 7, dyes and other coloring agents were introduced. It was pointed out that, for the most part, dye-fiber interactions involve ion exchange reactions between "small" molecules and particle surfaces (fibers, fines, filler). In this chapter we are going to move up in order of complexity and discuss the interaction of furnish components with polymers. A broad approach has been taken in this chapter and an attempt has been made to touch upon the most important aspects of dry strength additives. For more detailed discussions of specific topics, you are encouraged to read the literature references cited at the end of the chapter.

Before addressing the wet end chemistry subjects important to strength, it will be useful to review the general concepts associated with paper strength.

1. Introduction to the strength properties of paper

Paper has all the mechanical (strength) characteristics of common construction materials. Some of the specific mechanical properties important to the paper industry are listed in **Table 8-1.**

Table 8-1. Common mechanical (strength) properties of paper

- Tensile properties
- Tearing resistance
- Folding endurance
- Bending resistance
- Bursting strength
- Surface strength
- Internal bonding strength
- Compression resistance

Because of the special effect of water on paper strength, it is common to distinguish between the "dry" strength and "wet" strength properties of paper. The two subjects are obviously related, but we will adhere to tradition and treat wet strength as a separate topic in Chapter 9.

Paper strength is affected by many furnish and process variables. On the furnish side, longer softwood fibers produce stronger papers than shorter hardwood fibers. Fillers reduce strength. Alkaline pH conditions produce stronger papers than acid pH. On the process side, increased wet pressing and refining both increase paper strength. These furnish and process variables affect paper strength through their influence on the four basic factors listed in **Table 8-2**.

Table 8-2. Basic factors that influence paper strength

- Individual fiber strength
- Interfiber bond strength
- Number of interfiber bonds (bonded area)
- Distribution of fibers (sheet formation)

Of the above factors, it can be assumed that wet end chemistry will have little effect on individual fiber strength, which is a function of wood species, the pulping and bleaching processes employed to obtain the fibers, and the number of recycles a fiber has undergone.

The distribution of interfiber bonds is strongly affected by sheet formation (the "property" formation), which is influenced by a wide range of mechanical and chemical factors. In general, poor formation adversely affects strength. However, experience indicates that formation must become very poor before significant strength losses occur. Nevertheless, the papermaker always strives to achieve good formation because it strongly affects other important properties.

As mentioned in Chapter 1, wet end chemistry influences formation through its influence on flocculation and drainage. Overflocculation leads to a "lumpy," poorly formed sheet. Slow drainage causes the papermaker to increase headbox consistency which, in turn, makes it more difficult to achieve good formation.

In general, common papermaking practice tries to minimize the adverse effects of wet end chemistry on formation while simultaneously improving it (formation) through mechanical adjustments to the papermaking system. The remaining two factors in **Table 8-2**, interfiber bond strength and bonded area, can be strongly influenced by dry strength additives, as well as other papermaking processes.

2. Achieving paper strength through refining

The papermaker has several approaches available for increasing the strength and number of interfiber bonds. Three of the most important approaches are:

- Refining
- Utilization of dry strength additives
- Wet pressing

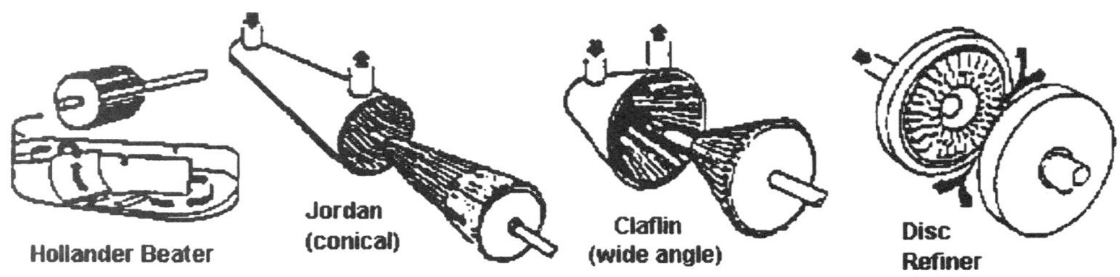

Figure 8-1. Equipment used in refining

Of the three, refining is the most frequently used approach.

Refining is the mechanical alteration of fibers to give them the characteristics required to be formed into sheets of paper or paperboard having the desired properties. This practice was begun in the middle 1700's with the invention of the Hollander Beater. Since that time, refining equipment has evolved and most mills now use disc refiners almost exclusively for this purpose. The general principles remain the same, however.

The Hollander Beater employs a beater roll that rotates in a tub containing the fiber slurry. A stationary bedplate is located beneath the beater roll. As the fibers circulate beneath the beater roll, they get subjected to squeezing and other mechanical forces between the rotating beater bars and the stationary bedplate bars. These forces disrupt internal and external fiber surfaces, leading to fiber cutting and fraying (fibrillation) and the imbibing of large amounts of water. The fibers swell and become more flexible and conformable so that when they come into contact with one another, more interfiber bonds having a greater bonded area are developed. **Fig. 8-1** contains schematic drawings of a Hollander Beater and its successors, conical refiners and disc refiners. Today, few Hollander Beaters are in use and most refining is done with disc refiners. The basic mechanisms apply, however.

a. Effects of refining on paper properties

The dramatic effects of refining on paper strength properties are illustrated in **Fig. 8-2.** There it can be seen that most strength properties increase with increased refining. Tearing resistance is the exception to this rule. Thus, if the customer desires both high tearing resistance and high tensile strength, it is difficult to achieve this need by refining alone.

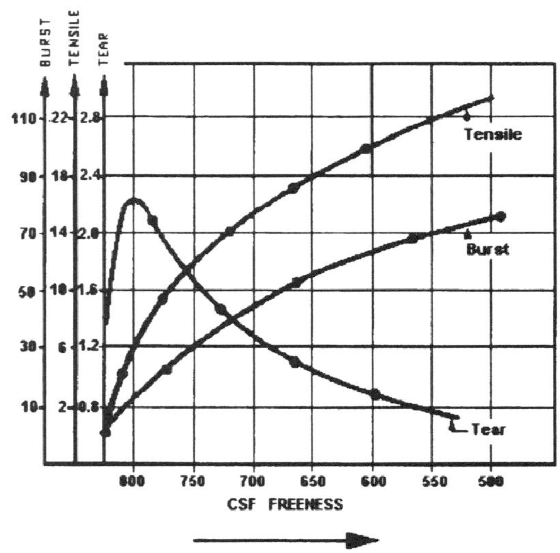

Figure 8-2. Effect of refining on strength properties. Increased refining causes freeness to decrease. Other strength properties, such as stiffness, surface strength, internal bonding, and folding endurance, also increase with increased refining.

b. Effect of refining on interfiber bonding

As stated above, refining affects paper properties primarily through its effect on interfiber bonding. **Fig. 8-3** illustrates just how dramatic this effect can be. In the figure it can be seen that fifty minutes of treatment in a laboratory beater greatly increases the extent to which fibers conform to one another at fiber crossings, leading to a denser, more highly bonded structure.

c. Other properties affected by fiber bonding

Refining affects many other properties in addition to strength. **Table 8-3** lists some of the more important ones.

It is clear that there is a limited number of property combinations that can be achieved through refining alone. The situation actually

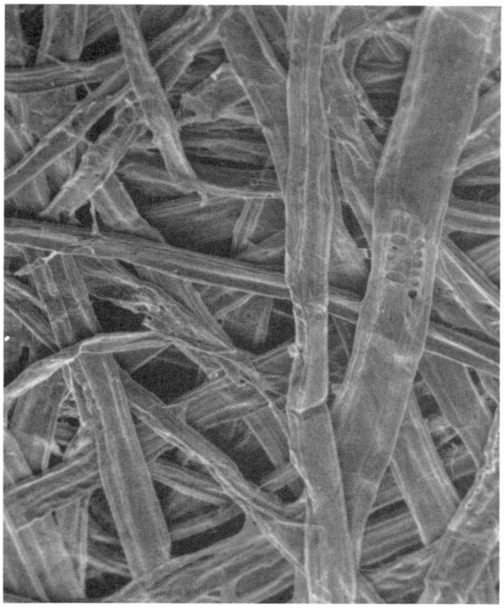

No Refining

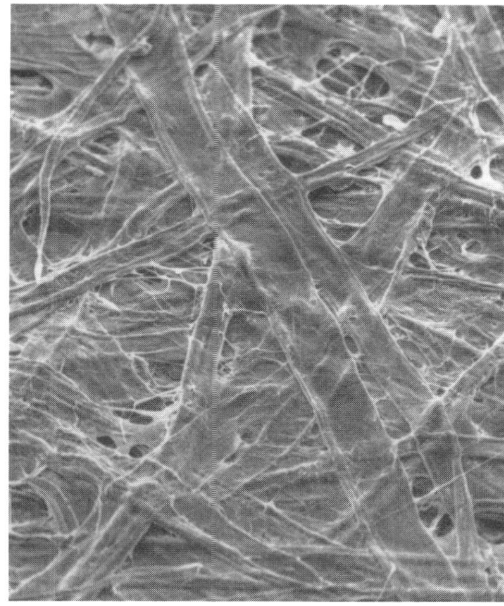

50 Minutes in Valley Beater

Figure 8-3. Effect of refining on interfiber bonding. Fifty minutes of treatment in a laboratory beater produces more conformable fibers that bond together to a greater extent in a formed sheet.

becomes more dramatic when slurry drainage performance, runnability, and cost are added to the equation.

One way to improve the situation is to added certain chemical additives to the furnish, called dry strength additives, that improve paper strength while maintaining other properties. In this manner, dry strength additives increase the combinations of properties available to the papermaker. **Fig. 8-4** provides a typical example of this.

Table 8-3. Effect of refining on properties other than strength

Property	Effect of Increased Refining
Opacity	Decreases
Roughness	Decreases
Permeability (Porosity)	Decreases
Dimensional stability	Decreases
Linting propensity	Decreases
Density	Increases
Formation	Improves
Brightness	Decreases

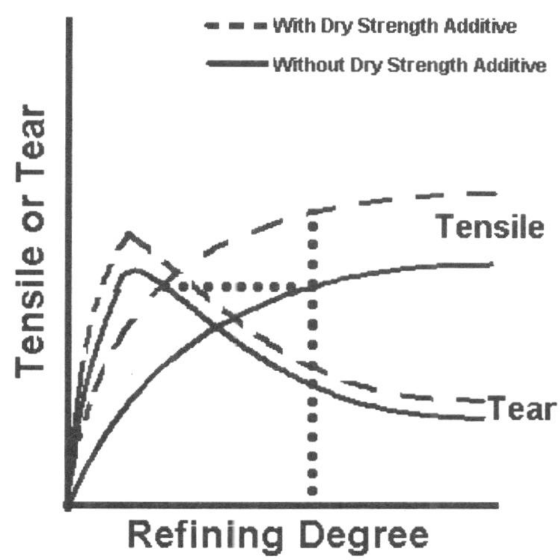

Figure 8-4. Effect of a dry strength additive on tearing resistance and tensile strength. In the figure it can be seen that, for a given refining level and tensile strength, it is possible to achieve a higher tearing resistance when a dry strength additive is utilized. Conversely, addition of a dry strength additive enables the papermaker to achieve higher tearing resistance at a given tensile strength.

3. Common dry strength additives

Many water soluble, hydrogen bonding polymers will act as dry strength additives. In fact, wood fibers contain their own natural dry strength additive in the form of hemicelluloses. It is well-known that removing hemicelluloses from wood fibers makes it more difficult to develop their bonding characteristics. An example of this is the difficulty of achieving interfiber bonding in paper produced from cotton fibers — which have no hemicelluloses.

Table 8-4 lists the three most common commercial types in use today, along with their approximate market shares (based on quantities used).

*Table 8-4. Common dry strength additives**

- Starch derivatives (ca. 95%)
- Gums (mannogalactans, ca. 2%)
- Synthetic dry strength agents (ca. 2%)

**Percentages based on quantities used*

Other substances, such as unmodified and anionic starches, soluble cellulose derivatives, wet strength resins, polyvinyl alcohol, and latex, account for the remaining one percent of dry strength additives. These materials are generally employed in specialty grades of paper.

a. Use of starch as a dry strength additive

Starch derivatives are the most common type of dry strength additive used in the paper industry today. The most recent edition of the TAPPI publication "Commercially Available Chemical Agents for Paper and Paperboard Manufacture" lists twenty-seven different starch dry strength products.[1]

Experience has shown that it is necessary to attach cationic substituents to the native starch structure in order to achieve the high retention levels required in papermaking. Native starch retention is quite low (<40%), which is undesirable from both the economic and environmental standpoints.

In addition to their applications as dry strength agents and retention aids, cationic starches are also used as emulsifying aids for synthetic sizing materials and as the cationic components of multiple-component retention systems. These subjects will be addressed in later chapters.

b. Some general information about starch

Starch is a white granular substance that occurs widely in the higher plants, frequently in the seed (as in corn and wheat), in various roots and tubers (tapioca and potato), or sometimes in the stem piths (sago). Its natural function is to provide a reserve food supply for the plant, sustaining it through winter dormancy and nurturing new plants during germination. Starch has been used in the manufacture of paper since the very early development of the industry, even long before the invention of the paper machine. The total spectrum of starch applications in the paper industry now includes wet end addition for filler and fines retention, paper dry strength improvement, surface sizing, pigment coating, and converting adhesive functions. The sum total of these applications represents more than two-thirds of the total US industrial starch consumption.

c. Chemical structure of starch

Starch is chemically similar to cellulose in that it is a polymer composed of glucopyranose units linked through carbons 1 and 4. **Fig. 8-5** contains a schematic drawing of the amylose starch molecule. The major difference between amylose and cellulose lies in the configuration of the glycosidic bond linking glucose units together. In cellulose, the bond connecting carbon 1 and the bridging oxygen is oriented in the same plane as the glucose ring (beta-link) and the resulting polymer molecule takes on an extended ribbon-like shape. In amylose, the bond connecting carbon 1 and the bridging oxygen is oriented perpendicular to the plane of the glucose ring (alpha-link) and the resulting polymer tends to take on a more coiled, helix-like shape. This difference in molecular structure between cellulose and starch leads to significant differences in their natural functions and in their industrial uses.

The structure depicted in **Fig. 8-5** is a linear molecule. In fact, the vast majority of starches occur as a mixture of linear and branched molecules. The branched fraction is called amylopectin. Amylopectin contains α-1,4 linked chains with branch points via α-1,6 links most commonly and occasionally through carbon 3. On average, there is one branch point for every 18-27 glucose units. **Fig. 8-6** illustrates the structure of amylopectin.

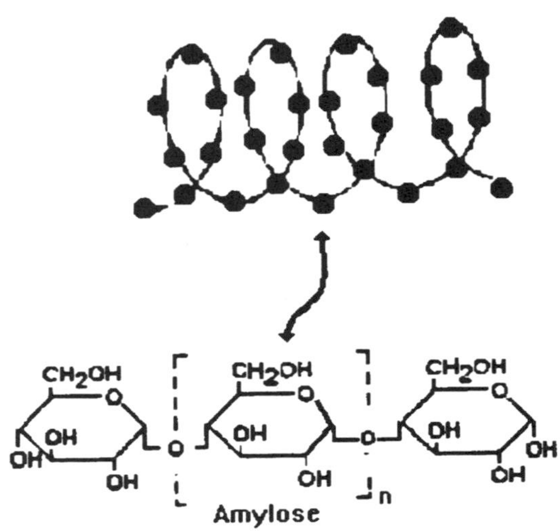

Figure 8-5. Molecular structure of the amylose component of starch. The regular structure of the molecule enables it to take on helical conformations.

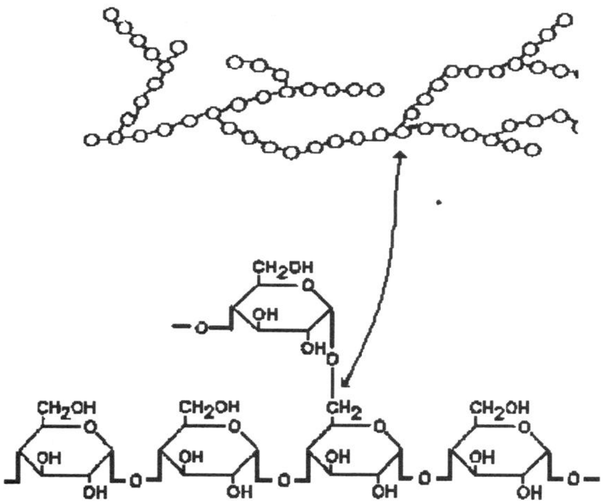

Figure 8-6. Molecular structure of amylopectin component of starch. The molecule contains α-1,4 linked glucose chains with α-1,6 branch points.

The amylose percentages for several common starch sources are listed in **Table 8-5**.

Table 8-5. Amylose contents of several common starch sources

Starch Source	Percent Amylose
Tapioca	17
Potato	20
Wheat	25
Dent corn	24
Waxy maize (corn)	1

The linear amylose molecules have a greater tendency to associate in crystalline aggregates (starch retrogradation) than the branched amylopectin molecules and so high amylopectin starches are preferred for wet end addition.

The molecular weight of unmodified starches depends on their source. For example, Harvey[2] reported average molecular weights of 3-3.5 x 10^6 for two samples of potato starch, while three samples of corn starch ranged from 0.8 x 10^6 to 1.8 x 10^6. These values correspond to degrees of polymerization ranging from 9,000 to 22,000. Within a given starch sample, the higher molecular weights are associated with the amylopectin fraction.

d. Granular structure of starch

Starch is composed of tiny cells or "granules," whose size and shape are characteristics of each botanical variety. The granules are quite small (5 micron-100 micron). It has been calculated that one pound of corn starch contains about three quarters of a trillion granules. Each granule possesses a highly complicated internal architecture. The starch molecules are laid down in concentric layers within a granule in an "onion-like" arrangement. It has been found that amylose molecules are associated in ordered crystalline regions in the granules, while amylopectin is found in amorphous areas. A characteristic feature is the appearance of a "Maltese cross" pattern on each granule when it is viewed in polarized light. The existence of this pattern is evidence that the structure of the granule is intact.

e. Starch dispersion

Unlike many synthetic polymers that are difficult to disperse, starch granules can be slurried easily in cold water without serious intergranular cohesion. To prepare starch for papermaking, the granules must be cooked to bring about either complete or partial disruption.

When starches are suspended in water and the temperature gradually increased, the granules suddenly swell enormously at the gelation temperature characteristic of each kind of starch. this occurs because the solvent power of the water increases with temperature and the granule is penetrated by the liquid causing it to swell. Concomitantly, molecularly dispersed amylose is leached from the swollen granule. The viscosity of the dispersion increases greatly as the swollen granules come into contact and form a continuum throughout the medium. Although the higher temperature reduces the viscosity of the fluid phase, this is more than offset by the increasing concentration of dissolved amylose and the increasing volume of swollen granules. A point is reached in the dispersion process at which the swollen granules begin to disintegrate and the viscosity then decreases. Subsequent cooling of the paste results in rapid viscosity increases and ultimately leads to gelling (set-back). Gelling is related to the amylose content and is a disadvantage in the handling of batches of cooked starch. The degree to which a starch exhibits gelling is related to the molecular weight of its amylose and reaches a maximum when the amylose contains 150-200 glucose units per polymer molecule. Anything above or below this value tends to decrease the propensity toward gel.

In steam-jet cooking the disruption of the granule is accomplished by a nearly instantaneous temperature increase to about 160°C accompanied by intensive mechanical shearing. Such dispersions differ from batch cooks in that all whole granules are gone, and, in the case of corn starch, the dispersions are remarkably resistant to gelling. Jet cooking can lead to a decrease in average molecular weight of the starch. Since the ability of starch to act as an effective interfiber bonding agent for paper is directly related to both the degree of molecular dispersion and the molecular weight, the balancing of those crucial factors in cooking is of considerable importance for optimizing performance.

f. Starch modifications

Unmodified starches are available that differ in properties such as amylopectin-to-amylose ratio, granule dimensions, gelatinization temperature, and the molecular weights of the component fractions. In addition to these variables, modified starches greatly extend the number of potential wet-end starch additives available for a given use. Much research has been carried out that has resulted in the development of new applications for starch in papermaking. Physical, chemical, and genetic modifications and combinations of these have increased the useful properties of starches. **Table 8-6** lists some of the methods applied.

Table 8-6. Starch modification methods

- Physical modifications
- Fractionation into amylose and amylopectin components
- Thermomechanical conversion
- Acid hydrolysis
- Chemical modifications
- Oxidation
- Derivatization
- Enzyme conversion

Thermomechanical conversion, acid hydrolysis, and enzyme conversion all lead to a decrease in the molecular weight of the starch and a corresponding decrease in the solution viscosity of starch solutions. This makes them more manageable and able to be pumped.

g. Cationic starches

Earlier in this chapter it was pointed out that in order to gain good starch retention on papermaking fibers, it is necessary to attach cationic substituents to the starch molecule. Cationic starches are applied in a wide range of paper grades because of their effect on strength, filler retention, and drainage. These starches have the advantage of having a high degree of dispersion and high retention by wood fibers.

Starches are cationized with either tertiary or quaternary amine groups. **Fig. 8-7** illustrates the structures of some common cationic groups that can be attached to starch. The most common tertiary reagent is diethylamine ethyl chloride hydrochloride. The most common quaternary reagent is epoxy propyl trimethyl ammonium chloride.

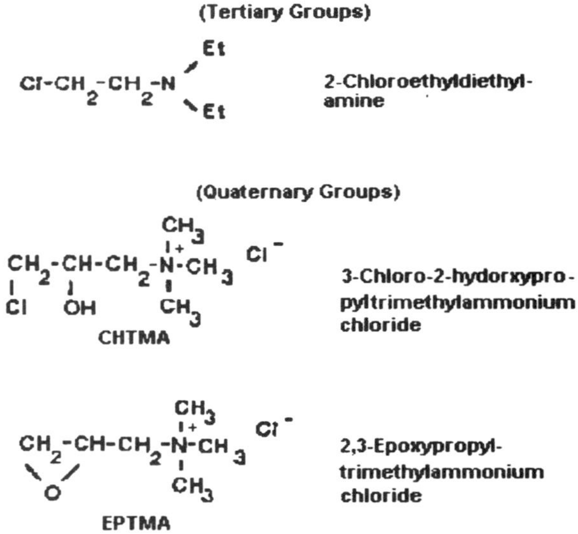

Figure 8-7. Some common reagents for imparting cationic character to starch

The cationic charge of the former group exhibits a degree of pH dependence, with decreasing charge at pH values above 7.0. Quaternary cationic groups retain their complete charge at all pH levels (**Fig. 8-8**).

Typical commercial wet end cationic starches have a degree of substitution (d.s.) of cationic groups ranging from 0.01 to 0.05. In other words, 1-5 of every 100 anhydroglucose units contain a cationic group. This represents a large amount of charge per molecule. For example, a starch of molecular weight 4 million and a d.s. of 0.01 will have approximately 250 cationic charges per molecule. A d.s. of 0.02 would contain ca. 500 charges per chain.

Cationization imparts high dispersability and solubility to starch. The internal cohesive forces of the granule are diminished and the starch cooks more rapidly than native starch. Cationic starches differ according to parent starch source, substituent groups, degree of substitution, distribution of cationic groups, and molecular size and shape.

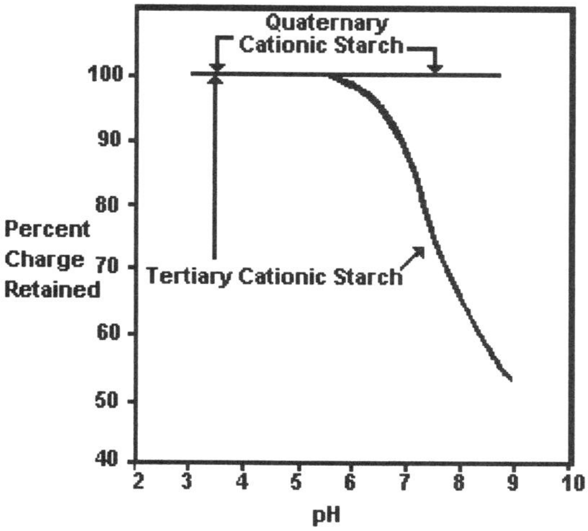

Figure 8-8. Effect of pH on the ionization of tertiary and quaternary cationic groups

Alternatives to cationic starches of fixed cationic charge density are the variable products available from continuous cooking and simultaneous cationization of starch by reactive chemicals, such as various amines and cationic resins. This practice is not widespread in the paper industry.

h. Amphoteric starches

Amphoteric starches contain both cationic and anionic groups. Amphoteric starches are useful when a furnish would become overcationized if a cationic starch were utilized. They are also capable of tying up cations, such as Ca^{+2} and Mg^{+2}.

When amphoteric starches are manufactured, a second reaction (after cationization) is carried out which may be alkaline etherification to produce sulfate or carboxyl groups, or an esterification to produce a phosphate derivative. The latter is most commonly used commercially for paper additives.

4. Mechanism of strength development

Dry strength additives, such as starch, are thought to act by several mechanisms. First, starch enhances interfiber bonding because its free glucose hydroxyls participate in hydrogen bonding with fiber surface cellulose molecules and supplement the number of hydrogen bonds that would naturally occur in the bonded area between two fibers. This is a "chemical hydration" of the fiber system. Refining hydrates the fiber, but in the process the fibers are generally weakened and shortened. It is for this reason that the decrease in tear strength that normally accompanies refining can be alleviated by the use of dry strength additives. The increase in bonding from chemical hydration does not affect individual fiber strength.

A second mechanism involves the improvement of sheet formation. This provides more uniformly distributed fiber-fiber bonds. A third, less important mechanism, involves the improved web consolidation that occurs when dry strength agents enhance fines retention and sheet dewatering.

5. Interactions between cationic starch and papermaking furnish components

In order for starch to participate in interfiber bonding, it must first adsorb onto fiber surfaces. We have already stated that for starch to be retained in high quantities it must have a cationic charge. This is logical since fibers and many other papermaking particles become anionic when suspended in water. Marton [3,4] has demonstrated that starch retention by fibers is intimately tied to their carboxyl content (hence, anionic nature). The higher the carboxyl content, the greater the amount of starch adsorbed. Other anionic furnish components also adsorb starch readily. **Fig. 8-9** illustrates typical results reported by Marton. [4]

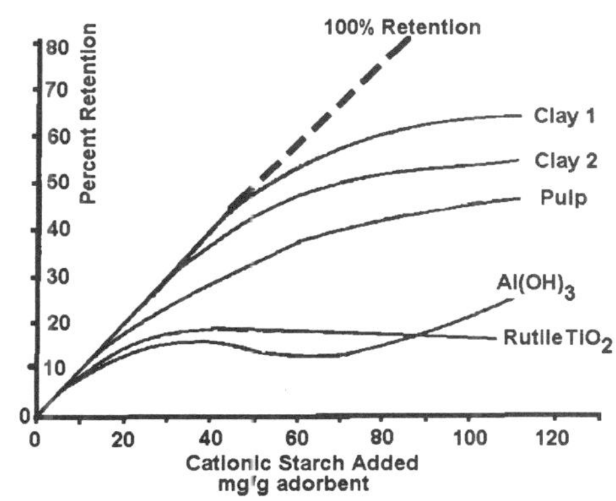

Figure 8-9. Adsorption of cationic starch by various furnish components pH 5 [4].

It has also been shown that cationic starch adsorption is essentially irreversible, with 85% of adsorbed starch remaining on the fiber with prolonged exposure to hot water.[3] It can only be redissolved by treatment with strong acid.

Starch does not distribute itself equally among all components when added to a mill furnish. In general, it tends to adsorb more on high specific surface area components. Thus, fiber fines and fillers will adsorb relatively more starch than fibers. The practical ramification of this is that less starch is available to enhance interfiber bonding when high levels of fines and fillers are present.

a. Effect of starch charge density and dosage levels on adsorption

Charge density and dosage levels play important roles in starch adsorption. As stated earlier, neutral natural starch retains to a very low degree on fibers since there is very little attraction between the two substances. Adding a cationic charge to the starch molecule creates attractive forces between starch and the anionic fiber surface and adsorption rates increase, as does the amount of starch that can be adsorbed.

As the degree of substitution of the starch increases, however, a point is reached where adsorption reaches a maximum and then declines.[5] This point is a combined function of charge density and dosage level. At low degrees of substitution (d.s. ca. 0.020), adsorption increases with increased charge density due to the increased attraction between the starch molecules and fiber surface. The adsorption maximum occurs when the charge density becomes so high that the fiber gets overcharged and the attraction between the starch molecules and the fiber surface declines. The higher the dosage rate, the sooner this point is reached. Papermakers, then, must pay close attention to both the dose level and starch charge density when using this material in order to achieve high efficiencies. This is especially true in view of the wide range of starch addition levels employed in papermaking (0.25%-2.5%).

The starch adsorption capacity of fibers can be increased by increasing the anionic character of the surface. For example, increasing the pH from 4 to 8 will greatly increase the starch adsorption capacity of the fibers. This observation is of relevance when papermakers convert from acid to alkaline papermaking.

b. Final remarks about starch adsorption

We have pointed out that fiber surface area, fiber surface charge, starch charge density, and starch addition levels play key roles in determining the adsorption (hence, retention) of starch by papermaking components. Other factors, like slurry consistency, mixing conditions, retention time, and competition from other additives will also play roles. Since rapid and complete retention of starch (or any dry strength additive) is the most important aspect of the efficient employment of these additives, it is critical that papermakers understand how these various factors act to enhance or degrade the final performance of the additive.

6. General considerations when using starch as a dry strength additive

It is clear that in order to use starch effectively, the papermaker must provide conditions that will promote the rapid and complete adsorption of starch on fibers (in preference to fines and fillers) and minimize adverse reactions with other additives or interfering substances. A balance must be gained between starch adsorption and charge density and addition level. If too much charge is added, it will still be possible to adsorb sufficient starch to obtain the desired strength response, but unadsorbed, dissolved starch will build up in the white water system and lead to environmental and other problems. **Table 8-7** lists some "rules-of-thumb" that apply to starch application.

Table 8-7. General rules for adding starch to a furnish

- Add at a point of good mixing.
- Avoid addition points that are close to other additives that may not have time to become well mixed prior to starch addition.
- Do not add starch close to an oppositely charged additive.
- Add after refining whenever possible.
- Allow sufficient time for the starch to adsorb onto the furnish components prior to sheet formation.
- Avoid exceedingly high addition levels. Anything above 2-2.5% will not be effective.

Finally, since starch adsorbs to different degrees on different furnish components, the adsorption is essentially irreversible, and starch adsorbed onto fibers provides the best strength gains, then selection of an addition point that maximizes the opportunity for adsorption onto fibers is particularly important.

7. Polyacrylamide dry strength resins

Polyacrylamides are also used as dry strength additives in addition to their widespread application as retention and drainage aids.[6] These water soluble polymers contain primary amide groups (**Fig. 8-10**) that can form hydrogen bonds with surface cellulose molecules in fibers and improve interfiber bonding.

Like cationic starch, polyacrylamide dry strength additives enable the papermaker to achieve combinations of paper properties that cannot be gotten through refining alone. For example, strength properties can be improved without affecting sheet bulk or appearance properties.

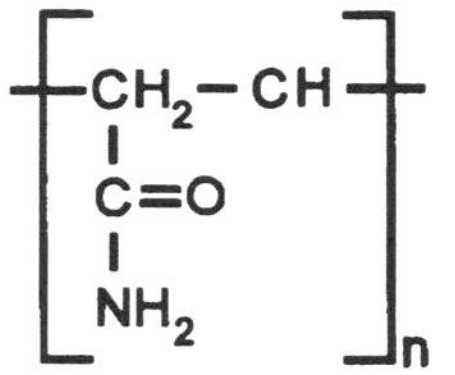

Figure 8-10. The chemical structure of polyacrylamide. The primary amide group can form many hydrogen bonds with cellulose molecules in the surface of fibers and enhance interfiber bonding.

a. Anionic and cationic polyacrylamides

The polyacrylamide structure shown in **Fig. 8-10** is a nonionic material that has very little attraction to papermaking fibers and cannot be retained to any significant degree when added to a furnish. Consequently, it is necessary to incorporate some kind of charged group in the polymer structure to make it useful in papermaking.

Anionic polyacrylamide — Anionic polyacrylamide was first produced back in the 1950s by copolymerizing acrylamide with acrylic acid.(To get acrylic acid, replace the amide group, $CONH_2$ in **Fig. 8-10** with COOH.) This introduced an ionizable carboxyl group into the polymer chain. The same result can be gotten through alkaline hydrolysis of the acrylamide amide group.

As we discussed in Chapter 3, carboxyl groups are weak acids and their extent of ionization depends upon pH conditions. Specifically, at pH levels found in acid papermaking (4.0-5.0), only a fraction of the carboxyls are ionized, and an anionic polyacrylamide is not fully charged under those conditions. As the pH is increased, the ionization equilibrium shifts and above pH 7.0 the group is essentially 100% ionized. Hence, under alkaline papermaking conditions (pH 8.0-8.5) anionic polyacrylamides are completely charged. (See **Fig. 3-7**, for the exact relationship between pH and the degree of ionization of carboxyl groups.)

Cationic polyacrylamide — Since anionic polyacrylamides are negatively charged, they are not directly attracted to papermaking fibers. A cationic substance, such as alum, must be used in conjunction with them to promote their retention. To avoid the need for a cationic promoter, it is possible to incorporate cationic groups directly into the polyacrylamide backbone. This has been done since the late 1960s and these materials have now become the most common form of dry strength polyacrylamide.

Cationic polyacrylamides are produced by copolymerization with cationic monomers or by modifying some of the amide groups to cationic groups. **Fig. 8-11** contains examples of typical cationic monomers employed for this purpose and **Fig. 8-12** illustrates a cationic polyacrylamide structure produced from the copolymerization of acrylamide and METAMS.

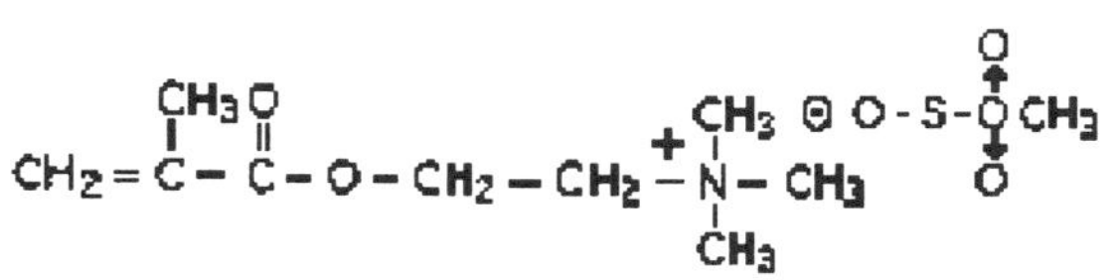

Methacryuloyloxyethyl trimethyl ammonium methosulfate (METAMS)

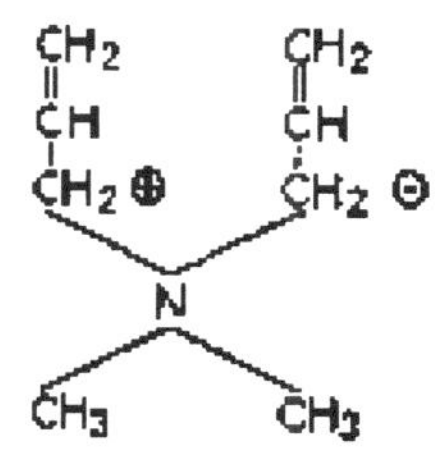

Dimethyldiallyl ammonium chloride (DMDAAC)

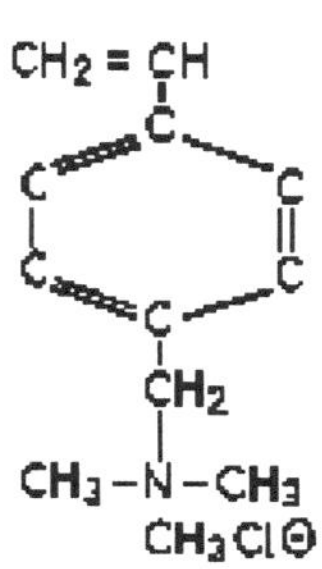

Vinyl benzyl trimethyl ammonium chloride (VBTAC)

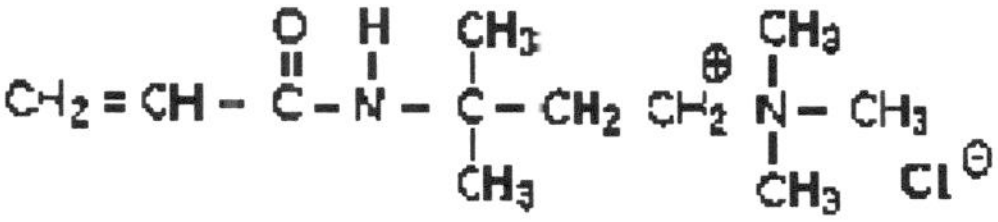

3-Acrylamido-3-methyl butyl trimethyl ammonium chloride (AMBTAC)

Figure 8-11. Typical cationic monomers used to produce cationic polyacrylamides for the paper industry

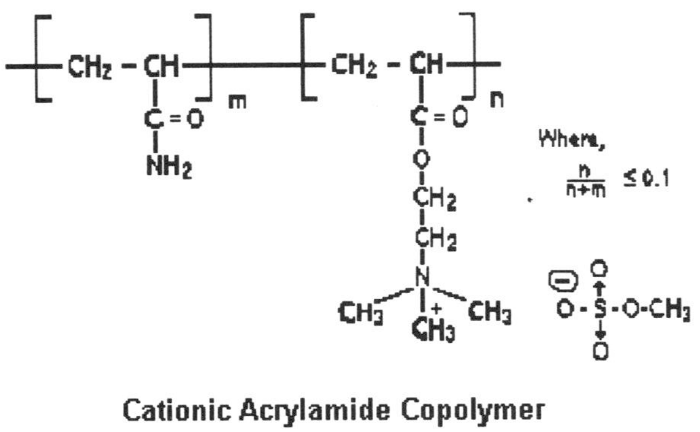

Figure 8-12. Cationic polyacrylamide produced by copolymerization of acrylamide and METAMS. Variable 'm' will be greater than or equal to ten times the value of 'n'.

When used as dry strength additives, typically 10% of the monomers will be charged (or have the capability of being charged). Unlike the anionic polymer, cationic polyacrylamides are completely ionized over the entire pH range of interest to papermakers.

Polyacrylamide-based dry strength additives typically have molecular weights between 100,000-500,000. This range is low enough that the polymers will not bridge between particles and cause flocculation, and high enough to retard migration of the polymer into fiber pores with its concomitant loss of activity.

b. Practical considerations in the use of polyacrylamide dry strength additives

Polyacrylamide dry strength polymers are supplied as ready-to-use aqueous solutions (10-20 percent) or as water soluble powders that must be dissolved prior to use. No other preparatory steps are necessary. They can be added to thick or thin stock. For best results, they must be added at a point of good mixing.

Addition rates of 0.1%-0.5% are adequate for most uses. Excessive use can overcationize the furnish and reduce the effectiveness of other cationic additives. All of the "rules" listed in **Table 8-7** for cationic starch apply equally well to cationic polyacrylamide dry strength resins.

8. Water soluble gums as dry strength additives [7]

Water soluble vegetable gums, such as locust bean gum and guar gum, began to be used as dry strength additives by the paper industry just prior to World War II. They represented some of the original "beater adhesives" used by the industry.

These highly hydrophilic polymers have chemical structures (**Fig. 8-13**) which are similar to cellulose, which enables them to participate in extensive hydrogen bonding with fiber surfaces. The natural materials are nonionic and are not retained by fibers to any extent. Conse-

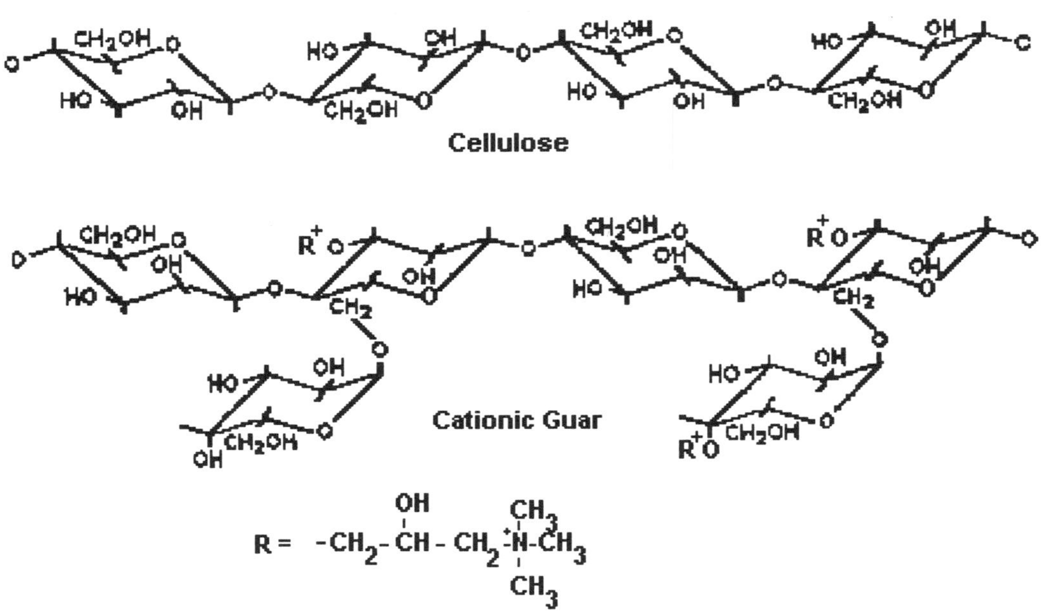

Figure 8-13. Molecular structure of guar gum. The main chain of the structure consists of b-1,4 linked mannose sugars and is structurally similar to cellulose. Single sugar a-1,6 linked galactose side chains also occur frequently.

quently, successful commercial products all have cationic groups attached to the main chain, which increases the attraction between a gum and fibers and results in improved polymer retention. The combination of high retention and effective dry strength enhancement makes it necessary to add only 0.1%-0.35% of the material in most instances.

9. Summary

This concludes our overview of the wet end chemistry of dry strength additives. In this chapter we reviewed the basic strength properties of paper and described how they are achieved. The role of dry strength additives in enabling the papermaker to obtain unique combinations of properties through the use of dry strength additives was described. The chemical structures of the most common dry strength additives were described and the critical wet end chemistry factors that determine their effectiveness were discussed. The critical role of polymer adsorption onto fiber surfaces was emphasized, as well as the practical wet end chemistry factors that operate.

10. References

[1] "Commercially Available Chemical Agents for Paper and Paperboard Manufacture." Fourth Edition, John Farewell, Editor. 83 pages. TAPPI PRESS, Atlanta, GA. 1990.

[2] Harvey, R. D. "The effect of wet end starch in neutral/alakaline papermaking systems." Proceedings of TAPPI Neutral/Alkaline Papermaking Short Course, Orlando, FL. October 16-18, 1990.

[3] Marton, J. and Marton, G., Tappi Journal, 59(12):121(1976).

[4] Marton, J. Tappi Journal, 63(4): 87(1980).

[5] Lindstrom, T. and Eklund, D. "Dry Strength," Chapter 3 in Paper Chemistry: An Introduction, DT Paper Science Publications, Grankulla Finland, 1991, p. 77.

[6] Reynolds, W. "Acrylamide based polymers for dry strength improvement of paper." Chapter 6 in Dry Strength Additives. E. W. Reynolds. TAPPI PRESS, Atlanta, GA, 1980, 188 pages.

[7] Werdouschegg, F. M. K. "Natural gums as dry strength additives." Chapter 3 in Dry Strength Additives, edited by W. F. Reynolds. TAPPI PRESS, Atlanta, GA, 1980, 188 pages.

[8] Marton, J. "Dry-strength additives." Chapter 5 in Paper Chemistry, edited by J. C. Roberts, Chapman & Hall, New York, 1991, pp. 63-75.

Chapter 9

Wet Strength Additives

1. Introduction

In this chapter, we will introduce the most important aspects of wet strength chemistry. Some excellent comprehensive reviews of wet strength have been published recently (See References [1-5]). Reference [2] contains a particularly extensive literature survey.

Wet strength additives, like dry strength additives, are polymeric materials that must be adsorbed by furnish components in order to be effective. In addition, all common wet strength resins must also undergo a curing step before they impart wet strength to paper. Thus, we add another layer of complexity in wet end chemistry interactions.

Paper is formed by filtering and thickening fibers and other particles from a water slurry and pressing and drying the formed web. When the sheet is consolidated in the press section, a few bonds form between fibers. However, the majority of the interfiber bonding occurs when the water is removed in the dryer section and surface tension forces bring the fibers into very close contact with one another.

Since hydrogen bonding plays a central role in interfiber bonding, and water molecules are capable of participating in extensive hydrogen bonding interactions, it is not surprising that paper loses most of its strength when it is brought into contact with water and fiber-fiber bonds are replaced by fiber-water bonds. Non-wet strength paper typically retains only 2-10% of its original dry strength when saturated with water. This property is a liability for a number of paper uses.

It is estimated that 5-10% of all paper and paperboard is given wet strength when it is manufactured. **Table 9-1** lists a number of paper grades that require some degree of paper strength in the presence of water.

2. Definition of wet strength paper

It is possible to produce papers that will retain more than 50% of their dry strength when saturated with water. Most wet strength grades will fall into the 20-40% category, however, and there is some general agreement that paper whose wet tensile strength is more than 15% of its dry tensile strength should be considered to be wet strength paper. (Wet strength is often expressed as the "wet-to-dry" ratio of a strength property. Tensile strength is a common strength property used fo this purpose.)

Wet strength papers are also classified in terms of their wet strength permanence. Non-wet strength paper loses its strength within seconds of being saturated with water. Some wet strength treatments will only slow down this strength loss and such papers are said to have "temporary" wet strength. Other additives give "permanent" wet strength in that under normal use conditions their wet strength properties remain intact indefinitely.

Table 9-1. Typical grades of paper that require wet strength

- Paper towels and other absorbent wipes
- Wrapping and bag papers, sack papers, carrier bags
- Folding boxboarc carrier stock
- Photographic paper, filter paper
- Milk carton raw stock
- Map papers
- Currency papers

3. Wet strength mechanisms

Dunlop-Jones points out that in order for paper to retain its strength when wet, it is necessary to take one or more of the following actions:[2]

- Add to, or strengthen, existing bonds
- Protect existing bonds
- Form bonds that are insensitive to water
- Produce a network of material that intermingles with fibers

Wet strength additives are believed to achieve the above phenomena by the following mechanisms.

Create a cross linked network within and around fibers that inhibits swelling and water sorption and protects existing interfiber hydrogen bonds. This mechanism is sometimes referred to as a *homo-cross linking* mechanism.

Create new, water resistant, interfiber bonds (covalent, acetal, hydrogen bonds) that cross-link fibers. This mechanism has been called the *reinforcement*, *new bond*, or *co-cross linking* mechanism. (Note that the two mechanisms are not mutually exclusive.

4. The need for curing wet strength additives

In Chapter 8 we discussed dry strength additives and pointed out that it was critical that they be retained on fibers with high efficiency.

If this is accomplished, then it can be expected that a good strength response will follow. This is not necessarily the case with many wet strength resins, where it is possible to have high retention and poor wet strength performance.

The reason for this is that most wet strength resins require a curing stage before they become effective. Even though curing does not occur in the paper machine wet end, curing is influenced strongly by the chemical characteristics of the sheet and, because of this, curing can be considered to be a wet end chemistry matter. We will address this matter below with reference to each common wet strength resin.

5. Wet strength resins

The earliest approaches to wet strengthening paper involved heating it to high temperatures or parchmentizing it in dilute sulfuric acid. Later, in the 1930s, it was discovered that certain water-soluble synthetic resins could impart wet strength to paper when added to the furnish and cured on the paper machine. Since that time, there has been a rapid growth in the use of wet strength resins and now more than $100 million worth of these materials are used every year by the US paper industry.

To be an effective wet strength resin, a material must have the following four properties:

- It must be polymeric and provide mechanical strength to protect fiber-fiber bonds against swelling and disruption.
- It must be cationic and provide the attraction to negative cellulose fibers needed for rapid and complete retention.
- It must be water-soluble or water-dispersible to insure uniform distribution throughout the furnish.
- It must be chemical network-forming (usually thermosetting) and provide paper with resistance against swelling by water.

While many different substances have been used as wet strength resins, the four materials listed in **Table 9-2** have found the most applications.

Table 9-2. Most commonly used wet strength resins

- Urea-Formaldehyde (UF)
- Melamine-Formaldehyde (MF)
- Polyamide-Epichlorohydrin (PAE)
- Polyamine-Epichlorohydrin

6. Urea-Formaldehyde Resin

a. Chemical structure

The first wet strength resins evolved from an early practice of impregnating paper with formaldehyde. Developments included the reaction of urea with formaldehyde to produce dimethylolurea (**Fig. 9-1**), which did impart wet strength, but whose low solubility made it impractical for wet end addition.

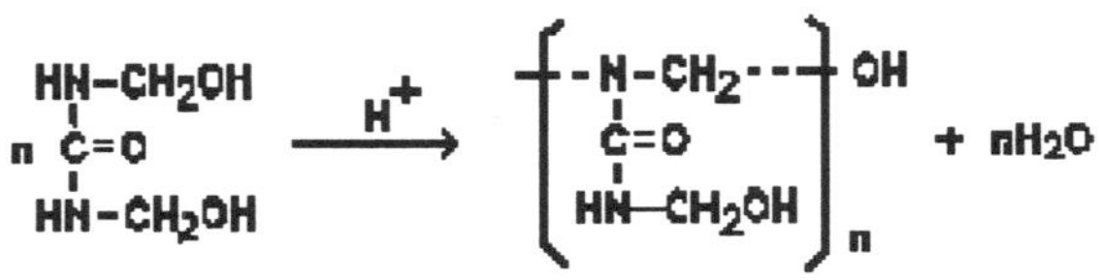

Figure 9-1. Reaction of urea and formaldehyde to produce dimethylolurea

Further research concentrated on producing a polymeric material. The primary reactions involved are illustrated in **Fig. 9-2**. The resin was still not soluble and this problem had to be overcome before urea-formaldehyde resins (UF resins) became viable wet strength additives.

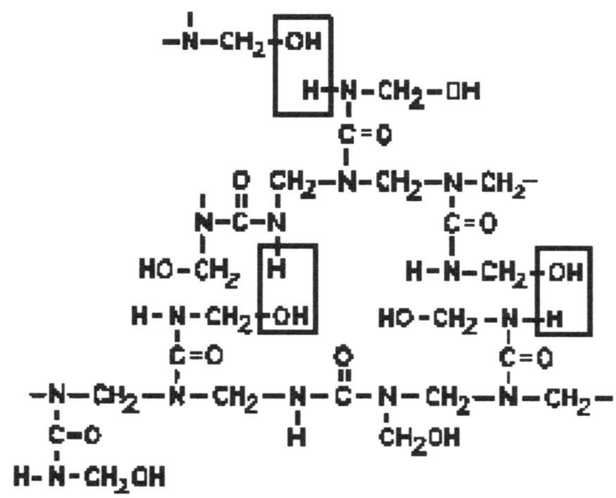

Figure 9-2. Cross-linking of dimethylolurea to form a resin that imparts wet strength to paper

A water-soluble, anionic material was next produced by incorporating sodium bisulfite or glycine in the product. This resin could be precipitated onto fibers by alum. Finally, it was discovered that incorporation of diethylenetriamine or other polyfunctional amine in the resin structure would produce a cationic polymer that was naturally attracted to pulp fibers and retained well.

UF resins are delivered to the mill as 25-35% solutions that have been polymerized to a small degree. They continue to polymerize on standing and must be used before they lose their effectiveness due to overpolymerization.

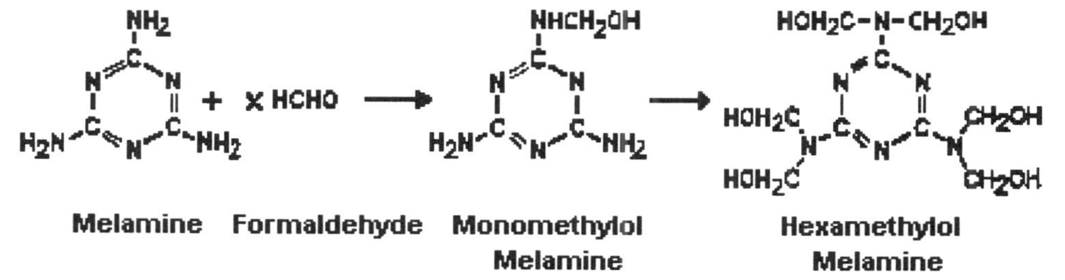

Figure 9-3. The reaction of melamine with formaldehyde to produce methylol melamines, the first step in the production of MF wet strength resin. Trimethylol melamines were found to be most effective.

b. Use of UF resins in the paper mill [3]

UF resins should be filtered and diluted to about one percent solids prior to addition. Since they are acid curing, alum or some other acidic material must be present. The headbox pH of a machine using UF resins must be between 4.0-4.5 and the pH of the produced paper must be in the same range.

Urea-formaldehyde resins are used in amounts of 0.5-3.0 percent, dry basis, depending on the wet strength requirements of the paper being made. It is usually added at the fan pump. When rosin size is used, it is essential that the alum-rosin reaction occur first, before the cationic UF resin has an opportunity to react with rosin to produce a foamy complex and deposits.

UF wet strength paper does not usually reach its final cured state on the paper machine and it must receive an accelerated cure (5-15 minutes at 104°C) prior to testing. It is important that a test specimen be completely saturated with water. This may require an extended soaking time or the use of a wetting agent if the paper is sized.

Fresh UF wet strength broke can be repulped without difficulty. Fully cured UF resin must be treated at pH 3-4 at 65°C or higher under repulping conditions in order to break down the resin.

Concentrated UF resin is highly cationic and will react strongly with any anionic material that it contacts. Dilute resin can tolerate mildly anionic materials but will react with strongly anionic materials. Both concentrated and dilute solutions polymerize rapidly at pH values much below 6.0.

Due to the health hazards associated with free formaldehye, the use of UF resins by the paper industry has been much reduced in recent years. Whenever they are used, strict adherence to rigorous safety standards is mandatory.

7. Melamine-formaldehyde resin [2,3]

a. Chemical structure

The second resin patented for use in the paper industry was melamine-formaldehyde (MF), which was first used in the early 1940s. The first step in the preparation of MF resin is the reaction of formaldehyde with melamine to produce methylol melamines (**Fig. 9-3**). Trimethylol melamine was found to be the most effective product which, when treated with hydrochloric acid, produces a highly cationic colloidal suspension that is very substantive to papermaking fibers. Commercial preparations containing a range of formaldehyde-to-melamine ratios are available.

Several reactions are possible, but it is believed that the formation of ether and methylene cross-links are important in the development of wet strength (**Fig. 9-4**). High temperature and low pH promote cross-linking reactions.

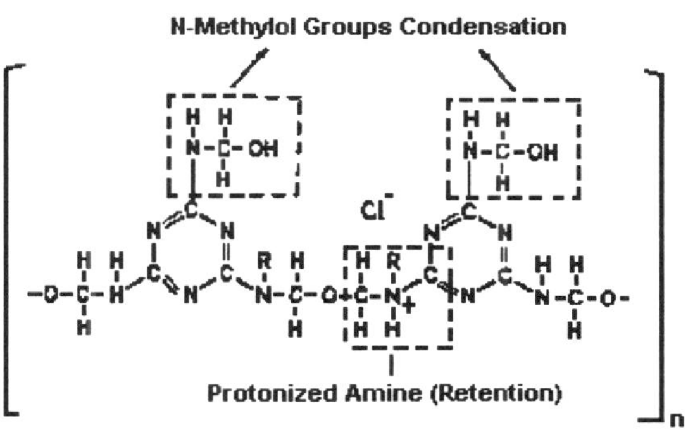

Figure 9-4. Potential cross-linking sites on the melamine formaldehyde colloid. Ether and methylene linkages are thought to be important in wet strength development.

b. Use of MF resins in the paper mill [3]

MF resins are delivered to the mill in two forms, a dry powder or a ready-to-use acid colloid form. The powder must be dissolved in dilute acid and aged under specific conditions before use. Both forms take on a strong cationic charge under acid conditions. MF resin makedown and handling details are described in Reference [3].

MF resin addition levels vary from less than 1 percent, based on dry fiber weight (e.g. toweling), to 5 percent (very high wet strength papers). For most grades, 1-3 percent provides satisfactory results.

MF resins should be added as close to the headbox as possible (one percent resin solution), while still allowing time for uniform distribution and adsorption. Avoid subjecting the material to refiners and Jordans. Optimum addition points vary widely from mill to mill.

When properly used, other additives such as sizing agents, alum, and starch do not interfere with MF resins. However, when incorrect addition sequences or addition point placements are employed, reactions can occur between MF resin and rosin or starch that result in machine deposits and spots in paper.

Stock pH plays a particularly crucial role in the resin cure attained on the paper machine. Within the normal papermaking pH range of 4.5-5.3, MF resin cures rapidly and should give adequate wet strength after normal drying on the paper machine. Machine pH's in the 5-6 range may result in incomplete cure. This may be checked by heating samples in an oven at about 120°C for 10 minutes and checking for wet strength improvement. If improvement is noted, then one should expect additional cure to occur as the paper is stored.

High sulfate levels in the stock (> 200 ppm) have been found to be detrimental to MF resin efficiency, although a small amount of sulfate (25-100 ppm) has been found to be beneficial.

MF resin treated paper should be repulped directly off the paper machine, whenever possible, since it is often not fully cured. High temperature (75-90°C) and low pH (2-3% alum) give the best results.

8. Neutral or alkaline curing polyamide wet strength resins

a. Chemical structures [1,2,5]

Approximately thirty years ago, wet strength polyamide (PAE) resins that could be used under neutral or alkaline papermaking conditions were invented. These materieals are now the most common wet strength additive. A multi-step synthesis sequence was followed, similar to that for the production of nylon. In the first step (**Fig. 9-5**) a dibasic acid is reacted with a triamine to give a polyamide.

Next, the polyamide is treated with epichlorohydrin to alkylate secondary amine groups and produce tertiary aminochlorohydrin groups (**Fig. 9-6**). These groups self-alkylate to form 3-hydroxyacetidinium groups, which are responsible for the cationic character and reactivity of the wet strength resin. The resultant product consists of a relatively low molecular weight, cationic polyamide (<100,000) having many reactive side groups.

PAE resins are believed to undergo at least two kinds of reactions that contribute to wet strength. One reaction involves the reaction of an azetidinium group in one molecule with a secondary amine group in another molecule, to produce a cross-link between the two molecules.

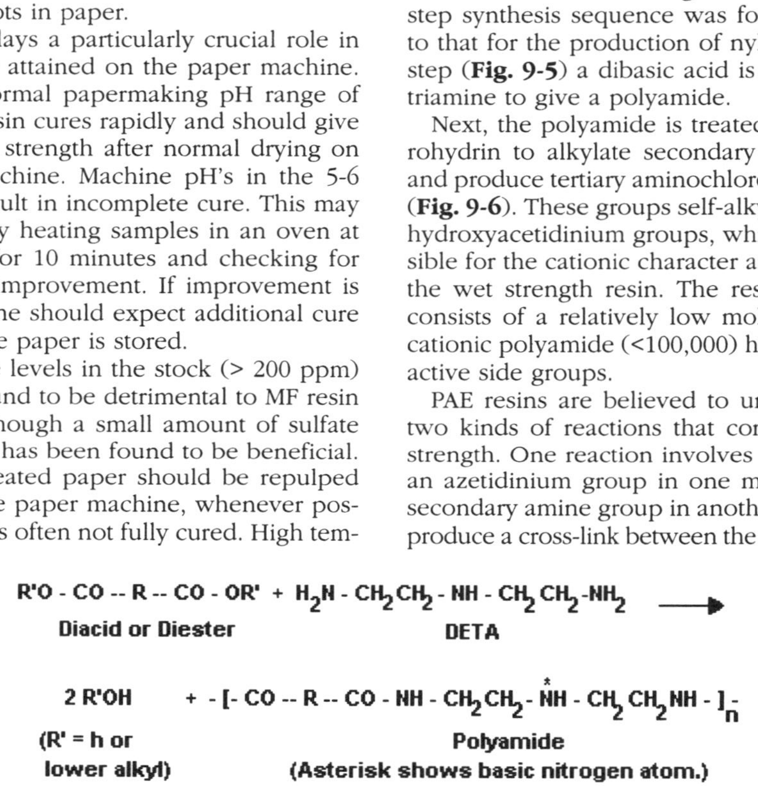

Figure 9-5. Preparation of polyamide-epichlorohydrin (PAE) wet strength resins. Step one: the synthesis of a polyamide from a dibasic acid and a triamine.

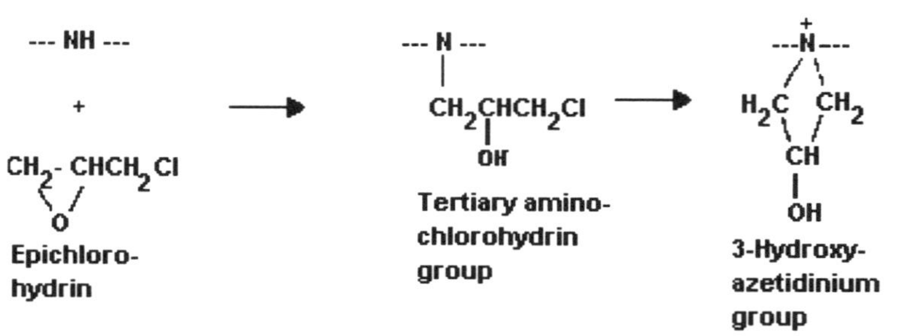

Figure 9-6. Preparation of polyamide-epichlorohydrin (PAE) wet strength resins. Step two: alkylation of the secondary amine groups in the polyamide produced in synthesis step one to produce tertiary aminocholorohydrin groups, which self-alkylate to form 3-hydroxyazetidinium groups which are responsible for the reactivity and cationic character of this wet strength resin.

In the second reaction, at least two azetidinium groups on a single resin molecule react with carboxyl groups on two different fibers to produce an interfiber cross-link.

b Use of polyamide epicholorohydrin resins in papermaking[3]

For stability purposes, PAE resins are acidified to a pH of 3.5-6 at the end of their preparation. Therefore, acid resistant pipes, pumps, and tanks must be employed. The resins are shipped at solids contents in the 12.5% to 33% range and can be handled by a variety of pumps. PAE resins are thermosetting, which means that they will polymerize to a water-insoluble condition by the action of heat alone. It is suggested that storage tank temperatures be kept below 30°C.

PAE resins can be used in headbox pH ranges of 5-9, however, the preferred range is 6-8. At lower pH's the reactivity of the resin is reduced. Typical PAE addition levels are 0.25% to 0.75%. They are often added at the fan pump, or earlier in the system if longer retention times are desired.

At the 25%-33% solids levels the resins are approaching their solubility limits, and will precipitate when mixed with practically any other substance. In addition, since they are highly cationic, PAE resins are incompatible with practically all types of anionic materials. Raising the pH above the normal storage value will accelerate the self cross-linking reactions and cause the resin solution to gel.

Dilute PAE solutions are compatible with some mildly anionic materials, but incompatible with strong anionics. Thus, it is important to keep PAE addition points on the paper machine well separated from anionic additive addition points. Rosin size is a good example. Rosin and PAE resins react to form foamy complexes and deposits. Raising the pH above 5 will gel dilute resin solutions.

PAE resins do not normally come to full cure on the paper machine and further polymerization occurs in the hot paper roll. Consequently, in order to predict the final level of wet strength that will be achieved, it is common practice to oven cure paper samples at 80°C for 30 minutes prior to testing.

Freshly made PAE wet strength paper can be readily repulped because it is not fully cured. Fully cured paper must be treated with more severe conditions. At the minimum, hot (>80°C) alkaline conditions (pH 11) are needed. In certain cases, it is necessary to employ an oxidizing agent, such as sodium hypochlorite, to break down the wet strength bonds.

9. Factors affecting the performance of wet strength resins

As with other cationic polymeric additives, many factors influence wet strength resin effectiveness. **Fig. 9-7** illustrates the four areas of interest: the resin, the fibers, the chemical

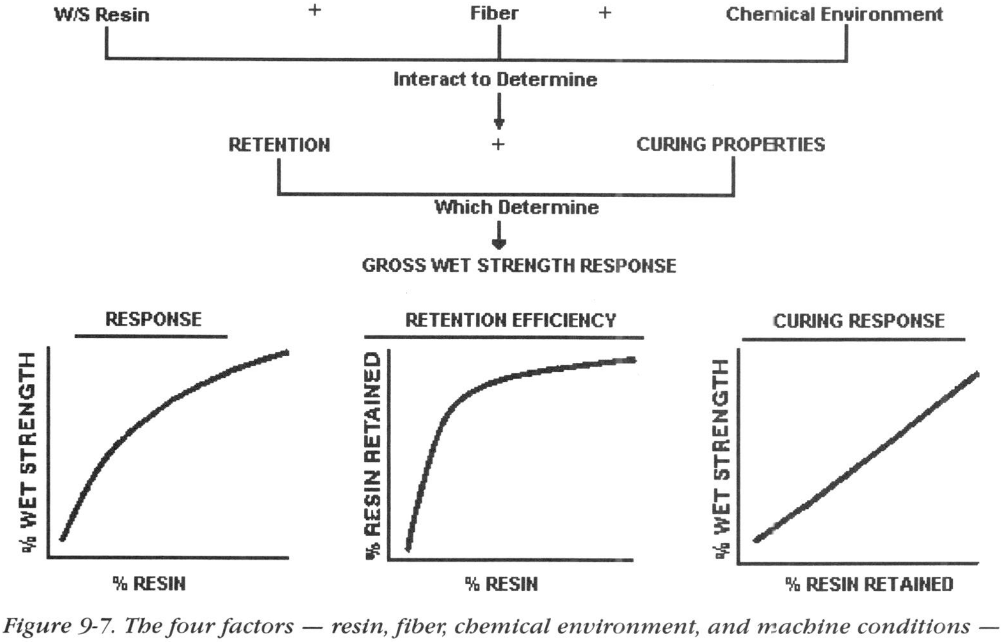

Figure 9-7. The four factors — resin, fiber, chemical environment, and machine conditions — that influence resin retention and curing and, in turn, produce wet strength.

environment, and the machine conditions. These interact to produce resin retention and curing which, in turn, produce the level of wet strength achieved.

Note that unlike other polymeric cationic additives, a high retention rate is insufficient to insure effective performance. The retained polymer must also be cured before a wet strength response is obtained.

10. Factors affecting the adsorption of wet strength resins by pulp fibers

Table 9-3 lists a number of important variables that influence resin retention by papermaking fibers.

Table 9-3. Variables important to the retention of wet strength resins by papermaking fibers [5]

- Fiber anionic sites
- Pulp consistency
- Contact time
- Resin concentration
- Pulp refining

a. Pulp carboxyl content

The carboxyl content of pulps affects resin performance because ionized carboxyl groups provide anionic adsorption sites for resin molecules. The higher the carboxyl group content, the more rapidly and more extensively will a pulp retain resin molecules. In general, unbleached softwood kraft pulps have the highest carboxyl contents, followed by bleached hardwood kraft and bleached softwood kraft, and finally, bleached softwood sulfite pulp. Wet strength resin retention follows the same order.

Wet strength resins exhibit typical Langmuir adsorption behavior, with the first resin added being completely adsorbed. As increasing amounts of resin are added, adsorption continues, but at a decreasing rate, due to impending saturation of the fiber anionic sites. The path of wet strength development parallels the adsorption behavior for the most part, with the initial resin having the most effect.

b. Pulp consistency and contact time

Both pulp consistency and contact time affect resin retention. The adsorption process is more rapid and complete at higher consistencies. This is presumably because at higher consistencies the polymer molecules have shorter distances to travel before colliding with a fiber surface than at lower consistencies. The data in **Table 9-4** supports this point.

Table 9-4. Effect of pulp consistency and contact time on resin retention [5]

Pulp Consistency	Contact Time	Resin Retained, Percent of Furnish	Tensile, Percent Wet/Dry
2.5%	20 sec	70	28
	5 sec	67	26
0.25%	20 sec	47	24
	5 sec	47	23
0.02%	20 sec	34	23
	5 sec	27	20

c. Resin dilution

To gain the best resin distribution, it is imperative that it be diluted at least 10:1 with fresh water. Fresh water is preferred because white water and other kinds of dilution water may contain anionic substances that will react with the resin and neutralize it.

d. Pulp refining

Refining significantly enhances the performance of PAE resins only at high resin addition levels because at high resin levels, a more highly refined stock will have greater surface area available for resin adsorption (i.e. higher resin capacity). At low resin levels, even lightly refined fibers will have sufficient surface area to adsorb all of added resin and no noticeable effect of refining on wet strength will be found.

11. Effect of the chemical environment on resin performance

In addition to the resin retention factors discussed above, the chemical environment plays a key role in determining wet strength resin performance (**Table 9-5**).

Table 9-5. Chemical environment variables that affect wet strength development

- Chlorine residuals
- pH
- Stock temperature
- Anionic contaminants

a. Effect of chlorine residuals on PAE resin performance

As was mentioned earlier, sodium hypochlorite is used often to break down PAE resin during broke reworking. Active chlorine present in papermaking stock (from repulpers or bleach plant carryover) will also react with PAE resin and reduce its effectiveness. **Fig. 9-8** illustrates this point. In such instances, the stock should be treated with an antichlor, such as sodium sulfite, prior to resin addition.

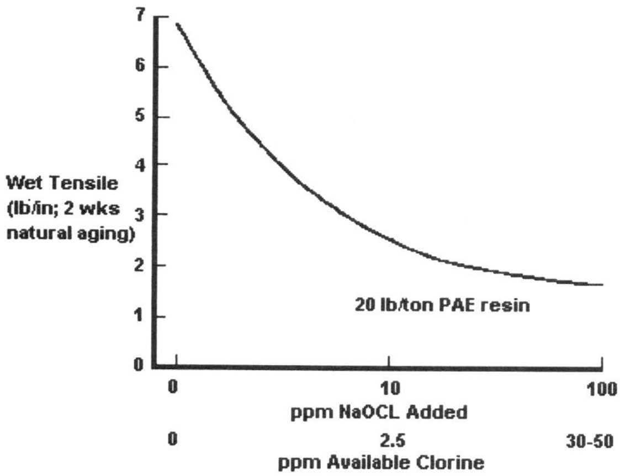

Figure 9-8. Effect of residual chlorine on wet strength

b. pH

PAE resins are effective over a pH range of 5-9. Below pH 5, resin effectiveness declines considerably for two reasons. First, pulp carboxyls are less ionized at low pH levels and become less attractive for the cationic resin. Second, at low pH levels, the secondary amine groups on the resin molecule become protonated and do not react readily in cross-linking reactions with azetidinium groups. **Fig. 9-9** illustrates the effect of pH on PAE resin performance.

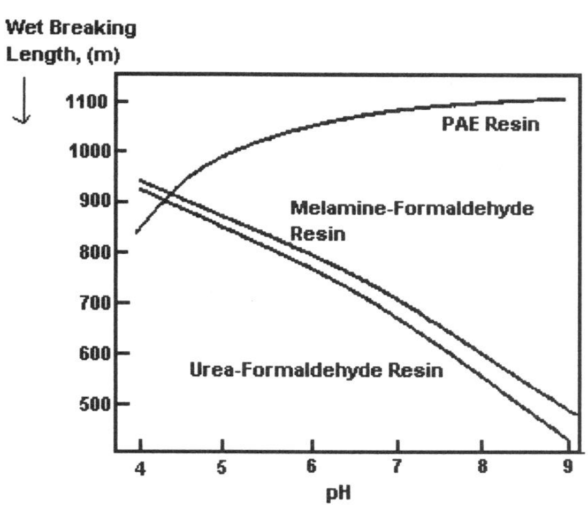

Figure 9-9. Effect of pH on wet strength development

UF and MF resins require acid pH values in order for wet strength to be developed by curing. If the sheet extracted pH is greater than about 4.5, then curing will be adversely affected.

c. Stock temperature and PAE performance

Prolonged exposure to high stock temperatures tends to reduce PAE resin efficiency. This may be due to the hydrolysis of azetidinium groups. This is a relatively slow reaction, but the resin should not be kept in contact with high temperature stock for long periods of time.

d. Interfering substances

Interfering substances such as lignin compounds, hemicelluloses, pulp mill carryover, bleach plant carryover, recycled papermaking additives, and anionic dispersants can react with cationic wet strength resins and interfere with their adsorption on fibers. Thorough washing or treatment of stock with a charge neutralizing substance (e.g. alum) prior to resin addition is necessary when high levels of interfering substances are present.

Sometimes cationic resins will precipitate small quantities of "color bodies" onto fibers and cause brightness reversion in the produced paper. This is usually a reaction between the cationic resin and an anionic color body. A similar reaction occurs between resin and anionic dyes. This can be a particularly serious problem in the manufacture of deep shade wet strength papers.

There is at least one instance where the reaction between cationic PAE and anionic polymers can be put to good use to enhance resin retention by fibers having low carboxyl content. In this case, anionic carboxymethyl cellulose (CMC) is reacted with PAE to form a weakly cationic complex (called a "symplex") that adsorbs onto the fiber surfaces. In essence, the CMC provides the carboxyl groups required to attract more PAE onto the fiber surface. **Fig. 9-10** illustrates the effect of adding CMC to PAE in a handsheet study.

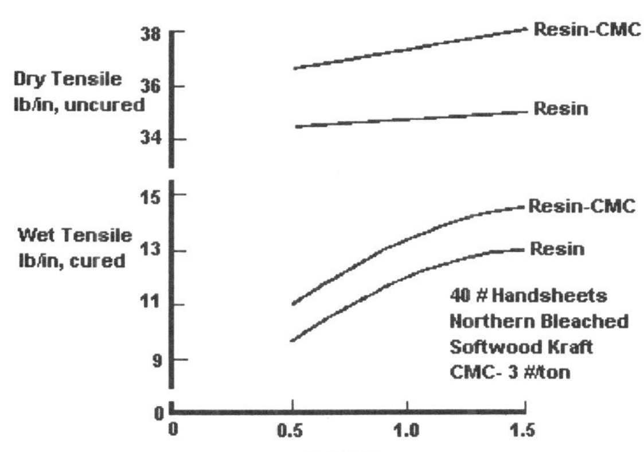

Figure 9-10. Effect of adding CMC to PAE in the preparation of wet strength paper

12. Effect of drying temperature on PAE curing

PAE resins are thermosetting and the drying conditions on the paper machine are critical to wet strength development. **Fig. 9-11** illustrates the effect of curing temperature on the rate and magnitude of wet strength development. The practical implications are that the drying program on a paper machine has a profound effect on the curing rate and final wet strength level attained. This means that a machine producing a given grade to the same final moisture content may nevertheless obtain different wet strength results if dryer section drying times and temperatures differ between grade runs.

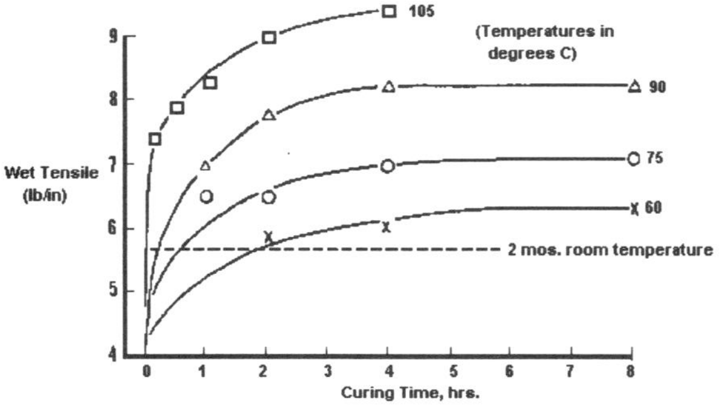

Figure 9-11. Effect of drying time and drying temperature on the development of wet strength

13. Summary

It is clear from the preceding discussion that wet strength wet end chemistry is complex and that the attainment of high wet strength levels and efficient resin utilizaton requires close attention to technical details. As is the case with cationic dry strength resins, rapid and efficient polymer retention (adsorption) is very important and system conditions and chemistry should be established to promote this goal. With wet strength resins, however, high retention does not always mean good wet strength performance because curing reactions must also occur.

14. References

[1] Espy, H. "Wet strength resins." Chapter IV in Pulp and Paper Manufacture, Volume 6: Stock Preparation. Editors R. W. Hagemeyer and D. W. Manson, Joint Textbook Committee of the Paper Industry, 1992, pp. 65-85.

[2] Dunlop-Jones, N. "Wet strength chemistry." Chapter 6 in Paper Chemistry. Edited by J. C. Roberts, Chapman and Hall, New York, 1991, pp. 76-96.

[3] Cates, R. "Wet strength resins." Chapter 9 in Chemical Processing Aids in Papermaking: A Practical Guide. Edited by K. Hipolit, TAPPI PRESS, Atlanta, 1992, pp. 129-148.

[4] Eklund, D. and Lindstrom, T. "Wet Strength." Chapter IV in Paper Chemistry: An Introduction. DT Paper Science Publications, Grankulla, Finland, 1991, pp. 89-105.

[5] Pahl, B. L. and Espy, H. H. "Advanced topics in papermaking with alkaline-curing polyamide wet strength resins." In Notes from 1987 Advanced Topics in Wet End Chemistry Seminar, October, 14-16, 1987, Memphis, TN. TAPPI PRESS, 1987, pp. 77-82.

Chapter 10

Fillers and Papermaking

1. Introduction

Mineral fillers have been utilized by papermakers since the very earliest days of hand papermaking when it was discovered that adding calcium sulfate to paper enhanced its printing and writing characteristics. As paper technology developed, fillers became essential components of many paper grades and these materials now represent, in terms of weight, the largest nonfibrous component of paper. Typical addition levels range from 3% to 30%.

For many years, low cost filler clay was the main papermaking filler and its primary function was to reduce furnish costs. Recently, papermakers have begun to focus more on the specific property improvements that can be gained from the use of a wider variety of filler pigments. This trend has been particularly accelerated by the large scale conversion from acid to alkaline papermaking that has occurred during the past decade in the US. **Table 10-1** lists some of the most common properties that can be improved by adding fillers to paper.

Table 10-1. Properties mproved by adding mineral pigments to paper

- Formation and sheet structure — fill in sheet voids
- Appearance properties — brightness, opacity, whiteness, gloss
- Texture and feel — smoothness
- Printability — smoothness, show-through, ink absorption
- Dimensional stab lity
- Cost savings — fiber substitution

In this chapter we will review the relationships among the properties of filler pigment, paper properties, and wet end chemistry. Further treatments of these subjects can be found in References [1-4].

2. Different types of mineral filler pigments

As indicated above, the papermaking has a wide range mineral filler materials available. **Table 10-2** lists some of the more common ones in use today.

Table 10-2. Different types of filler pigments

- Titanium dioxide (TiO_2)
  - Rutile
  - Anatase
- Kaolin clay
- Talc

(continued)

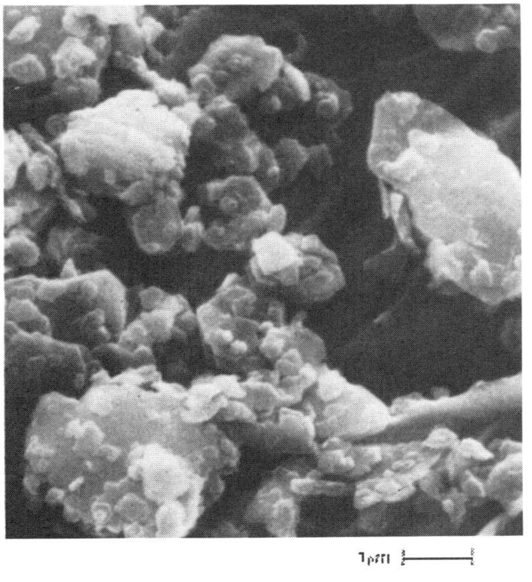

Figure 10-1. Filler clay electron micrograph. Most of the particles in the figure lie in the 1-2 micron (μ) range.

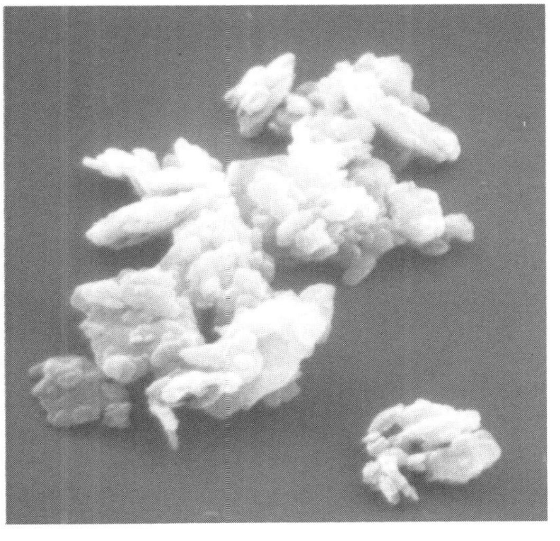

Figure 10-2. Calcined clay electron micrograph. Heating at high calcination temperatures drives water from within the clay lattice structure and produces a more aggregated and structured product.

(Table 10-2, continued)

- Calcium carbonate ($CaCO_3$)
  - Ground
  - Precipitated
- Aluminum trihydrate
- Silicas
- Plastic pigments

a. Kaolin clay fillers

For years, kaolin clays represented the largest single filler material utilized by US papermakers. This situation has changed recently with the growth in alkaline papermaking, but clays are still one of the most used filler materials, with current North American consumption on the order of 1 million tons per year.

Kaolin clay has a typical chemical composition of 39% Al_2O_3, 46% SiO_2, and 13% H_2O. Depending upon how they are processed, filler clays can be categorized as hydrous, fractionated, calcined, and/or structured. In general, the best brightness and opacity properties are achieved with clays that have undergone more beneficiation during their preparation processes (e.g. washing, delamination, degritting, bleaching, magnetic treatment, and leaching). These treatments result in higher brightess pigments having smaller particle sizes. Further treatments, such as calcination or treatment with high charge density cationic polymers, produce "structured" clays that have high suface areas and improved light scattering capabilities. **Figures 10-1** and **10-2** illustrate the plate-like structure of clay pigments.

Structured filler clays can also be prepared by treatment with high charge density cationic polymers under special conditions.

b. Calcium carbonate

Until recently, calcium carbonate pigments found their greatest use in US papermaking as coating ingredients. During the past decade, however, there has been a pronounced trend toward alkaline papermaking which has produced a significant growth in calicum carbonate filler utilization and these materials now rank with kaolin clays as the most common fillers. It is estimated that more than 1.2 million tons of calcium carbonate are now being used annually for this purpose.

Chemically, calcium carbonate is 98-100% $CaCO_3$. This apparently simple substance occurs in a wide variety of forms, however. **Table 10-3** lists some of the common types available to the papermaker.

Table 10-3. Types of calcium carbonate fillers available to the papermaker

- Natural
  - Chalk (calcite)
  - Limestone (calcite)*
- Precipitated
  - Scalenohedral (calcite)
  - Rhombic/prismatic (calcite)
  - Acicular (aragonite)
  - Spherical (calcite)
  - Cubic (calcite)

**Used as "ground" calcium carbonate*

Calcium carbonate occurs widely in the form of limestone. Depending upon the local geologic conditions, limestone deposits exist ion several different forms — marble and chalk being two examples. Of the two basic chemical types of limestone that exist, calcite and dolomite, the calcites are of the most significance to papermaking.

Calcium carbonate fillers fall into two different classifications, depending upon their process history — natural and precipitated. The natural calcium carbonates are produced by grinding limestone and are referred to as *ground calcium carbonate. Precipitated calcium carbonate* is produced by chemical reactions. The most common reaction used for this purpose is the reaction of slaked lime ($Ca(OH)_2$) with carbon dioxide gas (CO_2) to produce $CaCO_3$. This process is called the *carbonation process.*

The carbonation process is the basis for the on-site production of precipitated calcium carbonate at many paper mills. In these operations, the mill typically supplies water and CO_2 to the plant. These materials are reacted with quicklime lime (CaO) to produce precipitated $CaCO_3$. The pigment is pumped directly to the paper mill as a slurry (typically 20%).

Ground calcium carbonates usually have a large average particle size and a broader particle size distribution than the precipitated products. Within the category of precipitated calcium carbonates, different products can be produced by varying the precipitation conditions. These substances differ in particle size, shape, size distribution, and crystalline form. **Table 10-3** lists the various possiblilities. The most common form of precipitated calcium carbonate filler in use today is the scalenohedral calcite material. **Figs. 10-3** and **10-4** illustrate typical ground and precipitated calcium carbonate fillers.

c. Titanium dioxide

Titanium dioxide use in paper filler has declined somewhat in recent years due to its cost and environmental issues relating to its manufacture, but is still estimated to be approximately 115,000 tons per year in North America. It has the highest refractive index of any white pigment and, consequently, has the greatest specific opacifying power. This is an important asset in view of the continuing pressure to reduce paper basis weights, while retaining optical properties.

Natural deposits of titanium contain either 40-60% TiO_2 (ilmenite) or 90% TiO_2 (rutile). In most instances, the ilmenite ore is treated by one of two chemical processes to produce either rutile or anatase TiO_2 filler pigments of purity greater than 98% TiO_2. The manufacturing processes can be controlled closely, resulting in uniform pigment specifications. For example, the narrow particle size distribution of particles is illustrated in **Figure 10-5**.

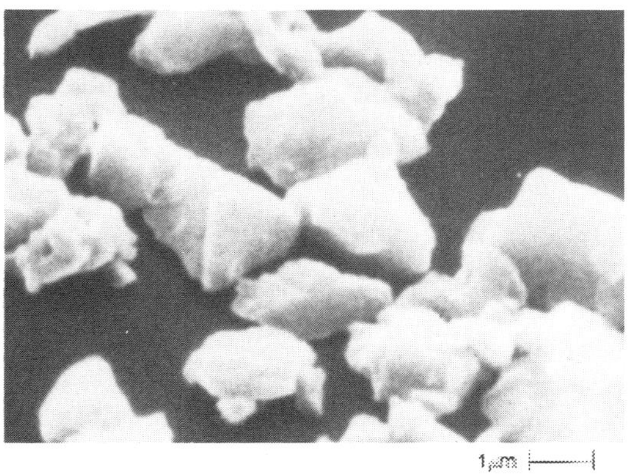

Figure 10-3. Ground $CaCO_3$ electron micrograph. This material is known as an ultra-fine ground limestone and has particle sizes in the one micron (μm) range.

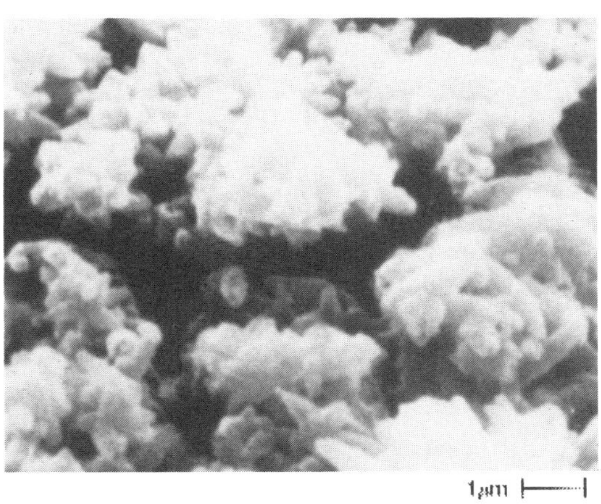

Figure 10-4. Scalenohedral precipitated $CaCO_3$ electron micrograph. The barrel-shaped particles can be clearly seen to have grown from single crystallization nuclei to produce the "rosette" structures seen in the micrograph.

Figure 10-5. Anatase TiO_2 electron micrograph. The spherically shaped pigment particles are clearly evident in the micrograph. The primary particles are approximately 0.3 μm in diameter.

Figure 10-6. Precipitated silica electron micrograph. The extremely fine ultimate particles can be seen in the micrograph. These particles are aggregated into larger, grape-like clusters that have very high specific surface areas, e.g. approximately 20 m^2/g.

d. Precipitated silica

Small quantities (ca. 30,000 tons per year in North America) of synthetic amorphous silica pigments are used in conjunction with other filler pigments to enhance brightness and opacity, for the most part. The hydrated silica form of this product consists of 78% SiO_2 and 5% CaO. **Fig. 10-6** illustrates one such material.

Precipitated silicas have extremely small, spherical ultimate particles. These particles are aggregated into grape-like clusters that have very high specific surface areas.

e. Aluminum trihydrate

Approximately 8,000 tons of aluminum trihydrate — 65% Al_2O_3, 34% H_2O — was consumed in paper filling in 1994 in North America. This specialty pigment contributes brightness, opacity, ink receptivity, and flame retardancy to paper.

Like calcium carbonate, there are two categories of aluminum trihydrates, native and synthetic. The native pigment is produced from very pure bauxite ore that undergoes beneficiation processes similar to ground calcium carbonate. **Fig. 10-7** contains an illustration of ground aluminum trihydrate.

Synthetic aluminum trihydrates are made by dissolving bauxite ore in sodium hydroxide. The solution is then hydrolyzed to precipitate alumina trihydrate in crystalline form.

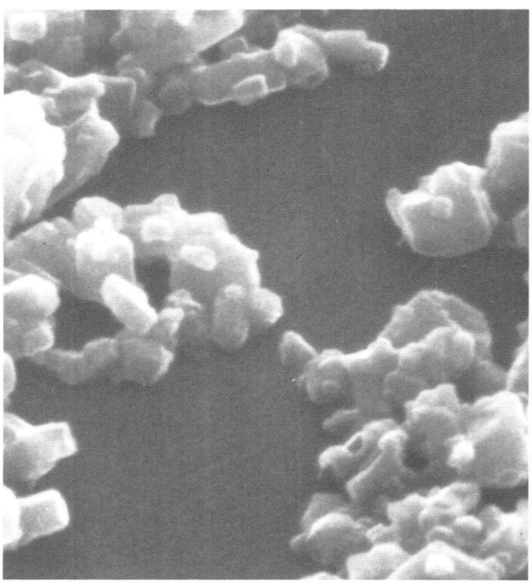

Figure 10-7. Ground aluminum trihydrate. The small, cubic-shaped pigment particles can be seen.

f. Talc

Talc (**Fig. 10-8**) is not utilized to any great extent as a filler by the US paper industry. However, due to its unique surface properties, considerable quantities — approximately 80,000 tons per year — are consumed in pitch control applications and much of that material is eventually retained in the produced paper.

Figure 10-8. Talc electron micrograph

3. Filler particle shape

As is obvious from the electron micrographs presented above, fillers take on a wide variety of particle shapes, which affects their behavior in several areas. For example, particle shape directly influences the way pigment aggregates are organized in the sheet (i.e. their "packing"). Spherically shaped TiO_2 and silica fillers tend to pack into tight aggregates that have many filler-to-filler interfaces and few filler-to-air interfaces. The platy clays and talcs do the same. These materials also can align themselves on fiber surfaces. Ground $CaCO_3$ fillers are irregularly shaped. Precipitated $CaCO_3$ can be produced in three different shapes — rhombohedral or barrel-shaped calcite crystals, acicular rods or needle-like aragonite crystals, and scalenohedral, rosette structures with ellipsoid-shaped calcite crystals. The latter two forms do not pack into tight aggregates, but form particle clusters that have significant void spaces and many pigment-to-air interfaces. This behavior significantly affects the light scattering attributes of the pigments — particularly at higher loading levels.

4. Fillers and appearance properties

a. Interaction of light and paper

When light strikes a piece of paper, several things can happen — as illustrated in **Fig. 10-9**. Some of the light is reflected directly from the surface of the sheet. When the angle of incidence equals the angle of reflection, this phenomenon is called *specular reflection*. The rest of the light enters the sheet, where it is either absorbed or scattered by the component particles. Some of the scattered light reemerges from the incident surface and is said to be *diffusely reflected*. The remaining portion of the light eventually emerges from the bottom side of the sheet. It is called *transmitted light*. Taken in sum, these phenomena — reflection, absorption, and transmission — define the color, brightness, opacity, and gloss properties of paper.

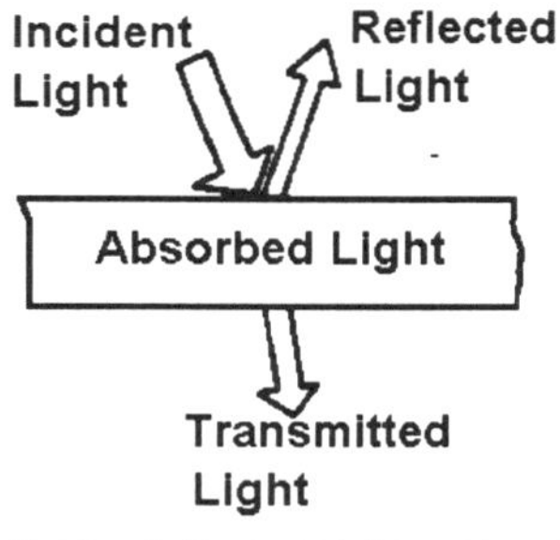

Incident Light = Reflected Light + Absorbed Light + Transmitted Light

Figure 10-9. Interactions between paper and incident light. Light is either reflected, absorbed, or transmitted. Since the web is comprised of light scattering particles, the light is said to be "diffusely" reflected and "diffusely" transmitted.

b. Relative scattering coefficient

The relative scattering coefficient of a pigment expresses the relative decrease in light intensity of transmitted light which has its cause in the reflection of light.

$$s = \lim_{w \to 0} \sum \frac{I_R \div I}{w}$$

where:

s = specific scattering coefficient, cm^2/g
I = incident light energy
I_R = reflected light energy
w = grammage of paper, g/cm^2

Table 10-4 lists some representative light scattering coefficients of paper fillers and papermaking fibers.

Table 10-4. Typical light scattering coefficients for fillers and fibers

Filler	Scattering Coefficient, cm^2/g
Filler clay	1100 — 1200
Calcined clay	2600 — 3000
Ground $CaCO_3$	1400 — 1700
Precipitated $CaCO_3$	2200 — 2700
Titanium dioxide	4500 — 6000
Bleached chemical pulps	220 — 350
Mechanical pulps	500 — 750

It is clear from the table that a wide range of light scattering abilities exist in mineral pigment fillers, with titanium dioxide being by far the best light scattering material. It is also evident that the mineral pigments scatter light to a much greater extent than pulp fibers.

5. Factors that affect reflectance and are related to wet end chemistry

There are several factors that influence light scattering and light reflectance in paper. Four are particularly important. They are:

- Particles whose sizes are half the wavelength of the incident light (i.e. a diameter of ca. 0.25 μ) scatter light the most extensively.
- The greater the refractive index of a particle, the greater its light scattering potential.
- The greater the number of optical interfaces available, the greater the potential for light scattering.
- The greater the difference in refractive index between two materials forming a light scattering interface, the greater the light scattering at the interface.

Table 10-5 presents some of the key filler properties that contribute to these four factors.

Table 10-5. Important properties of fillers

Filler	Average Particle Size, μ	Refractive Index	Brightness, %	Specific Surface Area*, m^2/g
Filler clay	0.2 — 2.0	1.57	78 — 90	10 — 25
Calcined clay	0.7 — 1.5	1.60	90 — 95	15 — 25
Ground $CaCO_3$	0.7 — 3.0	1.59	80 — 95	2 — 12
Precipitated $CaCO_3$	0.3 — 3.0	1.59	96 — 100	5 — 25
Titanium dioxide	0.2 — 0.4	2.55 (anatase) 2.76 (rutile)	98 — 100	7 — 12
Talc	1.5 — 10.0	———	85 — 90	9 — 20
Precipitated silica	0.1 — 3.0	1.55	93 — 99	45 — 75
Alumina trihydrate			97 — 100	6 — 8

**The specific surface areas for fillers are much higher than the values found for fibers, which are on the order of 1-2 m^2/g.*

As indicated by the four factors listed above, the refractive index differences at particle-particle interfaces play key roles in light scattering in paper. **Table 10-6** lists refractive indices for some of the materials that contact fillers in paper.

Table 10-7 presents the relative light reflectance that occurs at various pigment intefaces with air, fiber, and water. It is clear from the table that TiO_2 scatters light much more effectively than the other two pigments.

The table also indicates that when an interface is composed of materials having similar or the same refractive indices, very little or no light scattering occurs. From this, it is evident why the self-aggregation of filler particles is usually detrimental to their light scattering capacity per unit weight. When filler surfaces get closer than ca. 0.05 μ they come into "optical contact" and light scattering is no longer possible. This occurs when pigment particles agglomerate in a papermaking system. This is especially true for fillers like TiO_2, whose aggregated particles pack together very closely. This fact is illustrated below in **Fig. 10-10**.

Table 10-6. Other relevant refractive indices

Substance	Refractive Index
Air	1.00
Water	1.33
Cellulose	1.53
Starch	1.53
Paraffin	1.43

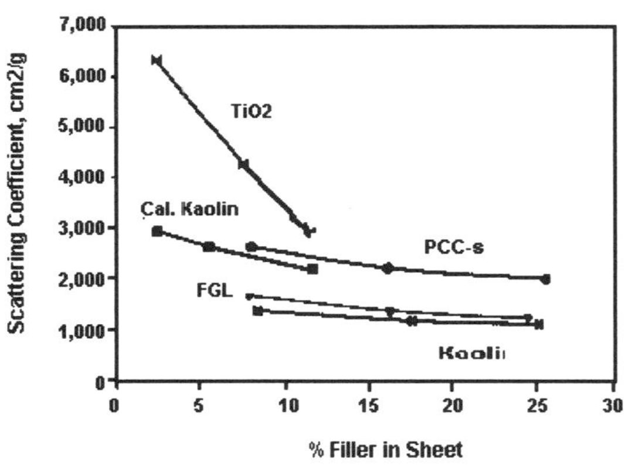

Figure 10-10. Influence of filler aggregation on light scattering. The usual methods for incorporating increasing amounts of filler in paper leads to filler aggregation. When filler surfaces get closer than ca. 0.05 μ they come into "optical contact" and light scattering is no longer possible and the light scattering coefficient declines. (Cal. Kaolin = calcined clay; PCC-s = precipitated calcium carbonate; FGL = ground calcium carbonate; Kaolin = kaolin filler clay)

Table 10-7. Relative percent light reflection at various pigment-media interfaces

Filler	Self	Air	Water	Fiber
Clay	0	4	0.5	<0.1
$CaCO_3$	0	5	1	<0.1
TiO_2 (Anatase)	0	19	10	7

6. Pigment surface charge

The electrostatic charge on the surface of filler particles plays important roles in the stability of pigment dispersions and in the the retention of filler in the paper web during papermaking. Pigments generally have zeta potentials in the -30 mV to +30 mV range , depending upon their preparation and treatment and the nature of the ionic content of the suspending medium. **Table 10-8** lists some of the agents that affect filler zeta potentials.

Table 10-8. Agents that alter pigment zeta potentials

- pH
- Inorganic salts
- Polymeric dispersants
- Polyelectrolytes (retention aids, coagulants)

Oxide minerals, such as titanium dioxide, have natural surfaces whose degree of ionization depends on the pH. They are usually negative at pH levels above 6. **Fig. 10-11** illustrates this point. Note that pH is not the only factor in the figure. The absolute magnitude of the measured zeta potential also depends on the concentration of other ions in the suspending water.

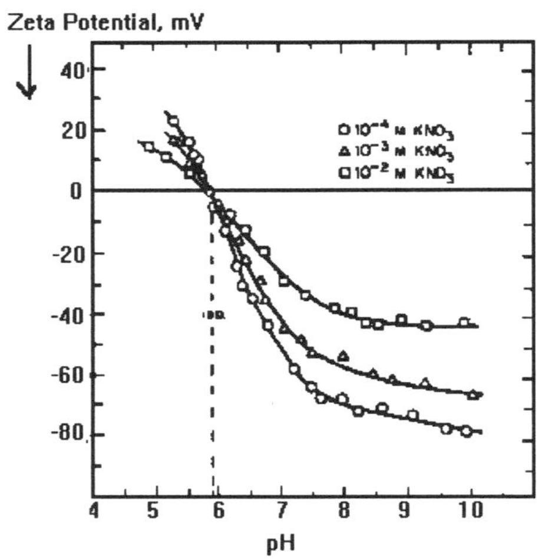

Figure 10-11. Variation of zeta potential of colloidal titanium dioxide as a function of pH in aqueous solutions of potassium nitrate at 25 degrees C[5]

Ions other than hydrogen ion also can act as potential determining ions. For example, Ca^{+2} concentration directly affects the zeta potential of suspended calcium carbonate, as demonstrated in **Fig. 10-12**.

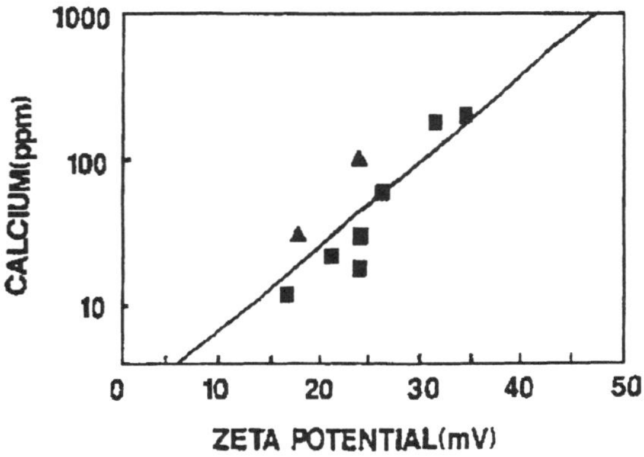

Figure 10-12. The relationship between the zeta potential of precipitated calcium carbonate and calcium ion concentration.[4]

Low molecular weight, anionic polyelectrolytes (polyacrylates, polyphosphates) are often employed in the preparation of pigment dispersions. These substances adsorb strongly onto particle surfaces, producing a high negative surface charge that leads to strong particle-particle repulsion and suspension stability. Other anionic substances, such as recycled size press starches, also adsorb onto filler particles. Similarly, cationic retention aids, coagulants, or alum, can also react with filler surfaces, imparting a positive charge to the particles.

It is clear that it is very difficut to predict the magnitude and sign of surface charge that filler particles will adopt when they are incorporated in papermaking systems. Consequently, the papermaker often must determine this information by making the kinds of charge measurements described earlier in Chapter 6. With this information, it is then possible to establish system conditions that will enhance filler retention and performance.

7. Preferential adsorption of additives by fine particles

Fillers and fines have very high specific surface areas relative to fibers. Consequently, they tend to adsorb greater quantities of additives per unit weight than fibers. **Table 10-9** illustrates this point.

This behavior has several important ramifications in papermaking. First, poor first pass filler retention leads to a buildup of filler in the white water loop and fresh cationic additives will readily adsorbed this excess material and be lost unless adjustments are made that improve first pass filler retention. Second, additives are often less efficient when they are adsorbed on filler surfaces than when they are located on fiber surfaces. An example of this phenomenon is the well-known inverse relationship between filler level and sizing. A third example is related to the unequal distribution of filler that exists through the thickness of the sheet and gives rise to paper property two-sidedness.

In order to minimize problems like these, the papermaker must be cognizant of the relative addition points of filler and other additives and constantly strive to achieve high first pass filler retention.

*Table 10-9. Relative adsorption intensity of some common wet end additives by fillers, fibers, and fiber fines**

Additive	Fibers	Filler Clay	Fines
Cationic Starch	1	4	5
Rosin Soap Size	1	4	16
Dispersed Rosin Size	1	16	20
Alum	1	2	3

**The relative adsorption intensities were derived from data presented in References [6 - 9].*

8. Adverse effects of filler on strength properties

The addition of filler adversely affects most strength properties by directly interfering with interfiber bonding. Particle size and particle shape play key roles in determining the severity of this effect for a given pigment. In general, the smaller the pigment particles, the greater their negative effect on strength. Spherically shaped filler particles, like TiO_2, and platelet particles, like clay, have greater effects than irregularly shaped or scalenohedral $CaCO_3$. **Fig. 10-13** illustrates the general behavior of pigments in this regard.

The exact effects are determined by the particle shape and size of the filler. In general, the smaller the particle size, the more adverse the effect. Hence, titanium dioxide, rhombohedral precipitated calcium carbonate, and the smaller scalenohedral precipitated calcium carbonates have strong effects on strength. Clay is particularly detrimental because of its plate-like structure. Ground calcium carbonate, with its larger particle size and "chunky" shape, is the least detrimental to strength.

One of the primary reasons for using cationic starch as a dry strength additive is to counteract the adverse effects of filler pigments on paper strength.

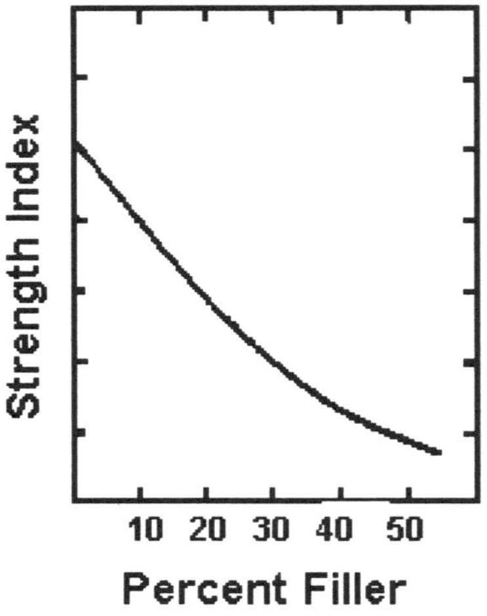

Figure 10-13. General adverse effect of filler on paper strength

9. Summary

This concludes our introduction to the properties and uses of filler pigments. The trend is toward increasing use of these materials in papermaking to achieve specific properties and to reduce furnish raw material cost. Since these particulate materials can have negative impacts on other papermaking additives, it is crucial that the papermaker use them judiciously and understand their fundamental nature and the various mechanisms by which they act.

10. References

[1] Bown, R. "Physical and chemical aspects of the use of fillers in paper." Chapter 10 in Paper Chemistry. Edited by J. C. Roberts, Chapman and Hall, USA, pp. 162-197(1991).

[2] R. A. Gill. "Fillers for papermaking." Chapter 4 in Applications of Wet End Paper Chemistry. Edited by C. O. Au and I. Thorn. Blackie Academic and Professional, London. pp. 54-75(1995).

[3] Eklund, D. and Lindstrom, T. "Fillers and pigments." Chapter IX in Paper Chemistry: An Introduction, DT Paper Science Publications, Grankulla, Finland, 223-263(1991).

[4] Gill, R. A. and Hagemeyer, R. W. "Fillers for paper". Chapter 2 in Pulp and Paper Manufacture, Third Edition. Edited by M. J. Kocurek, Vol. 6 , pp. 19-38(1992).

[5] Wiese, G. R. and Healy, T. W., J. Colloid Interace Sci. 51:247(1975).

[6] Marton, J. and Marton, T., Wet End Starch: Adsorption of Starch on Cellulosic Fibers, Tappi 59(12):121 (1976).

[7] Marton, J., The Role of Surface Chemistry in Fines Cationic Starch Interactions, Tappi 63(4):87 (1980).

[8] Marton, J. and Marton, T. Effect of Fillers on Rosin Sizing of Paper, Tappi J. 66(12):68 (1983).

[9] Marton, J. and Marton, T., Some New Principles to Optimize Rosin Sizing, Tappi J. 65(11):105 (1982).

Chapter 11

The Properties of Aluminum Compounds Important to Papermaking

1. Introduction

From its invention in approximately 105 A.D., papermaking was practiced under neutral or alkaline conditions. The tendency of inks to feather when applied to paper surfaces led to the development of tub sizing practices, with animal glue being one of the principal materials used for this purpose. It was found that addition of alum to the dilute size solution was beneficial to glue hardening and in preventing putrification.[1]

In time, alum was discovered to have other papermaking applications. It improved sheet formation and drainage, anchored rosin size to fibers, set the resinous components of wood and reduced pitch deposits, and so on. Some of the more common uses of alum are listed in **Table 11-1**. Of note is the fact that even though alum has been associated most strongly with acid papermaking conditions, current alkaline papermaking practice calls for its use, albeit in small, carefully controlled doses.

Table 11-1. Uses of alum in papermaking

- Retention aid (dyes, fillers, latexes, fiber fines)
- Drainage aid
- Formation aid
- Anionic trash charge neutralizer
- pH control agent
- Water purifying coagulant
- Pitch control agent
- Alkaline papermaking contaminant scavenger
- Rosin sizing agent
- Acid catalyst for wet strength resin cure

In addition to the advantages gained from alum use listed in the table above, there are also several notable disadvantages associated with alum. These are listed in **Table 11-2**.

The severe nature of some of the problems listed in the table have contributed to the recent trend toward alkaline papermaking in several paper grades. Nevertheless, many grades will continue to be made under acid conditions in the future and it is important that the wet end chemist have a good grasp of the basic aqueous chemistry of alum and aluminum compounds.

Table 11-2. Some adverse effects of alum in papermaking systems

- Alum-based systems, being acidic, are corrosive.
- Alum can degrade paper strength when used in excess.
- Alum-containing papers are impermanent.
- Calcium carbonate fillers cannot be easily used with alum furnishes.
- Alum deposits can be a serious problem in the wet end or a paper machine.

2. Different aluminum compounds used in papermaking

Papermakers have utilized four different types of aluminum compounds to accomplish the goals listed in **Table 11-1**. The first to be employed and the material that still receives the most attention is alum. The second compound to be utilized historically was sodium aluminate. More recently, polyaluminum chloride has been introduced as a high cationic charge density source of aluminum in papermaking. A fourth material, aluminum chloride, has been used in specialty technical paper grades where the sulfate ion is considered undesirable. We will discuss the first three compounds here, beginning with alum.

3. Papermaker's alum, $Al_2(SO_4)_3 \cdot 14H_2O$

a. Technical properties[2]

Papermakers' alum, as produced for the paper industry, has the formula $Al_2(SO_4)_3 \cdot 14H_2O$. The 14 water molecules of hydration differs from reagent grade aluminum sulfate, which has 18 water molecules associated with it. A typical way to characterize alum is by its *equivalent Al_2O_3 content*. While there is no actual Al_2O_3 present in alum, the convention provides a convenient way to compare the relative strengths of different solution concentrations. Dry alum contains 17.1% Al_2O_3. Commercial *liquid alum* solutions typically contain 48.5% $Al_2(SO_4)_3 \cdot 14H_2O$ with an equivalent content of 8.3% Al_2O_3. This concentration produces the minimum freezing temperature (-15°C). The following table, derived from the TAPPI Technical Information Sheet TIS 0506-

Table 11-3. Typical properties of commercial alum [2]

Property	Dry	Liquid
Chemical formula	$Al_2(SO_4)_3 \cdot 14H_2O$	48.5% solution
Molecular weight	594	—-
Al_2O_3 equivalent	17.1	8.3
Bulk density, ground (kg/m³)	1000-1100	
Specific gravity (@ 15.5°C)		1.335
Density (lb/gal, U.S.)		11.1
Density (kg/m³)		1330
lb alum/gal (U.S)		5.4
kg alum/m³		647

02, lists some of the technical properties of papermakers' alum.

A detailed discussion of the handling and metering considerations has been provided by Wortley. [3] Avery developed a method for quantitatively determining the amount of aluminum in a paper stock sample.[4] The procedure involves titration of an acidified sample of stock (filtered or unfiltered) with NaF. Endpoint determination is by fluoride-specific ion electrode. The method has been widely used.

b. The high charge density of aluminum ions

The many uses of aluminum in papermaking derives from its chemical properties, the most important of which are a result of the very high charge density of the aluminum ion — a product of its high (+3) charge and small ionic radius. **Table 11-4** compares Al^{+3} with other cations in this regard.

Table 11-4. Comparative ionic radii for inorganic cations

Cation	Radius, Angstroms
Na^+	0.95
K^+	1.33
Mg^{+2}	0.78
Ca^{+2}	0.99
Ba^{+2}	1.43
Al^{+3}	0.57
Cr^{+3}	0.65
Mn^{+4}	0.66
Fe^{+2}	0.67

The Al^{+3} ion can reduce its high charge density by forming complexes with six ligands, either neutral or anionic, that increase the size of the ion or reduce its net charge, respectively. Typical example ligands include H_2O, OH^-, SO_4^{-2}, $RCOO^-$, $H_2PO_4^-$.

c. Aluminum hydrolysis and the production of acidic solutions

The acidic nature of alum solutions is one of the substance's best-known characteristics. In fact, the pH of commercial alum solutions ranges from 3.25 (10% solution) to 2.4 (48.5%). The chemical reactions that produce these acid pH values fall under the heading of *hydrolysis reactions.* They can be described in terms of the complex formation introduced above. In this case, the active ligands are H_2O molecules and OH^- ions.

When an aluminum salt of a non-complex forming anion, such as Cl^-, is dissolved in water at pH below 3, the $Al(H2O)_6^{+3}$ is the predominant form of Al in solution. If the pH is raised so that the $[OH^-]$ increases, then ligand H_2O molecules get successively replaced by OH^- ions, which are stronger electron donors than H_2O. This process is often represented by the series of chemical equilibrium reactions shown in **Fig. 11-1**.

$$[Al(H_2O)_6]^{+3} + H_2O \rightleftharpoons [Al(H_2O)_5(OH)]^{+2} + H^+$$
$$[Al(H_2O)_5(OH)]^{+2} + H_2O \rightleftharpoons [Al(H_2O)_4(OH)_2]^{+1} + H^+$$
$$[Al(H_2O)_4(OH)_2]^{+1} + H_2O \rightleftharpoons [Al(H_2O)_3(OH)_3] + H^+$$
$$[Al(H_2O)_3(OH)_3] + H_2O \rightleftharpoons [Al(H_2O)_2(OH)_4]^{-1} + H^+$$

Figure 11-1. Chemical equilibrium equations involved in the aluminum hydrolysis equilibria that result in the production of acidic alum solutions.

Of course, even though the above reactions are written as if the aluminum ion is reacting with a water molecule, this is not really what happens. Instead, the water molecule is itself in equilibrium with OH^- and H^+. The Al ion complexes with an OH^-, upsetting the balance between OH^- and H^+ in the system and shifting the pH toward the more acidic side.

d. Formation of polynuclear aluminum species[5]

If the hydrolysis equilibria presented above were the only aluminum reactions occurring in aqueous solution, then the system would be fairly simple. Research has shown that this is not the case, however, because the mononuclear species depicted in **Fig. 11-1** are capable of complexing with themselves via a process called *olation*, which leads to the formation of complexes containing more than one aluminum. **Fig. 11-2** illustrates the formation of one such species. It is believed that this process is very important and that much of the aluminum is tied up in such complexes in the 4-5 pH range.

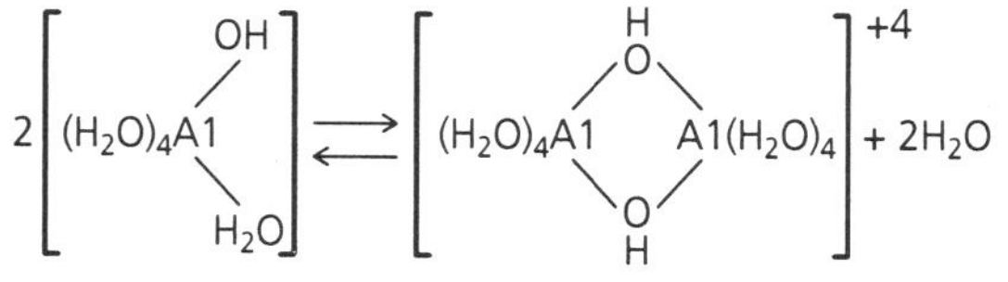

Figure 11-2. The olation process that leads to the formation of polynuclear aluminum species in aqueous solution

Extension of this process leads to the formation of complexes containing more than two aluminum atoms. Different researchers have hypothesized a number of different ones — $Al_{13}O_4(OH)_{21}^{+7}$ and $Al_8(OH)_{20}^{+4}$, for example. The exact composition of such complexes is uncertain because their formation is influenced by aging, temperature, concentration, and other system variables.

e. Equilibrium among aluminum species as a function of pH[5,6]

Fig. 11-3 illustrates the effect of pH on the aluminum species distribution over the normal pH range encountered in papermaking. Note that the polynuclear species occur at the pH conditions often employed in papermaking.

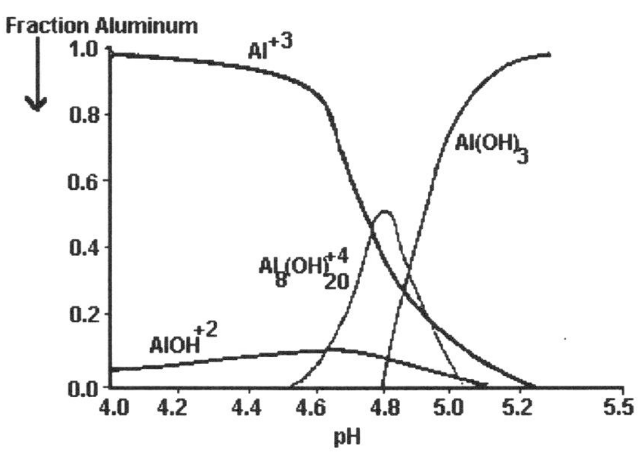

Figure 11-3. Equilibrium among different aluminum species as a function of pH under papermaking conditions[5,6]

The material studied was $AlCl_3$ and the solutions were allowed to equilibrate for very long times relative to papermaking time scales.

Fig. 11-3 only goes up to pH 5.5. Extending the pH range above 5.5 would find that $Al(OH)_3$ would dominate up to pH 8. The formation of $Al(OH)_4^-$ ion begins at pH 8, with a concomitant decrease in $Al(OH)_3$.

Continued increase in pH results in the disappearance of all $Al(OH)_3$ by pH 10 and the existence of a single Al species, $Al(OH)_4^-$[6]. At the low end of the pH spectrum, **Fig. 11-3** indicates that Al^{+3} is the predominant Al species up to pH 4.6, when it begins to decline in concentration and the polynuclear complex begins to form.

Many applications of alum rely upon its cationic charge. It is clear from **Fig. 11-3** that the different aluminum species have very different levels of charge. At the lowest pH values, Al^{+3} has a strong charge that begins to decrease as the pH is increased and $Al(OH)^{+2}$ forms. The situation changes again at about pH 4.6, where a +4 charged polynuclear species forms. This species reaches a maximum at about pH 4.8 and declines thereafter, disappearing by pH 5.0. $Al(OH)_3$, a gelatinous precipitate, begins to form at 4.8 and becomes essentially the sole species present above pH 5.0. Even though its molecular formula indicates that $Al(OH)_3$ is neutral, in fact it retains a cationic charge up to pH 8.5-9.0 due to the adsorption of cationic alum hydrolysis onto the precipitate.

A key feature of the equilibrium in **Fig. 11-3** is that it is concentration-dependent. If the concentration is increased, the three regions in **Fig. 11-3** still exist, but the curves shift. As the aluminum concentration is raised, the polynuclear species is formed at a lower pH and in increasing amounts. Also, the pH at which $Al(OH)_3$ precipitates out decreases.

Arnson has also postulated that the SO_4^{-2} ion in alum will have a similar effect on the equilibrium — i.e. the equilibria in **Fig. 11-3** would all shift to lower pH values. The polynuclear species formed would contain both OH and SO_4 ligands.

f. Adsorption of alum by papermaking fibers

The adsorption of alum by papermaking fibers is of great importance in papermaking. A number of workers have studied this subject and reported essentially similar results. [6,7,8] **Fig. 11-4** illustrates the observed effect of pH on aluminum adsorption by fibers. The aluminum concentrations in **Fig. 11-4** are similar to that found in many papermaking systems.

The adsorption of aluminum at both concentrations shows a sharp increase between pH 4 and 5. Reference to **Fig. 11-3** reveals that the adsorption increase occurs at the same pH as the beginning of polynuclear species formation.

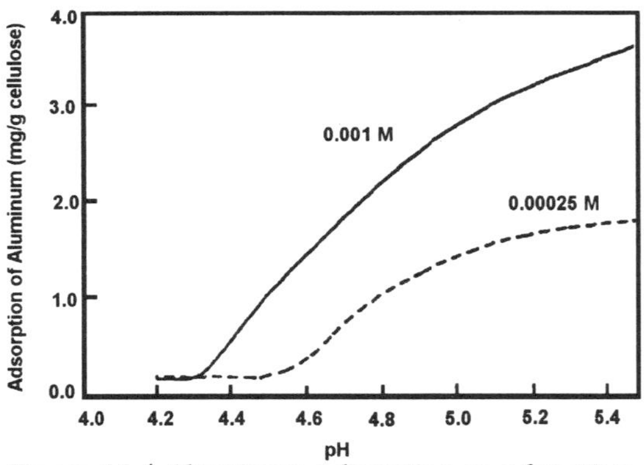

Figure 11-4. Aluminum adsorption as a function of pH and aluminum concentration for $AlCl_3$ [6]

It is clear from the above two figures that the different aluminum species exhibit quite different adsorption behavior. In the trivalent cation region, adsorption is low and independent of concentration. This is because the adsorption of Al^{+3} relies upon ion exchange with fiber carboxyl groups — which are present in limited quantity.

The polynuclear species demonstrates strong adsorption — probably due to the greater cationic charge (+4) and ability of the species to participate in hydrogen bonding interactions with fiber surfaces. Finally, increasing the pH above 5.0 further improves aluminum adsorption, which is now converting to the $Al(OH)_3$ form. This compound has a very low solubility in water and precipitates out. It has a cationic charge due to the presence of intermediate aluminum hydrolyzate products. Both factors, the charge and the low solubility, contribute to the strong adsorption of $Al(OH)_3$ by fibers. Once on the fiber surface, the material can hydrogen bond to it.

The observation that the higher concentration curve in **Fig. 11-4** turns upward at ca. pH 4.3, while the lower concentration curve does not increase until after pH 4.5 is consistent with the effect of concentration on polynuclear species formation seen in **Fig. 11-3**. Experiments comparing alum to $AlCl_3$ revealed the same shift to lower pH when the SO_4^{-2} ion was present.[6]

g. Influence of alum on fiber surface charge

The adsorption of cations by papermaking fibers always decreases with decreasing pH due to the protonization of hemicellulose carboxyl groups. This process is complete at about pH 2.8 and the fiber surface charge becomes zero. Conversely, increasing the pH creates negatively charged sites through carboxyl dissociation. It has been shown that a stoichiometric charge relationship exists between the quantity of adsorbed single cations and the degree of carboxyl dissociation. Therefore, charge reversal cannot be brought about by the adsorption of single cations. This is illustrated in **Fig. 11-5**,[9] where the effect of pH on the electrophoretic mobility of a microcrystalline cellulose/alum system is presented.

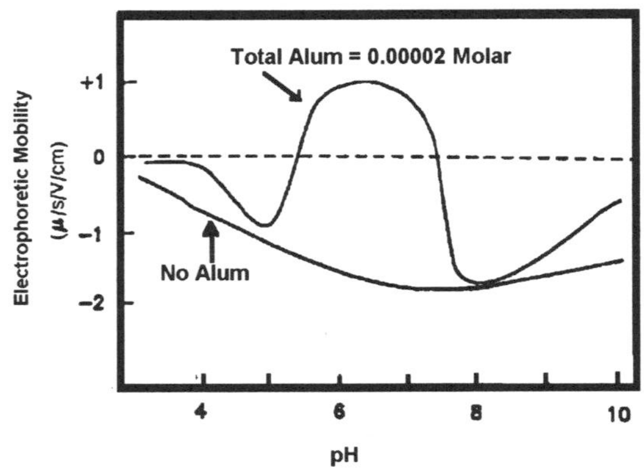

Figure 11-5. Electrophoretic mobility of microcrystalline cellulose as a function of pH — with and without alum [9]

No charge reversal occurs below pH 5.0, where much of the aluminum exists as Al^{+3} and mononuclear hydrolysis species. At higher pH values, polynuclear species that can adsorb by mechanisms other than simple ion exchange come into being along with $Al(OH)_3$ precipitate that contains occluded cationic species. Enough cationic charge is adsorbed to make the cellulose cationic. At higher pH's, the $Al(OH)_3$ converts to the negatively charged $Al(OH)_4^{-1}$, which cannot reverse the cellulose charge.

Since the aluminum species equilibria and adsorption behavior are concentration-dependent, the curves in **Fig. 11-5** will also depend on the alum concentration. At higher concentrations, one would expect charge reversal to occur at lower pH values since it is related to the polynuclear species and cationic $Al(OH)_3$, both of which will appear at lower pH for higher concentrations. (See the above discussion.)

h. Preferential adsorption of alum by fines

The above discussion of alum adsorption emphasized the roles played by the relative charges on aluminum species and fibers. In a papermaking furnish, the specific surface area of the particulate components also plays an important role. It is a general feature of all adsorption phenomena that the extent to which a particular furnish component participates is a strong function of its specific surface area. This is an important consideration in highly filled grades, groundwood papers, and other furnishes having high fines levels. Fines typically have specific areas 8-10 times higher than fibers. Consequently, adsorbing additives tend to partition themselves accordingly and a very high fraction of an additive will be found on the fines fraction of such furnishes. These additives are usually less efficient than those located on fibers and so the overall efficiency of the system declines. Worse yet, if the fines retention is low, then the additives can be lost completely. For a more complete discussion of this important subject, see References [10,11].

i. Practical implications of the aluminum/pH/concentration equilibrium

The apparently large number of polynuclear complexes that aluminum can form, coupled with their sensitivity to pH, concentration, temperature, and other factors, has lead to a certain amount of controversy regarding their exact stoichiometry and equilibria. Nevertheless, the equilibria presented in **Fig. 11-3** and adsorption behavior illustrated in **Fig. 11-4** represent useful "models" that provide a framework for thinking and reasoning about the wet end chemistry of alum-containing systems. And upon which wet end chemistry trouble-shooting and optimization strategies can be based.

Table 11-1 lists examples of papermaking applications of alum. It will be instructive to revisit some of them now to illustrate how the fundamentals discussed in the chapter come into play.

Dye retention - Many papermaking dyes are anionic and, consequently, will interact with a cationic material like alum. Two types of mechanisms are possible. First, cationic alum can react with anionic dyes to form a cationic precipitate that is attracted to fiber surfaces and retained there. This mechanism requires a pH in the 4.0-4.7 range, where the aluminum species will be highly charged.

The second mechanism involves occlusion of dye molecules within the gelatinous $Al(OH)_3$ precipitate. In this instance, one must work at a pH that produces $Al(OH)_3$ plus some cationic aluminum species that will contribute a net positive charge to the precept and make it attractive to fiber surfaces. Consequently, a pH in the 4.5-5.2 range should be preferred where both $Al(OH)_3$ and polynuclear species will exist.

Fillers and fiber fines retention; drainage aid - Alum acts as a retention aid through a coagulation mechanism where it reduces the zeta potential of particles to the point where aggregation can occur when they collide. Obviously then, one must operate at the 4.8-5.2 pH range where the cationically charged polynuclear and hydrolysis aluminum species predominate. These species exhibit the strongest adsorption on fiber surfaces and have the greatest ability to reduce the particle (fibers, fines, filler) zeta potential

Anionic trash, interfering substance neutralizer - Alum is a favorite material for this purpose due to its relatively low cost and potentially high charge density. Since the mechanism is a solution-based charge neutralization reaction, the pH should be in the low 4's to promote the formation of Al^{+3} and inhibit the formation of aluminum species that are strongly adsorbed by fibers, fines, and fillers.

pH control agent - This is a very clear cut case. Alum is used to control pH down to a range of 4.2-4.7. It does so by virtue of the hydrolysis reactions listed in **Fig. 11-1**. Consequently, the starting alum must be at a pH where there is adequate hydrolyzable aluminum ion present.

It should be emphasized that alum is not a very good pH control agent. It is a weak acid and is not effective in the face of high alkalinity. It is a better practice to decouple pH control and alum addition completely. Use an acid (e.g. H_2SO_4 or HCl) for pH control and meter alum according to functional needs (e.g. sizing, retention, dye fixation).

Pitch control agent - Natural wood contaminants (resin acids, fatty acids) can be precipitated onto fiber surfaces through reaction with cationic aluminum species. Consequently, to be effective, the pH and alum concentration must coincide with the formation of charged aluminum species.

Rosin sizing agent - This topic is discussed at length in Chapter 13. It does provide one of the clearest examples of the importance of the aluminum equilibrium, however, and so it is worth mentioning here. When using rosin soap size, alum provides the acid condition and the cationically charged ions which precipitate the rosin. In addition, alum provides a positive charge to the precipitate so that it will be attracted to the fiber surface. These reactions occur in suspension away from the fiber surface. Consequently, soap sizing is done at pH values in the 4.2-4.6 range, where the soluble, nonadsorbing, cationic aluminum species predominate.

Sizing with dispersed rosin size is a different story. There is only a limited reaction between alum and rosin in suspension. The papermaker must rely upon other mechanisms for rosin retention. The alum is retained on the fibers independently of the rosin and the two react in the dryer section. This mechanism calls for operating in the 4.8-5.5 pH range, where the aluminum species are strongly adsorbed by fibers.

These are just a few examples of how the papermaker can put fundamental knowledge about alum chemistry to practical use to produce a salable product. What about the adverse effects of alum listed in **Table 11-2**? Can they be minimized or eliminated by applying fundamental knowledge?

The first two considerations — the corrosiveness of alum papermaking systems and the impermanence of alum-containing papers — are pretty much a given. To be effective, alum must be used under acid conditions and these two effects follow. The third item in **Table 11-2** — the use of calcium carbonate filler in alum-based papermaking systems — is a result of the sensitivity of $CaCO_3$ to acid conditions. It is well-known that acid decomposes $CaCO_3$ to produce CaO and CO_2, a gas which creates foam. This fact makes it difficult to use alkaline paper broke in acid papermaking systems.

A related issue is the use of alum-rosin sizing under neutral papermaking conditions. Some papermakers desire to convert to $CaCO_3$ filler, but want to continue using a rosin sizing agent. It has been shown that, by using dispersed rosin size and employing tight process control, this is possible and some mills are currently operating this way.

The last item in **Table 11-2** relates to the excessive addition of alum to papermaking systems. Since papermakers add alum for so many different purposes, there has been a tendency to overuse it — leading to the advent of severe deposit problems. Alum deposition is a logical consequence when the concentration is increased beyond desirable limits in a papermaking system.

Figs. 11-3, 11-4 and the accompanying discussion provide an explanation of this phenomenon. Consider what happens to the aluminum species equilibrium when the concentration increases. As we stated earlier, the respective equilibria for the various species all shift to lower pH levels. In other words, $Al(OH)_3$ precipitates at a lower pH than at lower concentrations. Thus, even though you may be operating at a pH that would inhibit $Al(OH)_3$ formation under "normal" circumstances, excess alum addition shifts the precipitation pH downward and deposits occur.

There is another, related, alum deposit phenomenon called "pH shock" that many mills have experienced. In this instance, alum-containing white water is subjected to a sudden pH increase at some point — perhaps from a shower whose pH is greater than the white water. This local pH increase shifts the alum equilibrium toward $Al(OH)_3$ precipitation for a period of time that depends on the relative capacity of the system. In any event, a small quantity of $Al(OH)_3$ is produced that will deposit on the nearest surface or shear point. Such deposits commonly fill felts and forming fabrics and are very difficult to remove. A similar effect can be gotten from a temperature shock if the alum concentration is high enough. Refer to Reference [12] for an excellent analysis of a pH shock alum deposit case.

4. The use of sodium aluminate as an aluminum source in papermaking [13]

The $SO4^{-2}$ ion that accompanies the aluminum in alum can lead to severe deposit problems in the presence of Ba^{+2} and other metal cations that form insoluble sulfates. This situation often occurs in highly buffered systems that require high levels of alum to achieve a given pH. Partial replacement of alum by sodium aluminate, $NaAlO_2$, has been advocated as one way to address this problem.

The aluminate ion would appear in the aluminum species equilibria presented in **Fig. 11-3** once the pH had been increased to 8.0. Continued increase of the pH to 10 would see the conversion of practically all of the $Al(OH)_3$ into the $Al(OH)_4^{-1}$ ion, which can be changed to AlO_2^{-1} through the loss of two H_2O molecules.

Since aluminate is part of the aluminum/pH equilbrium scheme, adding $NaAlO_2$ to an alum solution does not interfere with the equilibrium and the aluminum level is increased without a concommitant increase in SO_4^{-2}. In spite of being theoretically attractive, the material has never been widely adopted by the paper industry for use as a chemical additive.

5. The use of polyaluminum chloride (PAC) as an aluminum source in papermaking [14–16]

In the discussion of alum applications above, it

was stated that cationic charge density and adsorption onto fibers and fines were important properties for effective alum functionality. It was also pointed out that the polynuclear aluminum species formed from alum in the 4.4-4.8 pH range carried the greatest cationic charge and also adsorbed strongly onto fibers. Unfortunately, when formed from alum, these species only exist over a narrow pH range. Further disadvantages of alum are the presence of SO_4^{-2} ion, which promotes the formation of $BaSO_4$ scale and the potential for $Al(OH)_3$ deposits when alum is overdosed or the pH is allowed to increase.

A material that avoids some of the above alum disadvantages and provides polynuclear aluminum species over a broader pH range is polyaluminumhydroxy chloride, which is also referred to simply as polyaluminum chloride, or PAC, in the paper industry.

PAC refers to a series of polyhydroxide chlorides of aluminum with the general formula "$Al_n(OH)_{3n-m}$," where *m* is determined by the degree of neutralization, with hydroxyl ions substituted for part of the chloride ions of aluminum chloride.

The degree of polymerization in PAC is sensitive to the hydroxyl ion concentration. For this reason, polyaluminum chloride is characterized by the *percent basicity*, which is defined as:

$$\text{Percent basicity} = 100 \times [OH^-]/\{3[Al^{+3}]\}$$

where $[OH^-]$ and $[Al^{+3}]$ are the molar concentrations of hydroxide and aluminum ions, respectively. PAC's used in papermaking typically have basicities of 25-75%.

It is has been shown that the nature of the polymerized aluminum species is strongly dependent on how the polyaluminum chloride is manufactured and the basicity. There are several manufacturing routes for the synthesis of PAC. For example:

- Aluminum chloride can be reacted with a base, such as NaOH, Na_2CO_3, NH_4OH or lime.
- Aluminum hydroxide can be reacted with HCl.
- Aluminum metal may be dissolved in aluminum chloride.

Many different polynuclear aluminum species have been reported to exist in PAC solutions. Between pH 2-5, $Al_7(OH)_{17}^{+4}$, $A_{18}(OH)_{20}^{+4}$, $Al_3(OH)_4^{+5}$, and $Al_{14}(OH)_{34}^{+8}$ have all been cited by various researchers. Above pH 5, the $Al_{13}O_4(OH)_{24}^{+7}$ structure is believed to be the most prevalent.

PACs have been used for a variety of papermaking applications, such as sizing (dispersed sizes under both acid and alkaline pH conditions), pitch control, white pitch control, deposit reduction, fines and filler retention, and interfering substance neutralization (acid and alkaline papermaking systems). In these uses, both charge neutralization and coagulation mechanisms are involved.

6. Summary

Alum has earned its title of "papermaker's friend," having been used for over 150 years. For much of this time, very little was known about the complex chemistry of the material or how to use it the most effectively. Fortunately, research has revealed much about the details of aluminum chemistry and it is now possible to employ this knowledge in a wide range of papermaking applications.

7. References

[1] Hagemeyer, R. W. "History of neutral/alkaline papermaking." in Course Notes from 1990 TAPPI Neutral/Alkaline Papermaking Short Course, Orlando, FL, October 16-18, TAPPI PRESS, 1-5(1990).

[2] "Papermakers' alum — properties". TIS 0506-02, TAPPI Technical Information Sheets, TAPPI PRESS, 1992.

[3] Wortley, B. H. 'Alum." Chapter 4 in Chemical Processing Aids in Papermaking: A Practical Guide. K. J. Hipolit, Editor. TAPPI PRESS, Atlanta, GA. 53-61(1992).

[4] Avery, L. P. "The quantitative alum analysis of a papermaking furnish and the effect of alum on retention." Tappi Journal. 62(2): 43(1979).

[5] Hayden, P. L. and Rubin, A. J., in Aqueous-Environmental Chemistry of Metals, A. J. Rubin, Editor, Ann Arbor Science, Ann Arbor, MI, 317(1976).

[6] Arnson, R. "The chemistry of aluminum salts in papermaking." Tappi Journal, 65(3): 125-130(1982).

[7] Strazdins, E. "Chemistry and application of rosin size." Chapter 1 in The Sizing of Paper. W. F. Reynolds, Editor, TAPPI PRESS, 1-31(1989).

[8] Jaycock, M. and Pearson, J. L., Svensk Papperstid. 78(5): 167(1975) and 78(8):289(1975).

[9] Lindstrom, T. and Eklund, D. "Aluminum and its chemistry in aqueous solutions." Chapter VI in Paper Chemistry: An Introduction. DT Paper Science Publications, Grankulla, Finland, 134-144(1991).

[10] Marton, J. and Marton, T. "Effect of fillers on rosin sizing." Tappi Journal, 66:12, 68-71(1983).

[11] Scott, W. E. "Fines management and control in wet end chemistry." Tappi Journal, 69:11, 30-34(1986).

[12] Springer, A., Porwal, S., and Procter, A. "Scale deposits on the fourdrinier wire of a fine paper machine." Tappi Jorunal, 63(6): 67(1980).

[13] Davison, R. W. "Internal Sizing." Chapter III in Pulp and Paper Manufacture, Vol. 6, Stock Preparation. R. W. Hagemeyer and D. W. Manson, Editors, Joint Textbook Committee of the Paper Industry, TAPPI/CPPA, 39-64(1992).

[14] Wortley, B. H. "Polyaluminum chloride and calcium carbonate application in papermaking." in Course Notes from 1990 TAPPI Neutral/Alkaline Papermaking Short Course, October 16-18, Orlando, FL, TAPPI PRESS, 119-124(1990).

[15] Crawford, R. A. and Flood, T. A. "Preliminary NMR study on the structure of polyaluminum chloride." in Proceedings of 1989 TAPPI Papermakers Conference, Washington, D.C. April 10-12, 55-57(1989).

[16] Rettinger, J. J. "New chemical technology gives improved sizing, retention and scale control." in Proceedings of TAPPI Papermakers Conference, Atlanta, GA, April 18-21, 477-482 (1993).

Chapter 12

Introduction to Internal Sizing

1. Introduction

a. Reasons for sizing paper

Papermaking fibers have a strong natural tendency to interact with water. This property is important to the development of strong interfiber hydrogen bonds during papermaking and is also the reason that paper loses its strength when saturated with water. This behavior is an advantage for certain absorbent paper grades such as towelling and tissue. Corrugating medium must also be highly absorbent to convert properly in the corrugating process. Such properties are not common for most paper grades, however. It is more typical for papers to interact only moderately with water during use. **Table 12-1** lists some common examples.

Table 12-1. Paper end uses and converting requirements needing liquid penetration resistance

- Liquid packaging
- Offset printing
- Wet/frozen foods packaging
- Gluability
- Water-based ink holdout
- Off-machine coating

Liquid penetration resistance is also important to many papermaking processes where paper is brought into contact with aqueous solutions, such as size pressing, coating, and calender water box applications. Finally, many papers (e.g. office papers, cereal boxes, cosmetic boxes, grocery bags) must be able to resist accidental wetting during use.

b. Two types of sizing processes

Processes that impart resistance to water penetation in paper are called sizing processes. There are two types of such processes. In one approach, called *surface sizing*, the paper web is formed and dried and a sizing agent is applied to the paper surface by a size press, coater, or calender box. In the second approach, called *internal sizing*, sizing additives are added to the papermaking furnish and get incorporated into the web when it is formed. Both processes affect the wettability of paper fibers, while surface sizing can also reduce the web pore size. **Table 12-2** lists some of the common materials used for these two purposes. This discussion will focus on internal sizing.

Table 12-2. Some typical materials used for internal and surface sizing

Internal Addition	Surface Treatment	
To Reduce Fiber Wettability	To Reduce Fiber Wettability	To Reduce Pore Radius
Emulsion size	Reactive size	Carboxymethyl cellulose
Reactive size	Wax emulsion	Stryene–Maleic Anhydride copolymers
Rosin size		Starches
Wax emulsion		
Stearates		
Fluorocarbons		

c. Some sizing terminology

Up to now, the term *resistance* has been used without definition. With respect to sizing, the term refers to characteristics that slow down or resist the passage of liquids through paper, i.e. sizing is a rate phenomenon. It is important to recognize that resistance properties are different from *barrier properties*. Barrier properties are attributes that absolutely prevent the passage of liquids through paper. Internal sizing produces resistance properties, while other processes, such as coating, saturation, or lamination are required to achieve barrier properties.

Papers are said to be waterleaf, slack sized, or hard sized, depending upon the their degree of sizing. **Table 12-3** contains definitions of these terms.

Table 12-3. Some common sizing terms

Waterleaf Papers: Very absorbent papers which have no water resistance (e.g. toweling, tissue, corrugating medium).

Slack Sized Papers: Papers that have intermediate levels of water resistance (e.g. offset printing papers, writing papers).

Hard Sized Papers: Very highly sized papers that have high liquid penetration resistance. (e.g. cup stock, milk carton raw stock).

Fugitive Sizing: The disappearance of liquid penetration resistance in a paper that exhibits this property shortly after it is manufactured (e.g. a loss of more than 25% of the original sizing test value within seven days of manufacture).

Self Sizing: The gradual development of water resistance in a paper that does not have this property immediately after being manufactured. This phenomenon can occur by the vapor phase redistribution of naturally occurring resin and fatty acids.

d. Chapter contents

As you might expect, since most papers receive some degree of internal sizing during manufacture, internal sizing is an extremely important topic in wet end chemistry. A complete discussion of the topic would be beyond the scope of this book and we will adopt a broad approach to the subject instead. Fortunately, excellent detailed treatments of sizing exist and you are referred to References [1-10] for further study.

2. Basic surface science considerations important to sizing

a. Introduction

We are interested in the rate and extent of liquid (usually water) penetration into paper under a variety of conditions. Such phenomena come under the general surface and colloid science heading of *capillarity*. An introduction to this subject may be found in any of several classic texts, such as those designated in References [11] and [12]. It is clear that the complex physical and chemical nature of a paper web makes it impossible to undertake a rigorous and complete theoretical analysis of the situation. Much can be learned about the interactions between paper and liquids by applying the fundamental principles in a general fashion. This has been done by several researchers. The treatments by Lindstrom and Swanson are recommended reading for anyone wishing to learn more about this subject.[4,13] The following discussion leans heavily upon the insights of these two authors.

b. Four ways that liquid penetrates paper[14]

It has been hypothesized that liquid moves through paper by the following three mechanisms:

- Capillary action through pores
- Diffusion through pores
- Diffusion through fiber solid fraction

It was also suggested that a fourth mechanism involving evaporation, vapor phase diffusion, and condensation could occur.

c. Capillary flow in pores: the Washburn Equation

Since a paper web can be considered to consist of a series of interconnected pores, it is natural that of the four mechanisms listed above, the most attention has been paid to capillary flow in pores. More than fifty years ago an equation, which has become known as the Washburn Equation, was developed to describe the penetration of liquid into fiber beds or packed powders. [15]

In its differential form, the Washburn Equation is the following:

$$dl/dt = r\gamma\cos\theta/(4\eta l)$$

where l = the capillary distance penetrated in time, t; r is the capillary radius; η is the liquid viscosity; γ is the liquid surface tension; and θ is the contact angle between the capillary walls and the liquid.

It is assumed that no significant external pressure exists on the penetrating liquid and that the penetrated structure does not change with time.

d. The contact angle and surface wetting

The Washburn Equation suggests that in order to achieve low capillary penetration, it is necessary to have a high contact angle and/or a small capillary radius. In papermaking, paper pores radius can be reduced by applying a film-forming substance at the size press. This is alluded to in **Table 12-2**. The contact angle is influenced by several factors.

When a nonvolatile pure liquid drop is placed on a clean, smooth, insoluble, plane solid surface, a balance of forces is established between the liquid and the surface. If the forces between the liquid and the solid are greater than the forces between molecules within the liquid, then the drop will spontaneously spread on the solid surface. This is referred to as *wetting* behavior.

If the forces reach an intermediate balance, then the liquid drop will establish a contact angle with the solid as shown in **Fig. 12-1**.

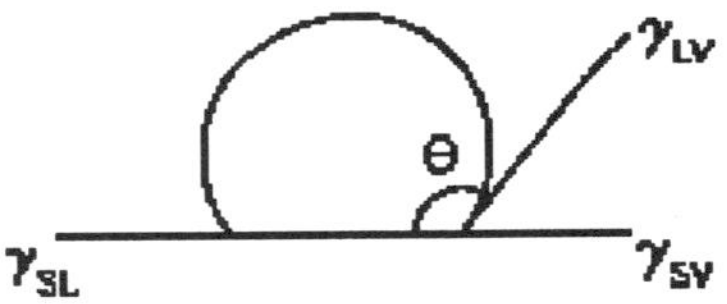

Figure 12-1. Formation of contact angle between a liquid drop and a solid surface, where γ_{LV} is the tension at the liquid-vapor interface, γ_{SV} is the tension at the solid-vapor interface, and γ_{SL} is the tension at the solid-liquid interface.

When the contact angle is less than 90°, the liquid is said to wet the surface. When the contact angle is greater than 90°, nonwetting is said to occur. **Fig. 12-2** summarizes the situation.

e. Relationship between contact angle and surface tension: Young's Equation

Tha balance of forces existing at the line of contact between the liquid and solid in **Fig. 12-1** is expressed by the Young Equation.

$$\cos\theta = (\gamma_{SV}-\gamma_{SL})/\gamma_{LV}$$

Reference to the Washburn and Young equations provides insight into how internal sizing acts to impart water resistance to paper. Sizing agents are hydrophobic and lower the surface energy (surface tension) of the fiber surfaces (γ_{SV}), thereby reducing the value of the numerator in the Young Equation. As ($\gamma_{SV}-\gamma_{SL}$) decreases, $\cos\theta$ decreases also and the dl/dt term in the Washburn Equation becomes smaller. Thus, the water penetration rate decreases — the objective of internal sizing. In the extreme, hard sizing ($\gamma_{SV}-\gamma_{SL}$) becomes less than zero and the capillary repels the penetrating liquid.

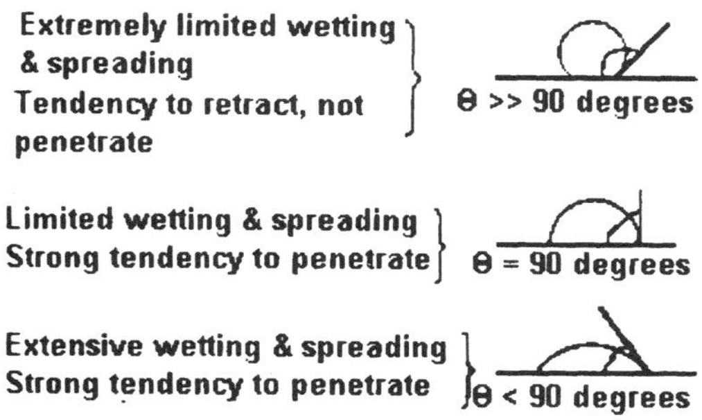

Figure 12-2. Relationship between contact angle and wetting.

f. Complicating factors

The above relationships were derived for pure liquids and clean, perfectly flat, stable surfaces. It is clear that none of these assumptions hold for paper — a rough, porous, swellable material having variable surface composition. All of these factors influence the measured contact angle and introduce uncertainty. It has also been observed that it makes a difference whether you measure an "advancing" or "receding" contact angle (contact angle hysteresis). In spite of these drawbacks in the theoretical treatment, the basic concepts hold and provide a good model upon which to base one's thinking.

g. The importance of external applied pressure

In most cases where paper is brought into contact with a liquid, (such as size pressing, coating, or printing) an external pressure exists on the liquid. Lindstrom has provided an approach that takes this fact into account. [4] His version of the Washburn Equation has the following form:

$$l^2 = [2r\gamma t\cos\theta + P_E r^2]/(4\eta)$$

where P_E is the external pressure applied to the liquid.

In this case, the relative importance of the surface chemistry capillarity effects are reduced and the capillary radius becomes more important. **Figs. 12-3** and **12-4**, taken from Reference [4], illustrate the strong interaction between surface chemistry and external pressure with respect to liquid penetration.

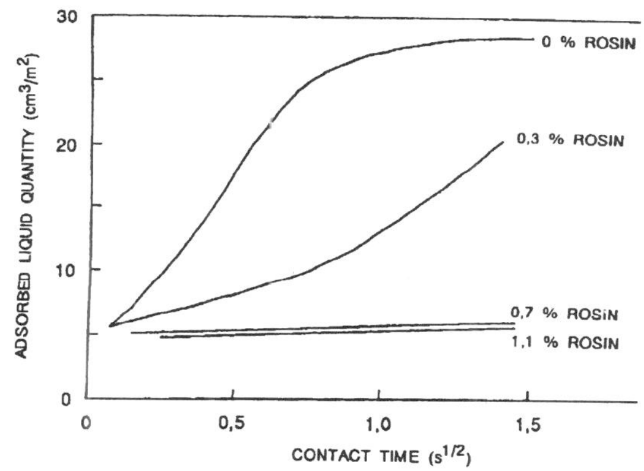

Figure 12-3. Liquid penetration in sized and unsized paper. No external pressure on the liquid. [4]

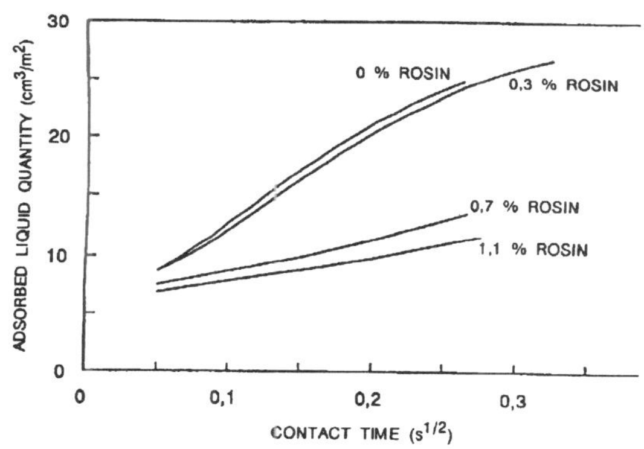

Figure 12-4. Liquid penetration in unsized and sized papers. External pressure of 0.5 atm. on liquid. [4]

Fig. 12-3 illustrates the strong effect of sizing on liquid penetration. At the highest sizing levels, practically no penetration occurs in the absence of an external pressure on the liquid. At sizing levels of 0.7% or higher, the only mechanism for liquid transport is diffusion.

Adding an external pressure to the capillary forces reduces the relative influence of the sizing and increases the overall liquid penetration at all levels of sizing. Lindstrom points out that under these conditions, it is likely that the primary role of a sizing agent is to protect fiber-fiber bonds from being disrupted by the penetrating water. [4]

h. Closing comments

These brief remarks have introduced some of the fundamental considerations important to the interactions between paper and liquids. You are encouraged to explore this topic further in order to gain a greater understanding of the basic surface science mechanisms important to sizing.

3. Measurement of degree of sizing

The papermaker carries out size testing for two purposes: (1) to meet customer specifications; (2) to control paper machine production rate. In the latter case, the testing is often related to size press and/or on-machine coater runnability.

There are a wide variety of sizing tests employed by the paper industry. **Table 12-4** lists several of the most common ones.

Most size testing methods fall into one of two categories. In one group of tests, the amount of water that a paper specimen can absorb in a given time period is measured. In the other group of tests, the time required for water to penetrate a specified distance into or through a paper specimen is determined. A few tests, such as the pen and ink and contact angle tests, do not fall into these two classifications.

Since liquids can penetrate paper by a variety of mechanisms, and different size tests emphasize one mechanism over another, the selection of an appropriate size test and penetrant is a complicated undertaking. This matter has been addressed at length by Swanson,[13] Kumler,[16] Gess,[1] and Scott and Abbott,[17] among others.

The path to follow is to select a test that best fits the use requirements of the paper being tested and then learn all that you can about the variables and conditions that affect the results achieved from the test. For example, the Hercules photosize size (HST) is one of the most common routine control tests used by papermakers today. However, HST results are influenced by the brightness, color, opacity, basis weight, and filler content of the paper being tested. The test should also be limited to papers which range in grammage from 50 g/m^2 to 250 g/m^2. Lightly sized paper can give erroneous results if the liquid penetration time is very short.

4. A few comments about the history of sizing in papermaking

Papermaking is an ancient technology, dating from about the year 105 A.D., when it is believed to have been invented in China. It is not known exactly when papermakers decided that paper needed to be treated for water resistance, but by the mid-1600's it was standard practice to tub size book and writing papers to prevent the feathering of inks. Animal glue was the principal sizing agent. Alum (potassium aluminum sulfate) was used to harden the applied glue. Earlier papers had been treated with starch solutions for surface glazing purposes. Alum rosin sizing entered the scene about 1807 and its introduction is usually attributed to Moritz Illig. Later, in 1876, papermaker's alum, $Al_2(SO_4)_3$, was introduced. The next period of intense sizing technology development began in the 1950's. Since that time, the various forms of rosin size

Table 12-4. Size testing methods

Name of Test	Property Measured	Types of Paper and Board
Water drop absorption	Time required for droplets of water to be absorbed by paper.	Heavy paper and paperboard, tissues, toweling.
Ink flotation test	Time required for writing ink to penetrate through a floating specimen and change the color of the upper surface.	Moderately sized paper.
Hercules photosize size test	Decrease in reflectance of the opposite side of a sheet of paper which has been covered with a given amount of ink.	Wide range of sized papers.
Cobb size test	Amount of water absorbed by a given area of paper in a specified period of time.	Paper and paperboard that do not become saturated too quickly.
Pen and ink feathering test	Extent of spreading of lines drawn on sample with a steel pen and ink.	Bonds and other writing papers.
Contact angle	Tangential angle from the horizontal which the base of a drop of liquid develops when carefully placed on a sheet of paper.	Any paper of board that does not absorb the drop too quickly for the measurement to be made.
Edge penetration test; edge wicking test	Lineal penetration of liquid in the in-pane dimension of paper.	Extruded papers, e.g. liquid packaging, photographic.

(paste, dispersed, fortified), alkyl ketene dimer size, and alkenyl succinic anhydride size have all been developed and adopted by papermakers.

In the US, most paper was sized by alum rosin systems until the 1980's when a trend to convert to alkaline papermaking began. Since then, most fine paper manufacture has changed over and these grades are now sized with either alkyl ketene dimer or alkenyl succinic anhydride sizes. Alum continues to find limited use in conjunction with these materials.

For more complete discussions of the history of sizing, refer to References [4], [18], [19].

5. Requirements of a good internal sizing agent

In our discussion up to this point, we have defined sizing as a process whereby a sizing agent is applied to papermaking fibers to lower their attraction for liquid water. Another way to look at this is say that *sizing makes a paper surface more hydrophobic by reducing its surface tension (or surface energy).* It is generally accepted that to be a good sizing agent, a material must meet the following criteria.

Table 12-5. Requirements of a good internal sizing agent

__In order to be a good sizing agent, a substance must:__

- Have high retention on fibers
- Be uniformly distributed on fiber surfaces
- Be able to produce hydrophobic fiber surfaces
- Adhere strongly to fibers
- Be chemically inert to the penetrants of interest

In addition to the above five criteria, it is also desirable that a sizing agent have no adverse effects on the papermaking process. We will briefly consider each of these criteria in general terms here and later, with respect to each of the specific sizing agents described.

a. High retention of size on fibers

It is obvious that unless a sizing agent is retained in the papermaking web during sheet formation, it cannot be effective. The primary retention mechanism employed in internal sizing is the *heterocoagulation* of size particles with fibers and fiber fines. (See Chapter 15 for a discussion of heterocoagulation and other aggregation mechanisms.)

In order to promote this phenomenon, it is necessary to impart a positive charge to size particles so that they will be attracted to, and retained by, negatively charged fibers and fines. As will be discussed below, common sizing agents are either anionic or noninonic in natural form. Consequently, it is necessary to use a cationic retention aid to promote size retention. Many different materials can serve this purpose, such as alum, cationic synthetic polymers, and cationic starch. In all instances, they must react with the neutral or anionic size particles to produce positively charged entites that will be attracted to fibers and retained by them.

b. Uniform distribution of size on fiber surfaces

For the most part, uniform distribution is promoted by multiple addition locations at the addition point, good addition point mixing, low size concentration, and low stock consistency. The charge difference between fibers and size needs to be balanced so that size particle retention rate is not too high or the mixing forces will not have time to be effective.

c. Ability to produce hydrophobic fiber surfaces

This property is an absolute requirement of sizing agents because, as was described above, sizing agents work by increasing the hydrophobicity of paper. Most common sizing molecules have a dual polar:nonpolar nature and are called *amphipathic* molecules. The nonpolar part — usually composed of C's and H's — provides the required hydrophobic character.

In addition to having hydrophobic character, it is necessary that a mechanism exist by which the sizing molecule will orient its hydrophobic part away from the fiber surface. In this manner, liquid molecules impinging on the paper surface will "see" the hydrophobic size molecules instead of the water-attracting fiber surface cellulose and hemicellulose molecules.

d. Ability to adhere strongly to the fiber surface

A fourth requirement is that the oriented size moleules must become anchored to the fiber surface so that they do not overturn when contacted by water. **Fig. 12-5** schematically illustrates this point.

e. Be chemically inert to the penetrant

This is a logical requirement. If the sizing agent reacts with the penetrant, then it could be converted to a nonhydrophobic form or be released from the fiber surface and lose its effectiveness. Fortunately, water is the principal penetrant of interest, and it is a relatively benign substance.

f. Adverse effects on the papermaking process

This requirement is listed separately in **Table 12-5** because, in practice, it cannot be met all of the time for any sizing agent. All common sizing agents have attributes that can lead to papermaking process problems under certain conditions. The goal is to minimize these problems through the informed use of sizing materials.

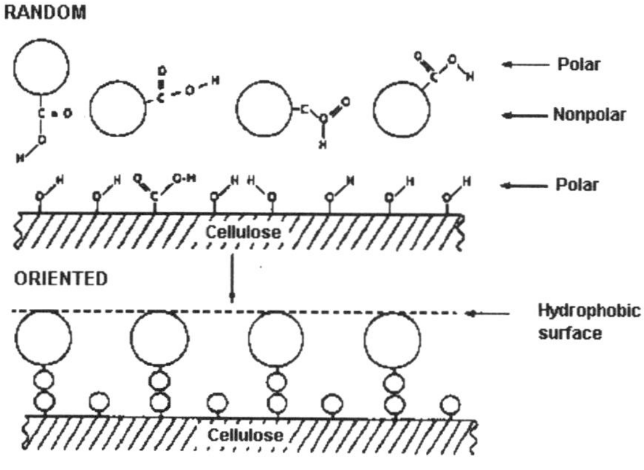

Figure 12-5. Schematic drawing that illustrates the orientation and anchoring of amphipathic (polar:nonpolar) size molecules on the fiber surface. When this occurs, impinging water droplets "see" a more hydrophobic surface than that provided by the fibers alone.

6. Summary

This concludes our introduction to internal sizing. In Chapters 13 and 14 we will consider alum-rosin sizing systems and synthetic sizing systems — the two main categories of sizing systems in use today.

7. References

[1] J. M. Gess. "The sizing of paper with rosin and alum at acid pH." Chapter 7 in Paper Chemistry, J. C. Roberts, Editor. Chapman and Hall, New York, 97-112[1991].

[2] J. C. Roberts. "Neutral and alkaline sizing." Chapter 8 in Paper Chemistry, J. C. Roberts, Editor. Chapman and Hall, New York, 114-131[1991].

[3] Otway, P. and Johnson, D. C. "Sizing in acid, neutral, and alkaline conditions." Chapter 9 in Applications of Wet End Paper Chemistry, C. O. Au and I. Thorn, Editors. Blackie Academic and Professional — Chapman and Hall, New York, 139-171[1995].

[4] Eklund, D. and Lindstrom, T. "Aluminum and its chemistry in aqueous solutions." Chapter VII in Paper Chemistry — An Introduction. DT Paper Science Publications, Grankulla, Finland, 134-141[1991].

[5] Eklund, D. and Lindstrom, T. "Water penetration and internal sizing." Chapter VIII in Paper Chemistry — An Introduction. DT Paper Science Publications, Grankulla, Finland, 192-214[1991].

[6] Sizing of Paper - Second Edition. W. F. Reynolds, Editor. TAPPI PRESS, Atlanta. 156 pp. 1989.

[7] Royce, R. Course Notes from 1995 TAPPI Introduction to Wet End Chemistry Short Course, February 15-17, 1995, Orange Beach, AL. TAPPI PRESS.

[8] Marton, J. and Marton, T. "Some new principles to optimize rosin sizing." 1982 TAPPI Papermakers Conference Proceedings, Atlanta, GA, p. 15.

[9] Wortley, B. H. "Alum." Chapter 4 in Chemical Processing Aids in Papermaking: A Practical Guide. K. J. Hipolit, Editor. TAPPI PRESS, 53-62[1992].

[10] Cates, R. E. "Internal sizing agents." Chapter 5 in Chemical Processing Aids in Papermaking: A Practical Guide. K. J. Hipolit, Editor. TAPPI PRESS, 63-80[1992].

[11] Ross, S. and Morrison, I. D. "Capillarity of pure liquids." Chapter II.A in Colloidal Systems and Interfaces. John Wiley and Sons, New York. 69-125(1988).

[12] Hiemenz, P. C. "Surface tension and contact angle." Chapter 6 in Principles of Colloid and Surface Chemistry. Marcel Dekker, New York. 287-352(1986).

[13] Swanson, J. W. "Mechanisms of paper wetting." Appendix in Sizing of Paper - Second Edition. W. F. Reynolds, Editor. TAPPI PRESS, Atlanta. 133-153(1989).

[14] Nissan, A. H. Proc. Tech. Sect. P.M.A., 30:96(1949).

[15] Washburn, E. W. "The dynamics of capillary flow." Phys. Rev.2:17, 273-283(1921).

[16] Kumler, R. W. "Testing paper and board for sizing." Chapter 7 in Sizing of Paper - Second Edition. W. F. Reynolds, Editor. TAPPI PRESS, Atlanta. 133-153(1989).

[17] Scott, W. E. and Abbott, J. C. "The barrier and resistance properties of paper." Chapter 7 in Properties of Paper: An Introduction. TAPPI PRESS, Atlanta, GA. 123-131(1995).

[18] Hagemeyer, R. W. "History of neutral/alkaline papermaking." in Course Notes from TAPPI Neutral/Alkaline Papermaking Short Course, Orlando, FL, 1-5(October 16-18, 1990).

[19] Hunter, D. Papermaking. The History and Tecnique of an Ancient Craft. Dover Publications, Inc., New York, 1978.

Chapter 13

Alum-Rosin Sizing

1. Introduction

In the previous chapter we were introduced to the need for developing a resistance to liquid penetration in many paper grades and the general approach, internal sizing, taken by papermakers to accomplish this task. In this chapter, we will focus on the alum-rosin sizing system. This system was the first internal sizing system invented and has been practiced universally for nearly 200 years.

Rosin sizing has the unique characteristic that it is based on a chemical substance that is produced by the tree itself. In this sense, then, it is a *natural* sizing process. Indeed, in some paper grades, it is possible to achieve the desired sizing level simply by adding alum to the furnish, thereby retaining some of the dispersed rosin present in the white water system.

Much research has been carried out and many publications have been written about alum-rosin sizing. It is our intention here to introduce the topic broadly and leave a detailed study of the subject to the reader. References [1-7] are recommended as excellent starting points for more in-depth reading.

2. How rosin imparts water resistance to paper

Before launching into our discussion of the different rosin sizes used in papermaking, it will be instructive to look at how rosin imparts water resistance to paper. In Chapter 12, **Table 12-5**, we listed the requirements of a good internal sizing agent. The table specifies that a good sizing agent must adhere strongly to fibers and be able to produce a hydrophobic fiber surface. To illustrate how these two requirements are met by rosin, we will consider a representative rosin size molecule, abietic acid, presented in **Fig. 13-1**.

The abietic acid molecule contains a bulky, nonpolar, tricyclic portion comprised solely of carbons and hydrogens and a polar carboxyl group (COOH). Molecules that have both polar and nonpolar characteristics are called *amphipathic* molecules.

The nonpolar portion is responsible for the molecule's ability to impart hydrophobicity to fibers. The polar carboxyl group plays an equally important role by participating in reactions with alum that anchor the rosin molecule to the fiber surface with its hydrophobic portion oriented outward. (See **Fig. 12-5** for a schematic illustration of this process.) In brief, then, this is how a rosin molecule acts as an internal sizing agent.

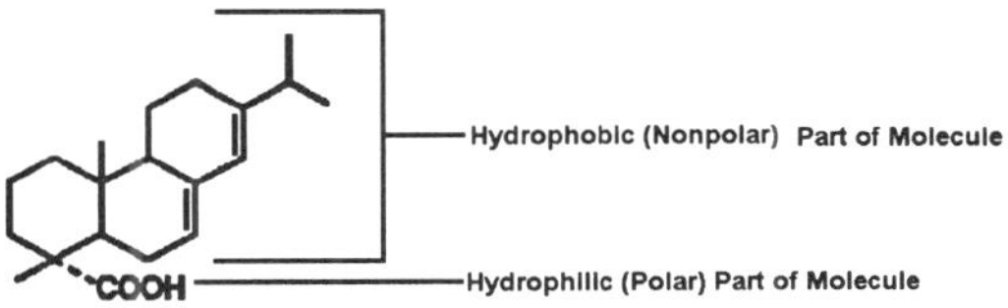

Abietic Acid

Figure 13-1. Molecular structure of abietic acid illustrating the polar and nonpolar parts of the molecule. Such structures are called "amphipathic." The nonpolar portion of the molecule imparts water resistance to paper. The polar carboxyl group participates in the process that attaches the molecule to the fiber surface.

We will discuss how rosin molecules meet the other requirements of a good internal size when we introduce the various forms of rosin used by papermakers below. Before that, we will briefly review some general information about the source and nature of rosin acids.

3. The source and characteristics of rosin acids

a. Introduction

When softwood trees used in papermaking are wounded so that their bark is ruptured, exposing inner wood tissue, a sticky, gummy material exudes from the wound and eventually seals it, thereby protecting the tree from further damage. The exudate is called *oleoresin* or *gum rosin* and is composed mainly of volatile terpenes and higher molecular weight resin acids. The resin acids form the sticky film that seals the wound.

Until the early 1900's, paper sizing was done with resin acids extracted from oleoresin. From about 1910 to 1940 resin acids extracted from pine stumps were used. Since the 1940's, tall oil derived from the kraft pulping process chemical recovery cycle has been the principal source of resin acids.

b. Chemical structure of resin acids

On average, about 5% of the weight of a softwood is comprised of materials, called extractives, that can be isolated by extraction with neutral organic solvents. Resin acids represent 50-80% of the total extractives, depending on the species involved. They consist of the family of structures presented in **Fig. 13-2**.

It is clear that the principal differences among the structures in **Fig. 13-2** are in the positions of their double bonds. The acids are relatively unstable and readily undergo isomerization upon heating and oxidation when exposed to air or sunlight. Abietic acid is usually the predominant acid. It is readily formed from levopimaric acid, and is also formed by prolonged heating of neoabietic and palustric acids.[8]

Due to this reactivity, it would be expected that oleoresin composition would differ from that in tall oil. **Table 13-1** illustrates this point. This compositional variability occurs among species and within a given species in different seasons.

Table 13-1. Composition of oleoresin and tall oil pinewood samples [8]

Acid	Gum Rosin	Tall Oil Rosin
Abietic	33	37
Palustric	5	9
Neoabietic	14	2
Dehyodroabietic	6	23
Isopimaric	18	11
Pimaric	5	3
Dihydroabietic	<1	2
Dihydropalustric	<1	2

4. A brief history of the rosin size products

Rosin sizing of paper dates from the very early 1800's. The resin acid forms used in the early days were not very efficient and, as a result, size suppliers worked to improve their products. The results produced improved sizing efficiency, material handling, size stability, and tolerance to a wider pH range. Gess [3] has traced the major developments in rosin size technology. **Table 13-2** is based on his remarks.

Table 13-2. Development of major rosin size product technology[3]

Approximate Time Period	Product
Early 1800's	First alum-rosin sizing with oleoresin rosin
1930's	First stable rosin acid dispersions
World War II to 1960's	Fortified rosins
1970's	Microparticulate dispersed sizes
1980's	Thin and extended soap sizes
1990's	Cationic dispersed rosin sizes

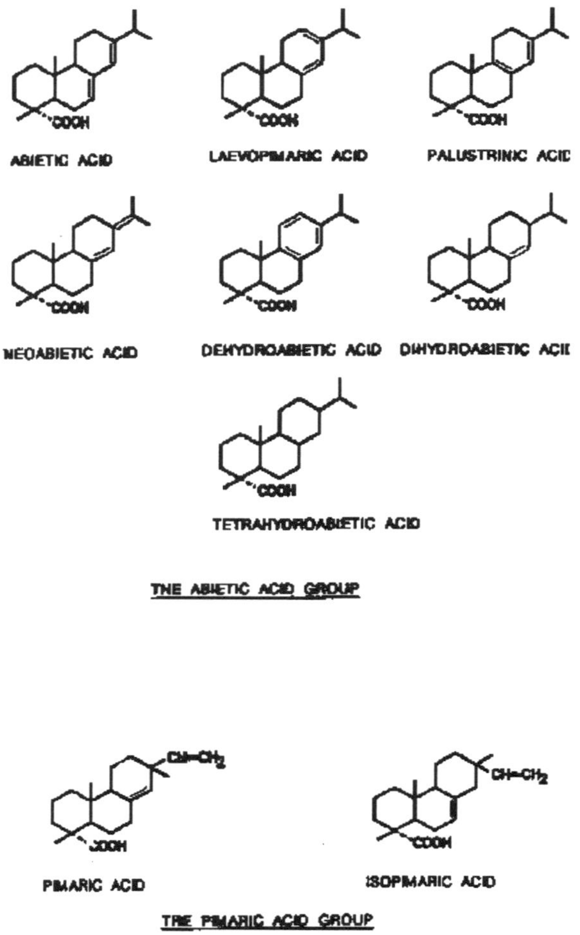

Figure 13-2. Resin acid structures from which rosin size is derived

5. Rosin sizing terminology

A considerable collection of rosin sizing technical jargon has developed over the years. In order to facilitate our discussion of different rosin sizes, some of the more important terms are defined in **Table 13.3**.

Table 13-3. Definition of important terminology associated with rosin sizing

Free rosin: Resin acids whose carboxyl groups are in the completely protonated form.

Dispersed size: High (95-100%) free rosin dispersion of resin acids. The dispersed size particles are stabilized by surfactants, starch, polymers, or proteins.

Rosin soap size: Rosin size composed of the sodium salts of resin acids. Formed by neutralizing (saponifying) resin acids with NaOH or some other sodium base, such as Na_2CO_3. Different forms of soap sizes are available, including paste, extended, and dry sizes.

Extended size: A 50% solids product consisting of a 50-50 mixture of completely neutralized resin acids and urea.

Fortified size: Rosin size that has been reacted with fumaric acid or maleic anhydride to form the Diels-Alder adduct. The reaction product contains extra carboxyl groups and produces more efficient sizing response than the unreacted resin acids. Used as a starting material for other size products.

Dry size: A 100% neutralized, dry rosin size product that dissolves easily in water.

Paste size: An 80% neutralized rosin size product marketed as a 70-80% solids paste.

6. Fortified rosin size

As a result of extensive experimentation with rosin sizes that occurred during the period following World War II, it was discovered that rosin sizing efficiency could be markedly improved by reacting the levopimaric acid component of rosin with maleic anhydride, fumaric acid, itaconic acid, or other acids that could undergo a Dials-Alder reaction with the material (**Fig. 13-3**).

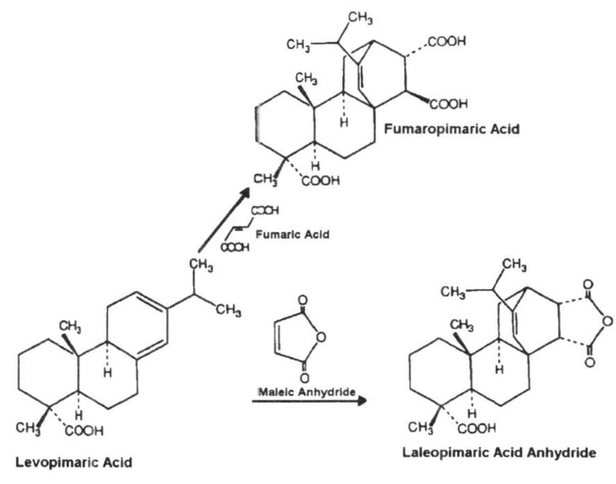

Figure 13-3. Preparation of fortified rosin by reaction of levopimaric acid fraction with fumaric acid or maleic anhydride. The resulting adduct improves sizing efficiency.

The reaction product contains three carboxyl groups and is more hydrophilic and anionic than the original levopimaric acid.

Empirically, it has been found that products containing approximately 20% of the adduct give the most efficient sizing.[3] The reason for this improvement has never been conclusively proven. It is hypothesized that fortified sizes, being more anionic, tend to form smaller particle sizes and are thereby distributed over the fiber surfaces more widely and uniformly. A second suggestion is that the extra carboxyl groups provide more reaction sites for alum and, hence, promote stronger anchoring of size molecules to the fiber surface. Whatever the reasons, fortification does lead to improved performance. Consequently, current rosin size products are all produced from fortified starting materials.

7. Commercial forms of rosin sizes in use today

There are two categories of rosin size in use today. The first, rosin soap size, is composed mainly of the sodium salt of free rosin acids. The second, dispersed rosin size, consists of essentially 100% free rosin acids. **Table 13-4** lists the common product types available for each category.

Table 13-4. Common rosin soap and dispersed size products used in papermaking

Rosin Soap Sizes	Dispersed Rosin Sizes
Paste	Anionic dispersed
Extended	Cationic
Dry	

Rosin soap sizes

The three types of rosin soap sizes have the following characteristics:

Paste size: Paste size contains about 80% of neutralized size and 20% of free rosin acid. It is sold as a 70-80% solids paste in water and requires special handling equipment in the mill. It is commonly diluted to ca. 5% solids prior to use on the paper machine.

Extended size: A 50% solids size that contains a 50:50 mixture of neutralized rosin and urea. It requires no special handling and dilution equipment in the mill.

Dry Size: A 100% neutralized rosin size that is dried. It is readily water soluble. It requires no special handling and dilution equipment in the mill.

8. Practical applications of soap size

a. Emulsification of paste size[2,6]

Paste sizes are viscous and do not flow at room temperature, but thin out and become pumpable at above 82°C. These materials, at 70-80% solids, are usually shipped in tank cars equipped with heating coils that will heat the size to a pumpable temperature. Low pressure steam should be used to prevent desiccation of the size on the heating coils, which reduces the heat transfer.

A two stage process is required to produce a stable, dilute size emulsion. In the first stage, the paste size, heated to 82°C, is mixed with hot water to produce a 12-16% solids primary emulsion. In the second stage, the hot primary emulsion is diluted with cold water and good agitation to 2-5% concentration.

The appearance of emulsified size at 3% solids depends upon the free rosin content and the amount of neutral, non-rosin components present. A properly emulsified rosin size is almost transparent and is stable for prolonged periods. If the dilute size is cloudy, it may be further destabilized by hydrolysis, CO_2 absorption from air, and water hardness (formation of insoluble Ca^{+2} soap). An unstable emulsion is strongly opaque and may cause deposits in storage tanks and coat the inside of dilute size lines. In areas of water hardness above 100 ppm, softened water must be used for emulsion preparation.

b. Preparation of extended rosin size

This material, which can have a solids level of 35-60%, has the viscosity of water and requires no special emulsification procedure in use. It can be diluted directly with cold water and stored at room temperature.

c. Preparation of dry rosin size

Dry rosin size is cold water-soluble and can be added directly to paper stock. It may also be predissolved in cold or warm water to give a 2-5% solution that can be added batchwise or continuously to the stock.

d. Addition of soap size to paper stock

Traditionally, rosin size was added to thick stock at the blend chest, machine chest, or the machine chest pump, at a pH of 7.0-7.5. Alum was then added to a final headbox pH of 4.4-4.6. Recently, it has become more common to add the alum first and then the rosin. This procedure, known as *reverse sizing*, insures that the rosin reacts with alum rather than other substances present in the stock (e.g. Ca^{+2}, alkalinity, interfering substances).

The exact size addition level employed will depend upon furnish conditions and desired sizing level. Most dosage levels range between 2-25 lbs/ton. Common alum-to-soap size ratios are 2.0-2.5.

9. Mechanism of sizing with soap sizes

a. Introduction

From early on, investigators have been interested in the fundamental mechanisms by which rosin sizing agents operate and a large number of papers have been written on the subject. In recent times, Strazdins and Marton have made particularly significant contributions to this area and you are encouraged to examine References [2] and [9] for a discussion of their findings.

According to the criteria laid down in Chapter 12, soap size must meet the following requirements to be a good size:

1. The rosin precipitate must be retained well on the fiber surfaces and be uniformly distributed.
2. The retained rosin must have a low surface energy and therefore a high hydrophobicity.
3. The wet rosin precipitate must be transformed to a stable low energy surface which remains unchanged when brought into contact with water.

Rosin soap size is thought to meet the above requirements by the following mechanism.

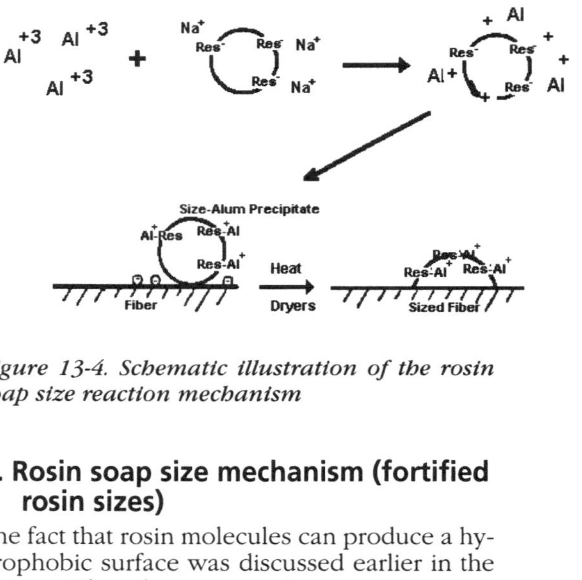

Figure 13-4. Schematic illustration of the rosin soap size reaction mechanism

b. Rosin soap size mechanism (fortified rosin sizes)

The fact that rosin molecules can produce a hydrophobic surface was discussed earlier in the chapter. The other two requirements are met as follows:

Retention — The neutralized rosin acid reacts with alum in the stock preparation system and on the paper machine wet end to form a positively charged alum-rosin precipitate that is attracted to the negatively charged fiber surface and is retained there.

Stable, hydrophobic surface — The aluminum, rosin, and fiber react in the dryer section to form a stable size complex that is attached to the fiber surface.

Fig. 13-4 illustrates this mechanism schematically.

The most efficient sizing occurs with rosin soap in pH ranges where alum exists primarily in the Al^{+3} and hydroxyaluminum complex forms. The formation of aluminum resinate occurs by reactions between neutralized rosin molecules and these aluminum species.

Referring back to the discussion of alum chemistry in Chapter 11, the Al^{+3} and several hydroxyaluminum forms predominate at pH values of 4.0-4.5 and normal alum concentrations. Higher pH values lead to formation of polyaluminum species and $Al(OH)_3$, which do not have the high cationic charges required to react effectively with sodium rosinate.

In practice, the situation is very complex because alum charge is influenced by a number of factors, including concentration, temperature, anion type and concentration, amount of shear in the system, and the level of alkalinity. Even so, mill studies have shown that maximum sizing and size retention is achieved if alum is 30-40% neutralized prior to reaction with rosin.[10]

Formation of cationic aluminum resinate is a most critical phase in rosin soap sizing. The precipitate formed from soap size consists of a mixture of mono- and di-resinate and some free rosin acid. The amount of free acid depends on the reaction conditions.

Soap sizes precipitate in large clusters, and their distribution on fibers is far from ideal. The size distribution suffers even more if the stock contains large amounts of Ca^{+2} or Mg^{+2} ions or recycled alum — substances that prematurely precipitate size as large aggregates.

In order to achieve the maximum effect from rosin size, it is necessary that the size precipitate melt (sinter) on the fiber surface, which promotes the best coverage and rosin orientation. Soap sizes have high melting points — 120-250°C — which depend on the composition of the precipitate and the amount of moisture present. If the proper moisture/temperature relationship is not attained during drying, then sintering may not occur at all and sizing efficiency will be correspondingly reduced. It has been found empirically that some sintering occurs if the moisture content is about 40% when the maximum temperature is applied. Therefore, a properly controlled paper drying program is critical for achieving maximum sizing with soap sizes.

It is clear that in order to achieve high rosin sizing it is necessary to exercise good process control. Since the alum cationic charge is a function of alum concentration and degree of neutralization, both the stock alkalinity and amount of alum present in the papermaking system must be carefully monitored.

Fig. 13-5 illustrates the importance of having the aluminum in the correct form when carrying out soap sizing. This essentially means controlling pH within the narrow range (4.0-4.5) where Al^{+3} and the highly charged mononuclear aluminum hydrolysis species exist.

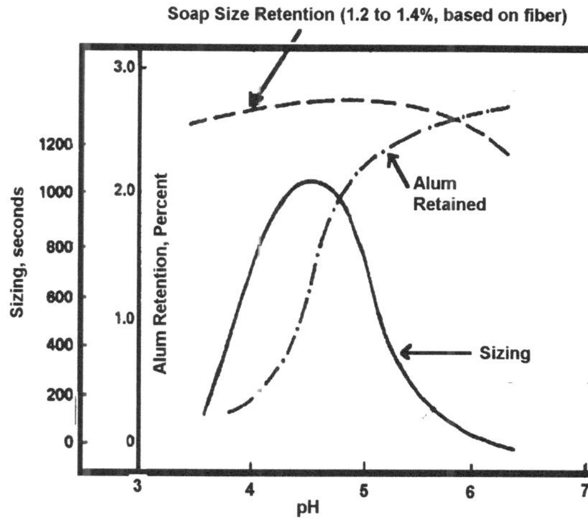

Figure 13-5. The influence of pH on alum and rosin soap size retention and sizing development [2]

Fig. 13-5 illustrates the importance of having the alum in the correct form so that the aluminum resinate precipitate can form and produce sizing. The decline in sizing occurs because the aluminum is in the form of $Al(OH)_3$ at the higher pH values and cannot react with the rosin soap to produce aluminum resinate. The gelatinous $Al(OH)_3$ can assist in retaining the rosin by inclusion in the aluminum precipitate. However, any rosin that does not get anchored to the fiber surface when the sheet dries will lead to sizing loss when the paper comes into contact with water.

c. Use of sodium aluminate in rosin soap sizing [1]

Sodium aluminate ($NaAlO_2$), an alkaline source of aluminum, can be used as a partial substitute for alum in rosin sizing. This practice enables the papermaker to achieve a higher aluminum concentration without increasing sulfate ion levels. This is a desirable practice since sulfate contributes to the formation of inorganic scale deposits in pipes and tanks. The absence of sulfate also means that the aluminum species cationic charge will be greater than for alum.

10. Anionic dispersed rosin sizes

Rosin soap sizes are being replaced by dispersed sizes due to the inflexible pH requirements of soap sizes, their relatively high alum and size level requirements, and the fact that soap sizes cannot easily achieve hard sizing. Anionic dispersed rosin sizes contain 95-100% free rosin acid emulsified in water at typically 35% solids and stabilized by anionic surfactants or proteins. The emulsion is a white, fluid product at pH 5.5-6.5.

Excessive pumping shear must be avoided in order to retain the small particle size required for efficient sizing. Prior to stock addition, the 35% solids material must be diluted to 2-5% with dilution water whose hardness is less than 50 ppm and whose alkalinity is less than 50 ppm. Softened water is often required.

Dilute dispersed rosin size is added at the machine chest pump or as late as the fan pump suction side. Alum addition can be simultaneous or before. The recommended headbox pH range is 4.6-5.3, although higher pH values have been run successfully. Dispersed sizes are more efficient than soap sizes and addition levels of 3-4 lbs/ton are typical. The small fraction of neutralized soap size present in dispersed size requires a less alum than soap size.

11. Cationic dispersed rosin sizes

Cationic dispersed rosin sizes are essentially 100% free rosin acids emulsified at 35% and stabilized with cationic polymers. Mill handling and addition practices are similar to anionic dispersed sizes, except that the preferred pH range is 4.6-5.3 for the cationic materials. Since no alum is needed for size retention, the total alum requirement is less than for anionic dispersed sizes.

12. Mechanism of sizing with dispersed rosin sizes

The mechanism by which dispersed size meets the requirements of a good size differs from soap size in several important aspects. First, the retention of dispersed rosin size by fibers is only lightly aided by alum. Instead, cationic starch or synthetic cationic polymers are used to promote the retention of anionic dispersed size, while cationic dispersed size is essentially self-retaining.

Second, it is critical that alum be present on fiber surfaces so that it can react with rosin molecules in the dryer section. Therefore, care must be taken that conditions are favorable for alum retention. This is accomplished by running at a pH range of 4.6-5.3, where the predominant aluminum species strongly adsorb onto fiber surfaces. A typical alum-to-dispersed size ratio is 1:1.

Third, in the dryer section, the retained dispersed size particles melt (sinter) and spread out on the fiber surface, where they come into contact with aluminum species and where reactions occur that bond the rosin to the fiber surface. This process is promoted by the lower melting point of dispersed sizes (70-110°C) compared to soap sizes. **Fig. 13-6** contains a schematic illustration of the entire mechanism.

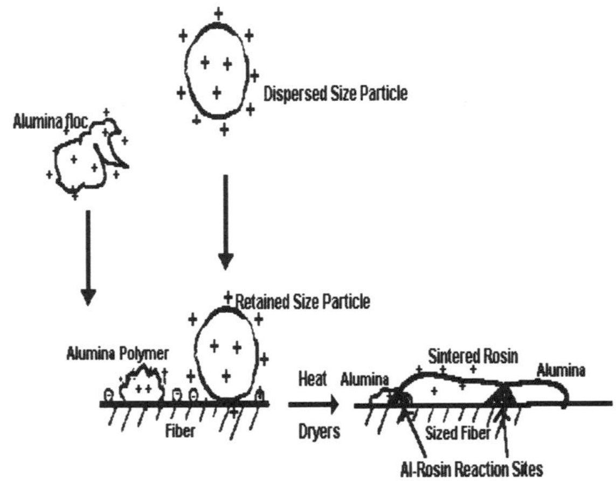

Figure 13-6. Schematic illustration of the sizing mechanism of dispersed sizes. The aluminum and rosin are independently retained on the fiber surfaces. The rosin melts and flows in the dryer section and comes into contact with the retained aluminum. The anchoring reaction then occurs.

13. Factors that influence the effectiveness of rosin sizing agents

a. Introduction

Rosin sizing is affected in a variety of ways by mill conditions. Strazdins has presented a comprehensive discussion of this subject.[2] We will be content with a brief summary here.

b. Mill conditions and sizing

Many mill factors affect sizing. **Table 13-5** lists some of the more important ones.

c. Effect of other additives on sizing

Table 13-6 indicates how other functional additives and chemical processing aids impact sizing.

Table 13-5. Mill factors that affect sizing

Factor	Sizing Effect
Stock temperature	High stock temperatures reduce the effective charge on aluminum species, thereby reducing sizing.
High SO_4^{-2}, citrate, hexameta phosphate ion levels	These anions coordinate with aluminum and lower the net cationic charge.
High Ca^{+2} ion levels (>100 ppm $CaCO_3$)	Ca^{+2} reacts with soap size to produce a calcium resinate precipitate that does not produce sizing. The overall effect is a reduction in sizing efficiency.
Pulp type	In general, highly bleached or purified pulps are much more difficult to bleach than unbleached pulps. Dispersed rosin size is less affected by pulp type than rosin soap size.
Refining	Refining increases the fines level and fiber fibrillation — both of which increase the specific surface area of the pulp and increase the amount of size needed to reach a given sizing level. Refining can also release hemicelluloses and other anionic materials (cationic demand) that react with alum and reduce its availability to react with size.

Table 13-6. Effect of other additives on sizing

• Wet strength resins • Retention aids • Drainage aids • Dry strength resins • Cationic starch	Whenever these substances are cationic, they tend to improve fines retention and thereby improve size retention. They have a beneficial impact on sizing.
Fillers	All fillers reduce sizing because they increase the specific surface area of a furnish. (See the discussion below.) Fillers can also have anionic dispersants that interfere with alum.
• Defoamers • Pitch dispersants • Wetting agents • Oils • Oil emulsions • Lubricants • Softening agents • Felt cleaners • Other cleaners • Oxidizers — when present at greater than 3-4 ppm	All interfere with sizing. A variety of mechanisms are involved.

d. Preferential adsorption of size components on fines

It was indicated in **Table 13-6** above that fillers have an adverse effect on sizing. The same can be said for fiber fines or any other substance that increases the specific surface area of the stock. This occurs because adsorbing materials, such as size, will partition themselves among furnish components according to their respective surface areas. Thus, stock components containing the highest surface area will adsorb the most additives. **Table 13-7**, derived from Marton's studies on the papermaking impact of fines indicates this fact.[11,12]

e. Size reversion (fugitive sizing)

Size reversion, or *fugitive sizing*, refers to the loss of sizing degree over time in a paper that originally had acceptable sizing. Size reversion almost always can be attributed to the presence of rosin size molecules that are not bonded to the fiber surface. When the paper surface is in contact with air, these unbound rosin molecules will orient themselves so that their hydrophobic parts extend outward, away from the surface, most of the time. However, the molecules are free to overturn and do so occasionally due to random thermal motion. If liquid water happens to be in contact with the paper surface at the time a rosin molecule overturns, then the rosin carboxyl (or carboxylate anion) will hydrogen bond with water molecules — creating an energetically favorable situation. With time, more and more rosin molecules undergo molecular overturning and the paper surface re-

Table 13-7. Relative adsorption intensities of various fines components in papermaking furnishes [11,12]

	Relative Adsorption Intensity		
Additive	Fibers	Filler Clay	Fines
Rosin soap size	1	4	5
Dispersed rosin size	1	4	16
Alum	1	2	3

gains its ability to interact with water, i.e. it "loses" its sizing.

In order to prevent the occurrence of fugitive sizing, the key is to control the sizing process so that all rosin molecules get bonded to fiber surfaces via reaction between carboxyl groups and aluminum. For soap sizing, this means that the aluminum must be present in sufficient concentration and cationic charge state to react with all of the resinate molecules present. For dispersed sizing, sufficient quantities of aluminum must be precipitated onto the fiber surface to react with the dispersed size in the dryer section. Both instances mandate close control over pH and alum addition and knowledge of the alkalinity and other factors that influence the state of aluminum in a papermaking system.

14. Trouble-shooting sizing problems

Despite attempts to maintain close control over the sizing process, papermakers are often called upon to trouble-shoot rosin sizing problems during day-to-day mill operations. It is important that any trouble-shooting scheme be based on a firm grasp of sizing fundamentals. Once that is established, then it is possible to take advantage of the vast collection of empirical experience that exists in the industry. As an example of this, you are referred to the excellent compilation of sizing problem case studies presented by Otway and Johnson in their recent publication.[13]

15. Summary

Alum-rosin sizing continues to play an important role in modern papermaking. Over the years, our fundamental understanding of the process has improved immensely and it is now possible maintain close control over sizing in the mill environment. Our purpose here has been to provide an introduction to the technical knowledge base upon which good sizing process control must be grounded and from which deeper investigations can be undertaken.

16. References

[1] Davison, R. W. "Internal Sizing." Chapter III in Pulp and Paper Manufacture, Vol. 6, Stock Preparation. R. W. Hagemeyer and D. W. Manson, Editors, Joint Textbook Committee of the Paper Industry, TAPPI/CPPA, 39-64(1992).

[2] Strazdins, E. "Chemistry and application of rosin size." Chapter 1 in Sizing of Paper - Second Edition. W. F. Reynolds, Editor. TAPPI PRESS, Atlanta. 1-31(1989).

[3] J. M. Gess. "The sizing of paper with rosin and alum at acid pH." Chapter 7 in Paper Chemistry, J. C. Roberts, Editor. Chapman and Hall, New York, 97-112[1991].

[4] Eklund, D. and Lindstrom, T. "Water penetration and internal sizing." Chapter VIII in Paper Chemistry — An Introduction. DT Paper Science Publications, Grankulla, Finland, 192-214[1991].

[5] Sizing of Paper — Second Edition. W. F. Reynolds, Editor. TAPPI PRESS, Atlanta. 156 pp. 1989.

[6] Cates, R. E. "Internal sizing agents." Chapter 5 in Chemical Processing Aids in Papermaking: A Practical Guide. K. J. Hipolit, Editor. TAPPI PRESS, 63-80[1992].

[7] Hunter, D. Papermaking. The History and Technique of an Ancient Craft. Dover Publications, Inc., New York, 1978.

[8] Buchanan, M. A. "Extraneous components of wood." Chapter 7 in The Chemistry of Wood, B. L. Browning, Editor, Robert E. Krieger Publishing Co., Huntington, NY. 314-367(1975).

[9] Marton, J. and Marton, T. Pulp and Paper Magazine Canada, 83:11, 26(1982).

[10] Cordier, D. and Bixler, H. J. Tappi Journal, 70:11, 99(1987).

[11] Marton, J. and Marton, T. "Effect of fillers on rosin sizing." Tappi Journal, 66:12, 68-71(1983).

[12] Scott, W. E. "Fines management and control in wet end chemistry." Tappi Journal, 69:11, 30-34(1986).

[13] Otway, P. and Johnson, D. C. "Sizing in acid, neutral, and alkaline conditions." Chapter 9 in Applications of Wet End Paper Chemistry, C. O. Au and I. Thorn, Editors. Blackie Academic and Professional — Chapman and Hall, New York, 139-171[1995].

Chapter 14

Sizing with Synthetic Sizing Materials

1. Introduction

a. Background developments

As we discussed in the Chapter 13, beginning in the early 1800's, water resistance was imparted to many papers by alum-rosin sizing, which could only be done efficiently under acid pH conditions. For various reasons, such as the desire to use calcium carbonate filler or the need to produce permanent paper, certain water resistant paper grades needed to be produced under alkaline conditions. Other grades must have very hard sizing in order to resist edge wicking in liquid packaging applications. Needs like these lead to the development of *synthetic* sizing agents. (The term "synthetic" is used to distinguish these materials from the natural product rosin size.)

The first synthetic size, *alkyl ketene dimer (AKD)*, appeared in the patent literature in 1953 and in the open literature in 1956. The other common synthetic sizing agent, *alkenyl succinic anhydride (ASA),* was first synthesized for use as a paper sizing agent in 1974.

From 1956 to the early 1980's, these materials were adopted for use in certain specialty grades where very hard sizing was needed, such as the liquid packaging grades, and by a few paper producers who wished to produce paper under alkaline pH conditions. However, overall application of these sizes by US papermakers was low. The situation changed in the middle 1980's when many fine paper manufacturers began to convert from acid to alkaline papermaking in order to take advantage of a new high opacity, high brightness, low cost precipitated calcium carbonate filler. Before 1986, less than 15% of US uncoated, free sheet production contained synthetic sizes. By 1994, the percentage had grown to over 90% .[1] Similar, albeit less dramatic, trends have also occurred in other paper grades.

b. ASA and AKD — the two principal synthetic sizing agents

While many synthetic organic molecules are capable of imparting water resistance to paper, only ASA and AKD have been widely adopted by the paper industry and there is approximately equal usage of the two materials today in US alkaline fine paper manufacture. In Europe and other papermaking regions, the relative use rates of AKD and ASA differ from 50:50, but they still remain the preeminent alkaline sizing materials. For this reason, we will limit our discussion to AKD and ASA. The references listed at the end of the chapter contain information about other synthetic sizing materials.

c. Chapter format

The chapter contains four sections. The first part reviews the requirements of a good size introduced in Chapter 11 and indicates how synthetic sizes meet these requirements. The next two parts discuss the chemical structure, reactions, handling, unique characteristics, and practical applications, and problems of AKD and ASA separately. Finally, section four summarizes the similarities and differences between the two materials.

Much of the material presented in the chapter draws heavily on the excellent available technical information pertaining to synthetic sizes. References [1-10] are examples of the extensive literature available to the wet end chemist.

2. Sizing requirements and synthetic sizes

The general requirements of a good internal sizing agent were listed in Chapter 12, and are presented again in **Table 14-1**.

Table 14-1. Requirements of a good internal sizing agent

In order to be a good sizing agent, a substance must:

- Have high retention on fibers
- Be uniformly distributed on fiber surfaces
- Be able to produce hydrophobic fiber surfaces
- Adhere strongly to fibers
- Be chemically inert to the penetrants of interest

Requirement one — Since ASA and AKD are nonionic molecules that have no innate attraction for anionic fibers, other substances, such as cationic starch and synthetic cationic polymers, must be used as size retention aids. Sometimes these agents are incorporated directly into the emulsion itself. Other times they are added separately. This practice is similar to that discussed in Chapter 13 for dispersed rosin sizes.

Requirement two — ASA and AKD occur as a water-insoluble solids or liquids at room temperature — depending upon the length and degree of unsaturation of their hydrocarbon chains. In order to get good distribution of these substances on fiber surfaces, the liquid or solid must first be subdivided into very small (submicron) particles. Dilute suspensions (emulsions) of these small particles are then added to the papermaking system with good mixing.

The second feature that contributes to good distribution is that retained size particles melt and spread over fiber surfaces — similar to dispersed rosin sizes.

Requirement three — ASA and AKD molecules contain relatively inert hydrocarbon chains that impart hydrophobic character to the materials and enable them to do the same for fiber surfaces.

Requirement four — The molecules contain reactive centers that react with fiber surface hydroxyl groups to form covalent linkages that are stable in the presence of water and anchor the size molecules to fibers.

Requirement five — ASA and AKD are capable of providing sizing against more aggressive liquids than rosin sizes. This is a result of their strong bonding to fiber surfaces and the ability of their hydrocarbon chains to affect large areas of the fiber surfaces.

A secondary effect of sizing in an alkaline environment is the impact it has on the refining process, i.e. energy transfer to the more highly swollen fiber. The resulting denser sheet provides an additional sizing advantage of its own.

3. Alkyl ketene dimer size

a. Synthesis and chemical structure

Alkyl ketene dimer size has the general chemical structure shown in **Fig. 14-1.**

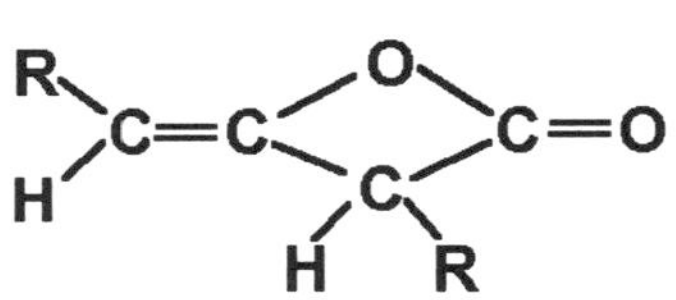

Figure 14-1. Chemical structure of alkyl ketene dimer sizes. '"R" refers to hydrocarbon chains containing 14-16 carbons, most commonly.

It is synthesized by dimerizing fatty acid chlorides. Commercial stearic acid (C_{16}) is the most common starting material for the synthesis. The resulting product has a melting point of about 50°C and is a waxy, water-insoluble solid at room temperature. AKD sizing efficiency against water has been found to increase with hydrocarbon chain length up to C_{14}. It levels off as the chain length is increased further.[2]

b. Reaction between AKD and fibers

There has been some argument among researchers in the past regarding the exact nature of the reaction between AKD and wood fibers.[2,3] It is most likely that the four-membered lactone ring in AKD reacts with cellulose and hemicellose hydroxyl groups to form β-keto ester linkages, as illustrated in **Fig. 14-2.**

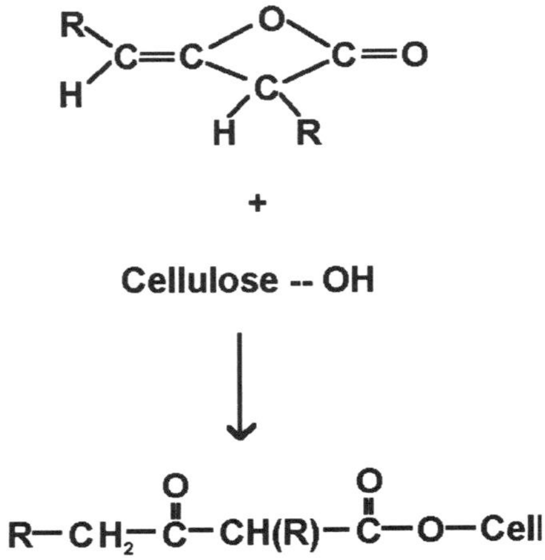

Figure 14-2. Reaction of alkyl ketene dimers with cellulose hydroxyl groups to produce a β-keto ester

A considerable body of data based on solvent extraction and sizing studies indicates that a fraction of AKD is irreversibly bound to the fiber surface and is chiefly responsible for the observed sizing. Direct evidence for the existence of β-keto ester linkages in AKD sized paper has been provided by Bottorff through the solid state[13] C NMR studies of filled and unfilled papers sized with AKD under papermaking conditions. [14]

The reaction between AKD and cellulose is very slow. It has been found that alkalinity will catalyze the reaction. Bicarbonate has been found to be particularly effective in this regard, as have polymers containing basic amine groups. [11] The latter are often added to AKD emulsions as "promoter resins."

Studies have shown that heat treatment is essential in AKD sizing. Roberts has suggested that one role of heating is to increase the diffusion and migration of size over fiber surfaces and thereby improve size distribution. [3]

Lindstrom pointed out the importance of water removal in the size spreading process. As long as size particles are surrounded by liquid water, they cannot spread because the high surface tension of water prevents this from happening. **Fig. 14-3** illustrates the relationship between AKD reaction and paper drying. [9] Paper must be dried to about 80% solids before

AKD particles are no longer surrounded by water. Once this is accomplished, size particles spread out and react with fiber surface cellulose and hemicellulose molecules.

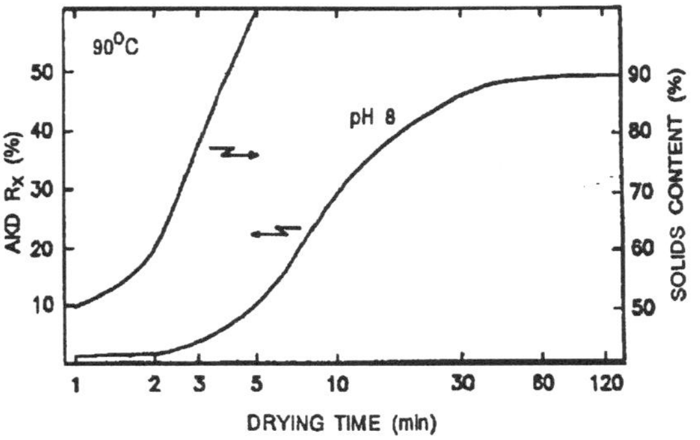

Figure 14-3. Relationship between the solids content of paper and the reaction between AKD and fiber cellulose [9]

One characteristic of AKD sizing is that significant sizing development usually occurs long after paper has been removed from the paper machine. The ultimate sizing level attained is a function of the drier temperatures and paper storage temperatures. **Fig. 14-4** illustrates this point.

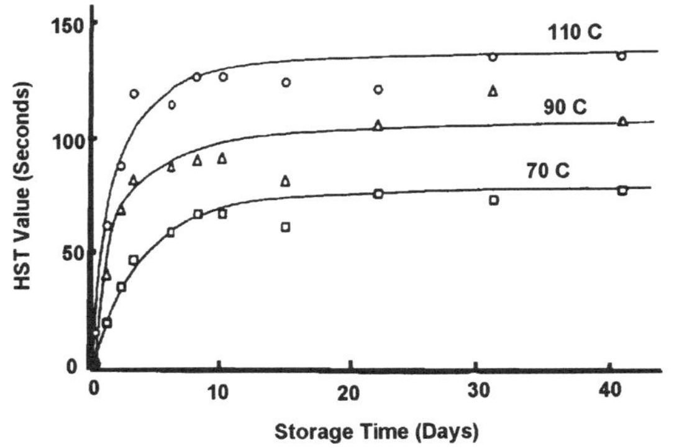

Figure 14-4. Effect of temperature and storage time on Hercules photosize test values of AKD sized paper [12]

c. Reaction of AKD with water (hydrolysis)

In addition to reacting with the hydroxyl groups on cellulose and hemicelluloses, AKD can also react with the hydroxyl groups in water. The reaction is analogous to that shown previously in **Fig. 14-2**, with water substituting for cellulose (**Fig. 14-5**). The b-keto acids produced can eliminate a CO_2 molecule to produce a ketone (palmitone).

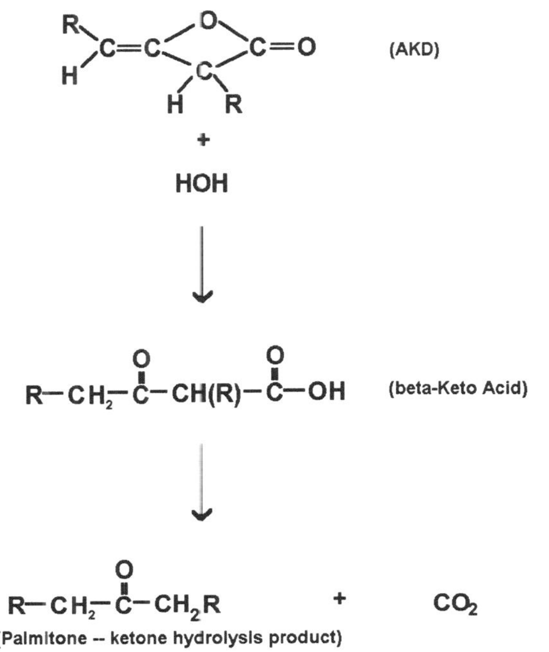

Figure 14-5. Hydrolysis of AKD. Potential reaction products are β-keto acids and their salts, and a ketone — palmitone.

In his investigation of AKD reactions by ^{13}C NMR, Bottorff reported palmitone to be the most common hydrolysis end-product, with β-keto acid occurring as an intermediate under certain conditions. [14]

Palmitone is a solid under most paper machine wet end conditions and does not cause serious deposit or runnability problems. It has been demonstrated that AKD hydrolysis products do not contribute to sizing[14] and their presence reduces the efficiency of an AKD emulsion. Fortunately, the hydrolysis reaction is slow and it is possible to produce AKD emulsions that are relatively stable over time periods on the order of tens of days at ambient temperatures. It can be held for months at temperatures just above freezing.

d. Relative reaction rates of hydrolysis and ester formation

AKD exhibits relatively low reaction rates with fibers under papermaking conditions. This is an advantage and a disadvantage. On the positive side, AKD hydrolysis occurs only to a limited extent in the paper machine wet end. On the negative side, it is difficult to achieve high on-machine sizing and substantial curing must take place off-machine.

Because of the low reaction rates, only a fraction of the retained AKD is covalently linked to fibers. The rest either exists as hydrolysis products or unreacted AKD. For example, Marton analyzed the composition of retained AKD in machine papers and found only 30-40% reacted AKD — with the remainder in the form of hydrolysis products.[19] Clearly, the relative hydrolysis and curing reaction rates are of importance to the papermaker.

Lindstrom and coworkers have studied the kinetics of ketene esterification,[15,16] while Marton has investigated the kinetics of AKD hydrolysis.[17,18] **Table 14-2** compares their findings.[18]

Table 14-2. Comparison of AKD hydrolysis and curing(esterification) kinetics (hrs units) at pH 8 [18]

°C	Esterification		Hydrolysis		k_{ester}/k_{hydr}
	k X 10^3	$t_{1/2}$	k X 10^3	$t_{1/2}$	
30	117	6	5	140	25
50	405	1.7	26	27	15
70	2160	0.3	200	3.5	10
Activation Energy (KJ/mole)	55		105	1:2	

Several conclusions can be drawn from the table. First, at room temperature, the esterification reaction is much faster than the hydrolysis reaction. As the temperature is increased, the two reaction rates become closer because the hydrolysis activation energy is twice that of esterification. At drier section temperatures, the two reaction rates will be comparable, which explains the high percentage of hydrolysis products found in machine-made papers by Marton.[19]

Two immediate practical implications can be drawn from the above data:

- AKD emulsions should be stored at low temperatures in order to reduce the hydrolysis reaction rate.
- Paper machine drier temperatures should be limited to produce the minimum acceptable AKD cure. This will produce the most favorable cure-to-hydrolysis ratio. Increasing the temperature will increase the cure reaction rate, but will increase the undesirable hydrolysis rate even more.

e. AKD emulsions[2,3,6]

As discussed earlier, synthetic sizes are prepared in the form of aqueous emulsions in order to facilitate their addition and distribution in the paper furnish. AKD emulsions contain AKD, cationic starch or cationic polymers, and other emulsion stabilizing additives. Solids levels are typically 6-15%, with AKD representing approximately 30% of the total solids. Particle sizes are 1-5 microns. The pH is 3.0-4.5. The emulsions have a shelf life of about one month at 24°C.

The hydrolysis reaction rate data presented in **Table 14-2** emphasizes the importance of emulsion storage temperature. It should be recognized that AKD hydrolysis occurs, even at room temperature, and emulsion storage times should be kept as short as possible. As an example of this, Marton found that a commercial AKD emulsion (9% AKD) stabilized with cationic starch lost about 15% of its reactive groups via hydrolysis during five weeks storage at room temperature.[17]

AKD is metered "as is" and diluted 10:1 before stock addition in order to obtain good distribution. It is added close to the paper machine (fan pump to pressure screen) in order to minimize the contact time with water. Addition levels range from 0.05% for slack-sized grades to 0.10% for hard-sized grades.

AKD retention occurs by a hetercoagulation mechanism where cationic emulsion particles are attracted to anionic fibers. The cationic charge can be incorporated into the emulsion when it is prepared, or it can be gotten through the addition of cationic retention aids, or both. High shear and long contact times tend to reduce AKD retention.

4. Effect of furnish and paper machine conditions on sizing with AKD

The rate of sizing development and efficiency of AKD utilization depends upon a number of machine and furnish parameters, such as pH, alkalinity, filler type and level, fiber fines level, fiber type, interfering substances, and dryer conditions.

a. Alkalinity and pH

The catalytic action of bicarbonate on the AKD reaction with cellulose has already been introduced. This is a significant factor in sizing development. **Table 14-3** supports this fact.[1]

AKD can be used over a pH range of 6-9, but it is clearly most effective at levels of 8-9. The profound effect of alkalinity is demonstrated by the great increase in sizing level accompanying the increase in alkalinity from 10 ppm to 150 ppm. Practical experience indicates that alkalinity levels of 150-250 ppm result in the best AKD performance.[2] It may be necessary to add $NaHCO_3$ or Na_2CO_3 to some furnishes in order to achieve these levels.

Table 14-3. Effect of pH and alkalinity on sizing development [1]

Headbox pH	Off-Machine Sizing (HST, sec)
	Total alkalinity<10 ppm
6	1
7	1
8	22
9	95
	Total alkalinity 150 ppm
6	—
7	69
8	366
9	389

b. Fillers

Fillers adversely affect the sizing performance of all sizes. Two mechanisms are responsible. First, since fillers have higher specific surface area than fibers, sizing agents tend to adsorb on fillers to a relatively higher degree than on fibers. (This phenomenon was discussed for rosin in Chapter 13.) AKD size attached to filler does not contribute significantly to sizing and, consequently, additional size must be added to make up for this deficiency. There have been recent attempts by one PCC filler supplier to pretreat the filler surface to reduce its size demand.[23]

The second mechanism by which fillers affect sizing efficiency is through their effect on first pass size retention. Research has shown that some of the size that is initially adsorbed by filler can be redistributed to fiber surfaces. However, if the filler is not retained, then this size gets lost with the filler. Therefore, high first pass filler retention is very important to efficient size utilization.

Calcium carbonate filler is thought to play a pivotal role in alkaline paper size reversion. This will be discussed in greater detail below.

c. Fiber fines

Increased furnish fines levels has a similar effect on sizing as fillers, except that size adsorbed on fiber fines can react and become effective in sizing. However, the high specific surface area of fines means that more size is required to achieve a given degree of sizing. This is illustrated in **Fig. 14-6**.[13]

d. Fiber type

Pulp sizability is a function of many variables, including carboxyl content, specific surface area, and accessibility of surface hydroxyl groups. Complete knowledge of these attributes is rarely known about a given pulp, but some generalizations have been developed. Unbleached pulps are easier to size than bleached pulps. High alpha-cellulose pulps are extremely difficult to bleach and require a much higher AKD addition level to achieve a given degree of sizing. Acid sulfite pulps are less responsive than kraft pulps. It is possible to use electrokinetic charge measurements to estimate the sizing response of a given pulp.

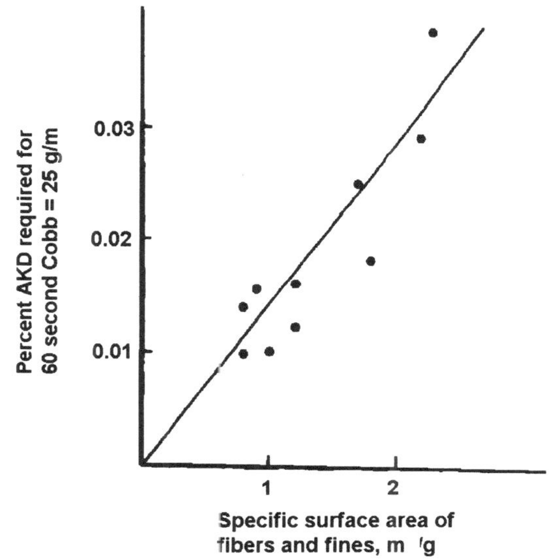

Figure 14-6. Effect of pulp specific surface area on the amount of AKD required to achieve a 60 second Cobb size value of 25g/m² [13]

e. Other additives

AKD sizes are impacted both adversely and positively by other additives in the papermaking system. **Tables 14-4** and **14-5** contain typical examples of both categories.

Table 14-4. Additives that are harmful to sizing

- Defoamers (wet end, size press, calendar)
- Surfactants (pitch dispersants, wetting agents)
- Oils, oil emulsions, and lubricants
- Softening agents
- Felt cleaners and other cleaners
- Oxidizers (>3-4 ppm)
- Anionic dyes and other anionic materials

A variety of mechanisms are involved, ranging from changing the surface energy of paper to make it more wettable to interference with size retention and anchoring.

Table 14-5. Additives that are beneficial to sizing

- Wet strength resins
- Retention aids
- Drainage and drying aids
- Dry strength resins
- Cationic starch

These additives are all cationic materials that act primarily by improving size, fines, and/or filler retention.

Alum does not enhance the effectiveness of AKD sizes directly. In fact, there is evidence that alum is detrimental in this regard. Alum is used in alkaline papermaking for the neutralization of interfering substances and can improve AKD sizing by this, indirect route. It is important that the alum and AKD addition points are well separated and that alum overdosing is avoided.

5. Size reversion (fugitive sizing)

Size reversion, or *fugitive sizing*, refers to the loss of sizing degree over time in a paper that originally had acceptable sizing. These terms were introduced in Chapter 13 for alum rosin sizing where it was pointed out that size reversion almost always can be attributed to the presence of rosin size molecules that are not bonded to the fiber surface well. When liquid water contacts them, they overturn so that their polar carboxyl groups interact with water molecules. When sufficient rosin molecules have overturned, the paper surface regains an ability to interact with water, i.e. it "loses" its sizing.

As the quantity of AKD sized paper has grown in the US, there has been an increase in the incidence of size reversion problems. Unlike rosin sizing, where size reversion occurs relatively rapidly, size reversion can take weeks to occur in AKD sized papers, which has made it more difficult to recognize as a widespread problem. Another consideration is that size reversion has never been a serious problem in Europe, where AKD sizing has been practiced for decades. Once the magnitude of the situation became apparent, a number of investigations were undertaken to determine the cause of the problem and possible solutions.[7,14,20–22] The following picture has emerged.

While reversion has been reported to occur in a wide variety of alkaline papers, papers containing precipitated calcium carbonate (PCC) seem to be by far the most susceptible. **Table 14-6** lists some other "themes" that have developed pertaining to reversion.

The facts in the above table suggest that reversion occurs because the AKD on the filler, which initially provides sizing, hydrolyzes to the nonsizing palmitone over time. The hydrolysis reaction is promoted by increased pH and increased alkalinity. The residual $Ca(OH)_2$ associated with PCC is highly alkaline and has been shown to catalyze the hydrolysis of AKD. Bottorff [14] has reported the direct observation of this reaction with ^{13}C NMR spectroscopy. Bottorff also found that increasing the AKD:PCC ratio reduced the rate and extent of the AKD hydrolysis — which could explain why papermakers have been able to overcome reversion in many cases by increased AKD addition levels. Another approach being investigated is the production of modified PCC fillers that have a reduced demand for AKD.[23]

Table 14-6. Factors that have been found to affect sizing reversion in AKD papers

Factor	Comments
PCC particle size and specific area	Smaller particle size, higher surface area PCC fillers exhibit greater size reversion.
Alkalinity	Higher alkalinity is associated with greater reversion.
pH	Higher pH is associated with greater reversion.
Residual $Ca(OH)_2$ in PCC	$Ca(OH)_2$ appears to promote reversion.
Other fillers	Fillers, such as clay and TiO_2, do not appear to promote reversion except in high alkalinity systems.
AKD dosage level	Size reversion can be minimized or stopped by increasing the AKD dosage level.
AKD retained on PCC filler particles retained on	AKD retained on PCC filler particles can apparently give good sizing for a period of time without being bonded to the particle surface.
AKD molecules on PCC filler particles revert to palmitone with time	Unbonded AKD on PCC filler particles gets hydrolyzed to the nonsizing palmitone with time.

6. Other potential papermaking problems related to AKD [2,3,7]

Papermakers have also experienced other problems when converting from acid to alkaline papermaking. Some examples of problems directly associated with AKD sizing include:

Wet press picking - This problem often occurs in the early stages of conversion to AKD sizing and alkaline papermaking when rosin-containing broke and stock is still in the system. It has not proven to be a major obstacle to conversion in most instances.

Insufficient sizing ahead of the size press - Only moderate sizing is required to achieve good size press runnability and by controlling pH, alkalinity, and drying within required ranges, this should be possible with cationic AKD sizes.

Paper slipperiness - AKD sized paper often has a lower coefficient of friction than rosin sized paper due to the fact that AKD is a waxy material. If high AKD levels are used in order to achieve high sizing levels, then the paper surface can become too slippery for certain converting and handling processes. There is also evidence that the AKD ketone hydrolysis product contributes to this problem.

7. Alkenyl succinic anhydride

a. Introduction [4]

Alkenyl succinic anhydride (ASA) is the other popular synthetic size used in alkaline papermaking. ASA has been preferred in several applications because of its greater reactivity, which produces higher on-machine sizing (>80% of total sizing) and better size press holdout. ASA is used most often in fine papers. It is also used in gypsum board and for a few other applications. It does not provide the hard sizing required by liquid packaging grades.

b. Synthesis and chemical structure of ASA [3,4]

ASA synthesis is a two-stage process that begins with the random isomerization of the double bond position in an *alpha*-olefin. Next, the internal olefin is reacted with maleic anhydride to form the final product. **Fig. 14-7** summarizes the reactions.

The starting alpha-olefin is usually in the C16-C20 range. It may be either linear or branched. The ASA product is a yellow to amber liquid at room temperature. Starting olefin chains containing more than 20 carbons produce solid ASA products that are less suitable for emulsification.

c. Reaction between ASA and fibers

Unlike AKD, whose reaction with fibers has been studied extensively to determine whether or not ester linkages actually occur, practically no published results relative to this subject are available for ASA. This is probably because ASA reaction rates are much greater than AKD reaction rates and the concept of the formation of ester linkages with cellulose under papermaking conditions is more readily accepted (**Fig. 14-8**). For instance, it has been found that sizing will develop in air-dried, ASA-sized handsheets without heat treatment.[3] Like AKD, a fraction of the ASA in sized papers is resistant to solvent extraction, indicating very strong bonding to the fibers. Unlike AKD, a much higher proportion of the retained ASA reacts

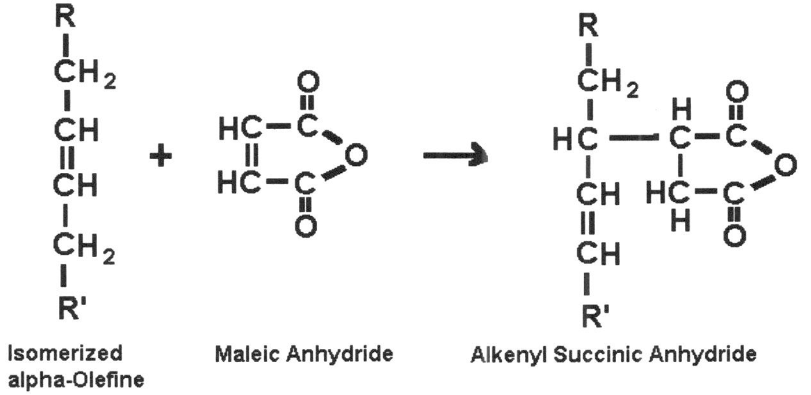

Figure 14-7. Synthesis of alkenyl succinic anhydride by addition of maleic anhydride to an internal olefin

with the fibers under paper machine conditions and less is available for extraction.

To summarize, ASA reacts with fibers to form ester bonds with accessible cellulose and hemicellulose hydroxyl groups. These linkages enable the molecule to orient its hydrocarbon side chains out from the sheet surface and impart a hydrophobic nature to it.

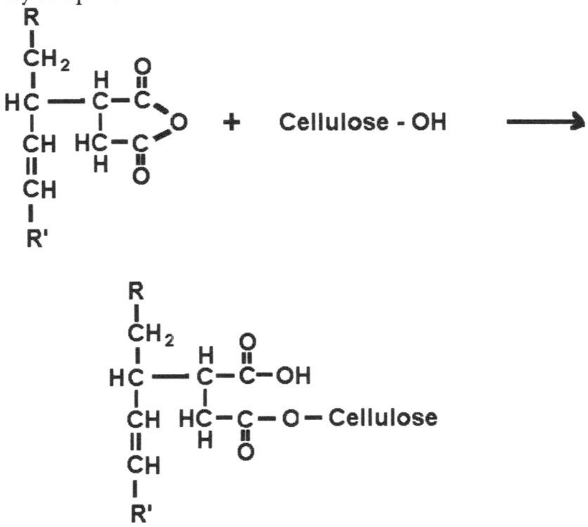

Figure 14-8. Formation of ester linkage between ASA and fiber surface. The linkage anchors the ASA to the fiber and enables it to orient its hydrophobic side chains away from the surface and produce sizing. The existence of such a linkage is inferred from the observed high reaction rate and the direct observation of ester links in AKD, which reacts much more slowly with cellulose.

d. Reaction between ASA and water (hydrolysis)

Fig. 14-8 illustrates the reaction between ASA and water that competes with ester linkage formation.

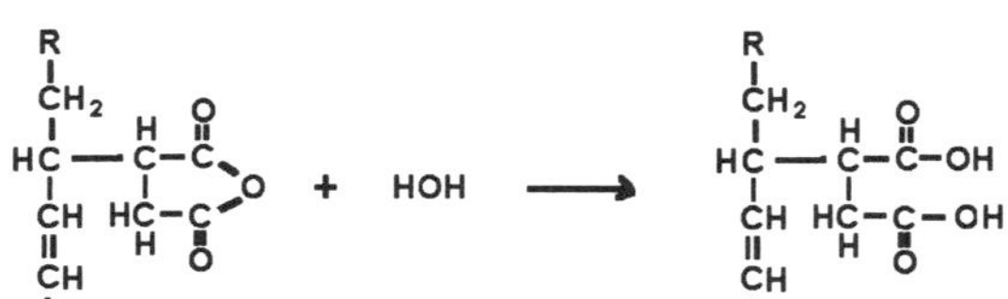

Figure 14-9. Hydrolysis of ASA to produce a diacid hydrolyzate product. The hydrolyzate is a liquid at room temperature. It forms tacky salts with Ca^{+2} and Mg^{+2} and acts as a desizing agent.

Like the esterification reaction, hydrolysis is rapid. Roberts reported that samples of tetradecenyl succinic anhydride would completely hydrolyze in 30 hours at 25°C and a starting pH of 7.2.[3] The reaction kinetics are affected by temperature and pH, as illustrated in **Figs. 14-10** and **14-11**.

In order to limit ASA hydrolysis, it is necessary to maintain low temperature and pH during ASA emulsification and short storage times prior to using the emulsion on the paper machine. On the paper machine, it is imperative that high first pass size and fines retention be achieved to prevent any substantial recirculation of unretained ASA in the white water loop where it will hydrolyze and form the diacid.

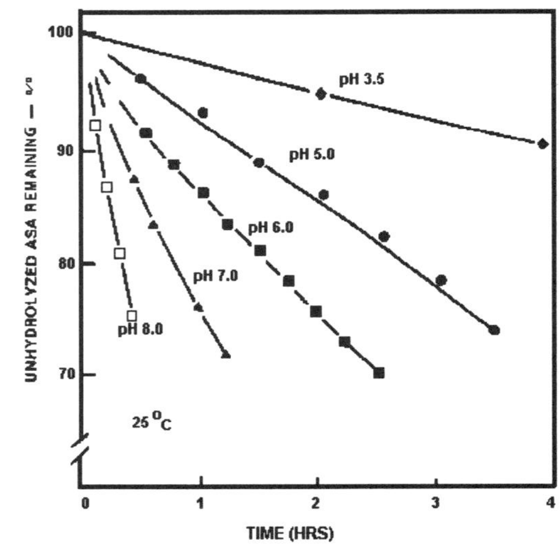

Figure 14-10. Effect of pH on the hydrolysis rate of ASA at 25°C [4]

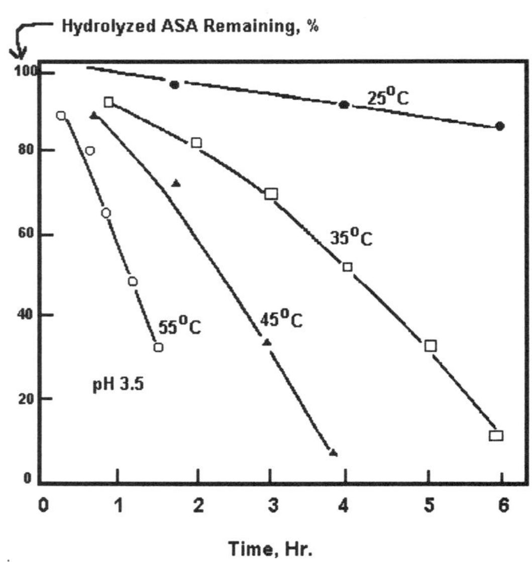

Figure 14-11. The effect of temperature on the hydrolysis rate of ASA at an initial pH of 3.5 [4]

In addition to reducing the sizing efficiency of ASA by reducing the amount of active agent present, hydrolysis has two other adverse side effects. First, the diacid hydrolysis product has been shown to inhibit sizing.[3] It has been suggested that the hydrocarbon chains on the diacid interact with similar chains on attached ASA molecules. This enables the two polar carboxyl groups on the hydrolyzate molecule to orient outward and be available to interact with water.

Second, the Mg^{+2} and Ca^{+2} salts of the diacid are extremely tacky substances that contribute to deposit formation and picking. Press picking has been found to be especially problematical with ASA. Because of this, it is absolutely imperative that the first pass retention of ASA be high so that the material is not exposed to white water during multiple passes around the system.

e. ASA emulsions [3,4,6]

Like AKD, ASA must be converted into an emulsion in order to get good distribution on fiber surfaces. High shear emulsification is most commonly used. ASA is passed through a high shear turbine pump in the presence of a protective colloid cationic starch (most often) or synthetic cationic polymer. The optimum starch solids to ASA ratio is 3-to-1.[4]

Emulsion particle size ranges between 0.5-3.0μ, with 1-2μ being optimum. Smaller particles would be more effective sizes, but are difficult to produce and undergo higher hydrolysis rates. Larger particle sizes produce unstable emulsions that tend to break under papermaking conditions and produce deposits. Emulsion quality is a function of the amount of energy input by the emulsifier, the starch attributes, the starch:ASA ratio, and the presence of other components, such as surfactants and synthetic promoters. Emulsion turbidity measurement has been suggested as a simple method of monitoring ASA emulsion quality in the mill.[24]

Whenever possible, emulsions should not be stored prior to use in order to avoid ASA hydrolysis. They should be prepared and immediately added to the furnish. This will help to avoid runnability problems associated with hydrolysis deposits. If storage cannot be avoided, then the temperature should be as low as possible and the pH should be in the 3.5-4.5 range.

ASA is typically added at the fan pump inlet, stock chests, refiner outlets, or pressure screen inlet. Minimal contact time with water is necessary in order to minimize hydrolysis. Typical use rates are 0.1-0.15% (2-3 lb/ton), based on the dry weight of the pulp.

f. Use of alum with ASA

It has become standard practice to add alum to the papermaking system in ASA sized grades at the 0.5% level. The aluminum ion forms an aluminum succinate salt with hydrolyzed ASA and reduces its tendency to form sticky deposits. It has also been shown that addition of a small amount of alum (20% of ASA weight) directly to the ASA during emulsification is a very effective method of enhancing ASA efficiency.[24]

8. Effects of furnish and paper machine conditions on sizing with ASA

The effects of furnish and paper machine conditions on ASA sizing are similar to those previously discussed for AKD. High first pass furnish retention is important for AKD sizing, but critical for ASA due to the severe problems associated with ASA hydrolysis. Fiber fines and fillers have the same effects on ASA and AKD, as do the other additive listed in **Tables 14-4** and **14-5**. The beneficial effect of alum to ASA sizing was discussed above. Alum, does not play as pivotal a role with AKD. ASA is effective from pH 5-10, a broader range than AKD. The reaction between and ASA and fibers is not affected by alkalinity to the same extent as AKD.

Drier temperatures are not as critical to ASA curing as for AKD curing. It has been pointed out previously that ASA curing can be gotten by air-drying. However, if the goal is to achieve sizing before the size press, then obviously it is important that the sheet be dried enough to promote the desired extent of curing. On the other hand, if the drier temperatures are too high, then ASA will steam distill out of the sheet and collect as an oily mess in the drier hoods. Any number of problems can be envisaged as a result of this occurrence!

9. Size reversion with ASA

Size reversion has not been reported to be as great a problem in ASA sized sheets as for AKD. This could be due to the greater ASA reactivity, which should lead to less unreacted size in the sheet. When reversion does occur, two mechanisms are possible:

The ester linkage bonding ASA to the fiber surface gets hydrolyzed and the newly freed size molecules are able to overturn and reduce the sheet's water resistance.[21]

Unreacted ASA residing on fiber surfaces hydrolyzes and is converted to the diacid, which is a desizing agent and reduces the sheet's water resistance.[3]

Table 14-7. Comparison of ASA and AKD attributes

Attribute	AKD	ASA
Reaction rate	Slow to moderate with fibers and water	Very fast with fibers and water
Physical state	Solid product and hydrolyzate	Liquid product and hydrolyzate
Hydrolyzate reactivity	Unreactive	Causes deposit problems and is a desizing agent
As received	Emulsion	Oil that requires in-mill emulsification
Impact on precision converting grades	Critical that addition rate be kept below levels that cause slippery paper	Addition rates not as critical because slip problems not prevalent
Effective pH range	Neutral-to-alkaline pH levels	pH range 5-9
Hard-sized grades	Able to do without significant problems	Difficult without deposit problems due to high dose levels

10. Comparison of AKD and ASA sizes

Now that we have discussed some of the most important attributes of ASA and AKD sizing agents, it will be useful to summarize with a comparison table (**Table 14-6**).

It is clear from the table that both sizes have their advantages and disadvantages. The decision as to which size to select is very much a matter of a mill's specific paper property requirements, technical capabilities, previous experience, and preferences.

11. Comparison of alum-rosin vs synthetic sizing response curves [10]

One significant difference between alum-rosin sizing and sizing with synthetics pertains to their relative sizing response times illustrated in **Fig. 14-12**. In the figure, the rosin system gives an immediate sizing response and continues to do so as the addition level is increased. Water resistance is imparted at all levels of rosin addition. Conversely, the synthetic sizes have a clearly defined minimum critical addition level before any water resistance is achieved. Above this size level, the HST increases sharply until a maximum is reached — which is greater for AKD than for ASA.

The figure also indicates that when compared at equal addition levels, synthetic sizes exhibit greater sizing efficiency than rosin sizes and can achieve higher absolute sizing levels.

It is often said that rosin sizing systems are more "forgiving" than synthetic sizing systems. **Fig. 14-12** indicates why. Synthetic sizes tend toward an all-or-nothing behavior. Small changes in synthetic sizes addition level can produce large changes in water resistance. Because of this, synthetic sizing systems are more difficult to control than rosin systems.

12. Summary

This concludes our discussion of the wet end chemistry of internal sizing and water resistance. It is clearly one of the most important subjects in wet end chemistry. Once you have mastered

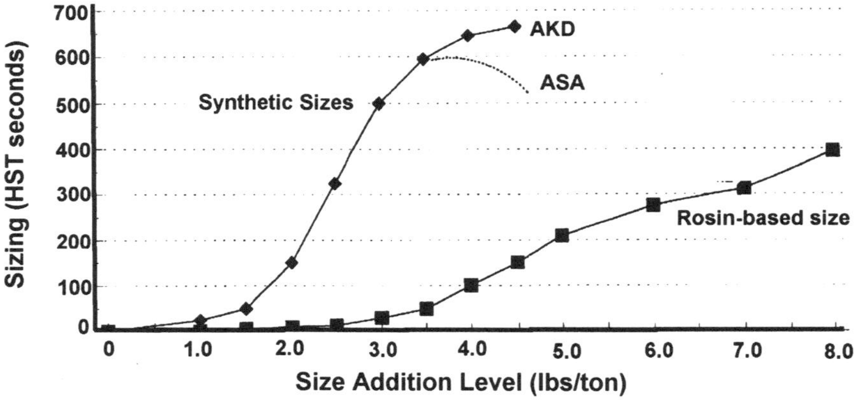

Figure 14-12. Comparison of sizing response for synthetic and rosin-based sizes [1]

the basic concepts discussed here, you are encouraged to undertake more advanced study of one or more of the chapter references. There is much to be gained from informed application of the knowledge contained there.

13. References

[1] Royce, R. "Internal sizing." in Course Notes from 1995 Introduction to Wet End Chemistry Short Course, Orange Beach, AL, February 15-17, 1995.

[2] Cates, R. E., Dumas, D. H., Evans, D. B. "Alkyl ketene dimer sizes." in The Sizing of Paper, Second Edition, W. F. Reynolds, Editor, TAPPI PRESS, 33-50(1989).

[3] Roberts, J. C. "Neutral and Alkaline Sizing." Chapter 8 in Paper Chemistry, Roberts, J. C., Editor, Chapman and Hall, 114-131(1991).

[4] Farley, C. E. and Wasser, R. B. "Sizing with alkenyl succinic anhydride." in The Sizing of Paper, Second Edition, W. F. Reynold, Editor, TAPPI PRESS, 51-61(1989).

[5] Davison, R. W. "Internal sizing." Chapter IV in Pulp and Paper Manufacture, Vol. 6, Stock Preparation, Hagemeyer, R. W. and Manson, D. W., Editors, Joint Textbook Committee of the Paper Industry, TAPPI and CPPA, 39-64(1992).

[6] Cates, R. E. "Internal sizing agents." Chapter 5 in Chemical Processing Aids in Papermaking: A Practical Guide, Hipolit, K. J., Editor, 63-80(1992).

[7] Otway, P. and Johnson, D. C. "Sizing in acid, neutral, and alkaline conditions." Chapter 9 in Applications of Wet End Chemistry, Au, C.O. and Thorn, I., Editors, Blackie Academic and Professional, 139-171(1995).

[8] Roberts, J. C. "Overview and theories of neutral and alkaline sizing." Proceedings of 1990 Neutral/Alkaline Papermaking Short Course, Orlando, FL, 55-62(1990).

[9] Lindstrom, T. and Eklund, D. "Water penetration and internal sizing." Chapter VIII in Paper Chemistry: An Introduction, DT Paper Science Publications, 192-222(1991).

[10] Crouse, B. and Wimer, D. "Alkaline sizing: an overview." Tappi Journal, 74(7):152 (1991).

[11] Lindstrom, T. and Sodeberg, G. Nordic Pulp and Paper Res. J. 2, 39-45(1986).

[12] Roberts, J. C. and Yajun, Z., Proceedings of PIRA Symposium on Neutral Papermaking, Stratford, England (1990).

[13] Lindstrom, T., Soderberg, G. and Obrien, H. Nordic Pulp and Paper J., 1: 26-42(1986); 2: 31-45(1986).

[14] K. J. Bottorff. 'AKD sizing mechanism: a more definitive description." Tappi J. 77(4): 105-116(1994).

[15] Lindstrom, T. and O'Brian, H., Nord. Pulp Paper Res. J. 1(1):34(1986).

[16] Lindstrom, T. and Soderberg, G., Nord. Pulp Paper Res. J. 1(2):39(1986).

[17] Marton, J. Tappi J. 73(11):139(1990).

[18] Marton, J. "Kinetic aspects in furnish interactions: sizing as an example." in Proceedings of 1995 TAPPI Papermakers Conference, Chicago IL, April 23-26, 97-99 (1995).

[19] Marton, J. Tappi J. 74(8): 187(1991).

[20] Moyers, B. "Diagnostic sizing-loss problem solving in alkaline systems." Tappi Journal, 75(1): 111-115(1992).

[21] Novak, R. W. and Rende, D. S. "Size reversion in alkaline papermaking." Tappi Journal, 76(8): 117-120(1993).

[22] Bartz, W. J., Darroch, M. D., and Kurrle, F. L. "Alkyl ketene dimer sizing efficiency and reversion in calcium carbonate filled papers." Tappi Journal 77(12): 139-148(1994).

[23] Gill, R. A. "Modified PCC fillers to reduce sizing demand and improve runnability." in Proceedings of 1990 TAPPI Papermakers Conference , TAPPI PRESS, Atlanta, GA 255-267(1990).

[24] Chen, G. C. I. "Optimizing an alkenyl succinic anhydride sizing process" in Course Notes from 1987 Advanced Topics in Wet End Chemistry Short Course, TAPPI PRESS, Memphis TN, October 14-16, 63-69(1987).

Chapter 15

Colloidal Aspects of Retention and Drainage

1. Introduction

Now that we have reviewed the characteristics of the major substances commonly found in papermaking furnishes, we will turn to a discussion of how wet-end components interact with one another immediately prior to, and during, sheet formation. The topics of interest are chemical additive adsorption, fiber flocculation, fines flocculation, and drainage. Dye, wet strength resin, and dry strength agent adsorption by fibers and fines have already been addressed in previous chapters. In this chapter, we will focus mainly on subjects related to the coagulation and flocculation of fibers, fines, and other furnish particles.

Fiber flocculation primarily influences the mass uniformity (formation) of the web. Fines flocculation influences both paper properties and the runnability of the paper machine. Since chemical additives are highly adsorbed on fines and filler particles, additive retention is also affected by fines and filler flocculation. The drainage processes occurring in the paper machine forming section are important to paper machine productivity.

It should be understood that the achievement of good sheet formation and high drainage rates always involves both hydrodynamic and chemical considerations. Consequently, the papermaker should insure that the mechanical configuration of a paper machine is in good order before attempting to work with its papermaking chemistry. An excellent description of the influence of mechanical factors on formation, drainage, and retention is given by Marton.[1]

In addition to the paper property and machine performance attributes listed above, colloidal interactions among furnish components also affect the cleanliness of the machine and its environmental impact. Furnish components that are not incorporated into the paper web, but remain suspended in recirculating white water, often form deposits or are lost in mill effluent. This is costly and to be avoided whenever possible.

The aggregation of furnish components to produce the paper web is an extremely important and complex subject that touches upon many areas of wet end chemistry. Maintaining close control over this aspect of papermaking is vital to producing a high quality product in an efficient manner.

Chapter topics are organized in the following order:

- Terminology associated with retention
- Materials and mechanisms important to retention
- Common retention systems
- Impact of interfering substances and fines level
- Impact of first-pass retention on paper properties
- The role of wet end chemistry in drainage

The issue of retention process control is addressed in Chapter 16.

2. Fines retention

a. Introduction

As discussed earlier in Chapter 3, fines are defined as any particulate material that passes through a 200-mesh screen. This definition encompasses a wide variety of substances found in a papermaking furnish; e.g. fiber fines and fillers. Fines influence the structural, strength and optical properties of paper. Poor fines retention adversely affects the runnability, cleanliness, and chemical efficiency of a paper machine.

b. Retention definitions

Fines retention will be treated in some detail from the paper machine practice standpoint in Chapter 16. Here, we will be content simply to define what we mean by retention. **Table 15-1** contains the two definitions of greatest interest to the papermaker.

Table 15-1. Retention definitions

Overall Retention: The ratio of the amount of material that is sent to the wet end of a paper machine compared to the amount that goes onto the reel at the dry end of the machine. (Typical values are 90-95%.)

First Pass Retention: The ratio of the amount of material that leaves the headbox slice compared to the amount of material that is contained in the paper web leaving the couch roll. (Typical values are 20-90%.)

It is common to refer to the first pass retention of particular fractions of a headbox furnish as well as the entire furnish (e.g. "first pass fines retention," "first pass fiber retention," "first pass ash retention").

c. Primary retention mechanisms: mechanical entrapment and colloidal aggregation

Fines are incorporated in the sheet by two mechanisms: *mechanical entrapment* and *colloidal aggregation*. The relative contribution of the two mechanisms in a given situation depends upon many factors, such as basis weight, machine speed, headbox consistency, forming section configuration, filler level, and mechanical pulp content. Many recent industry trends have tended to reduce the effectiveness of retention by mechanical entrapment and increase the importance of colloidal aggregation. Because of this, we will focus our attention in this chapter mainly on the colloidal aspects of retention.

3. Mechanical entrapment or filtration of fines by the forming web

Mechanical entrapment involves catching the fines in the interstices of the web formed by fibers deposited on the moving fabric. In the very beginning of the process, only the long fiber fraction of the furnish is captured by the forming fabric — whose openings are much larger than the fines particles (**Fig. 15-1**).

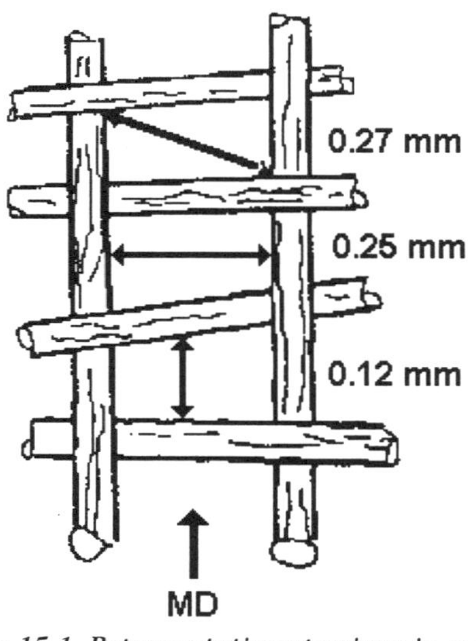

Figure 15-1. Representative openings in a single layer forming fabric. Note that the opening dimensions are much larger than the dimensions of fines particles. Consequently, the forming fabric will not capture the fine particles of a furnish.

A certain mat thickness must be formed before substantial entrapment occurs.[2] As the fiber mat builds, more fines are caught in the denser network which provides more available points of entrapment. This process results in a rough surface on the wire side of the web and a smoother, denser structure on the top side (felt side). The papermaker has only a limited control over this retention mechanism, which is a strong function of basis weight. Good formation promotes better retention by providing a more homogeneous mat to filter the fines. Likewise, the larger the flocs formed by colloidal mechanisms, the greater is the probability that they will be filtered out by the forming web.

The situation is further complicated on the paper machine because water removal occurs by the pulsed drainage action of forming section water removal elements. After each pressure pulse the forming mat expands and absorbs water and fines from the furnish suspension above and below it. Then, when the next vacuum pulse occurs, the mat is compressed quickly and the fines move in the direction of water flow. The net result is reduced fines retention in the web. The greater the pressure pulses and the lower the consistency, the lower will be the ultimate fines retention.

Mechanical entrapment or filtration is only effective for the long fiber fraction of a furnish. In order to increase the magnitude and uniformity of fines retention in a forming paper web, it is necessary either to attach the fine particles directly to the fiber surfaces so that the fines will be retained along with the fibers or form fines aggregates that are large enough to be efficiently filtered out by the forming web. Both of these alternative objectives can be accomplished through the application of colloidal aggregation phenomena.

4. Introduction to the colloidal aggregation of fines

Colloidal aggregation is the predominant mechanism employed for fines retention in papermaking and involves the formation of flocs composed of only fines, and flocs containing both fibers and fines. In the latter instance, the mechanism involves the adsorption of fines onto fiber surfaces. These fines are then immobilized and incorporated into the mat network along with fibers on the paper machine forming section.

Fiber-fiber flocculation is an undesirable phenomenon that should be prevented as long as possible in the papermaking process. In the case of fines, the situation is more complex. The papermaker strives to promote flocculation between fines and fibers, while at the same time discouraging the formation of flocs containing only fines so that the fines will be well distributed throughout the web. This can only be accomplished if a large percentage of the fines are attached directly to fibers prior to sheet formation. If large flocs containing only fines occur, then these fines will be retained primarily by mechanical entrapment, which can only occur after a certain percentage of the sheet has been formed. The first-formed portion of web will contain fewer fines, giving rise to greater sheet "two-sidedness."

A second reason why uniform fines distribution is desirable is that chemical additives tend to concentrate on fines and in order to achieve good distribution of chemical additives it is necessary that the fines be well distributed throughout the web.

5. Laboratory evaluation of colloidal interactions in a papermaking furnish

Fines flocculation occurs in stock preparation, the approach flow system, and the headbox. Once formed, a floc's ability to resist disruption by shear and turbulence in a paper machine headbox and forming section is determining factor in its retention. If flocs are sensitive to these hydrodynamic forces, they will break down, pass through the forming fabric, and accumulate in the white water.

The paper machine is too expensive to be used for routine or extensive exploratory retention experiments. Because of this, papermakers employ laboratory investigations to evaluate retention additives and elucidate the effects of various chemical parameters on first-pass retention. The former activity is usually done as a preliminary step to running machine trials with retention aids, while the latter is associated with trouble-shooting and process improvement studies.

A variety of devices have been employed in laboratory retention studies. Since it is impossible to simulate the conditions on the wet end of a paper machine with high accuracy, most persons have settled for a procedure that will enable them to rank retention aids according to their effectiveness in the laboratory test and also provide input as to whether or not they can be expected to fail completely on the machine. In most cases, it can be stated with fair certainty that any retention aid that is completely ineffective in a laboratory test will also fail under mill conditions. **Table 15-2** lists several of the devices that have been employed in laboratory retention studies.

Table 15-2. Devices used in laboratory retention studies

1. Static settling testers
2. Free drainage testers
3. Handsheet formers
4. Britt dynamic drainage jar

Of the devices listed in **Table 15-1**, the one receiving the most acceptance has been the Britt Dynamic Drainage Jar. [3] In addition to the original version, a number of variations of the device have been developed.[4,5]

The Britt Jar enables direct measurement of the colloidal retention of solids under turbulent conditions independent of fiber mat formation. The measurement of colloidal retention is of practical and fundamental importance because the paper stock system is known to be in a dynamic state at the headbox slice and initial forming areas of the paper machine. In addition, as was discussed earlier, fines retention is a strong function of colloidal flocculation phenomena.

The basic Britt Jar (**Fig. 15-2**) consists of a cylindrical stock sample-holding container that screws into a base designed to hold a perforated plate or sample of paper machine forming fabric. The base has a hole in the bottom into which is inserted a large, one-hole stopper. A small-diameter tube is inserted into the rubber stopper to a point flush with the top of the stopper. A propeller for stirring the stock sample is inserted into the container. The propeller is driven by a variable speed motor capable of operating in the range of 0 to 3000 rpm. Reference [6] describes the use of the Britt Jar in detail.

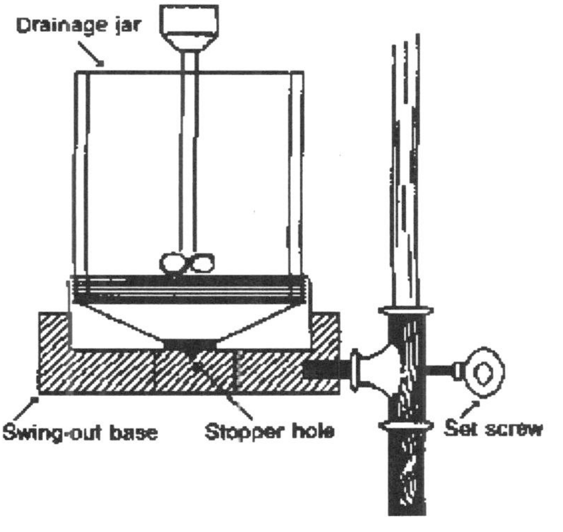

Figure 15-2. Diagram of the Britt Dynamic Drainage Jar

Since its introduction in the early 1970's, a number of different uses have been developed for the Britt Jar. The most important of these are listed in **Table 15-3**. A good introduction to the detailed procedures associated with the uses in the table can be found in TAPPI Method T 261.[3]

Table 15-3 . Uses of the Britt Jar

Fines Screening

- Characterization of stock sample fines content
- Evaluation of first-pass fines retention

Investigation of shear resistance

- Additive evaluation
- Britt Jar "calibration" against a paper machine

Study of variables that influence retention

- Retention aid studies
- Investigation of retention interferents
- Investigation of mechanical factors and first-pass retention

6. Chemical additives employed to promote the colloidal interactions of fines

a. Introduction

The chemical additives employed to promote aggregation interactions among furnish particles are classified as *chemical processing aids.* Chemical processing aids are additives that do not directly affect paper properties (as opposed to functional additives). Instead, they either assist the effectiveness of functional additives, or promote the cleanliness and runnability of the paper machine.

Excellent surveys of the different materials employed to promote the aggregation of papermaking furnish components have recently been published.[7–10] **Table 15-4** lists the three major categories of these substances used in the paper industry. Two of the categories represent polymeric materials that usually have ionic groups incorporated into their molecular structures. Such substances are called *polyelectrolytes.* Reference [11] provides a modern, comprehensive treatment of all aspects of polyelectrolytes.

Table 15-4. Types of Retention/Drainage Aids

- Inorganic salts (alum, $CaCl_2$)
- Natural polymers
- Synthetic polymers

The most important inorganic retention aids are alum, polyaluminum chloride, and calcium chloride — with alum being by far the most significant material. These aluminum-based compounds were discussed in Chapter 8. It was pointed out there that the Al^{+3} form of aluminum was not the best form for retention. Rather, various cationic polyaluminum species exhibit the best particle charging capability and also the best fines retention performance.

Calcium chloride, $CaCl_2$, has been used occasionally as a retention/drainage aid in linerboard manufacture. The Ca^{+2} ion is the active agent in this system.

The silicas and bentonite clays are inorganic pigment materials finding increasing use as components in *microparticle retention systems.* These systems utilize very small inorganic particles in combination with polymers to improve drainage, formation, and retention. They are described below.

b. Natural organic polymers

Cationic starch is by far the most important retention aid that is based on a natural polymer. Cationic guar gum is also used for this purpose to a small extent. Both materials were discussed in Chapter 11. It should be pointed out that the traditional use of these materials has been dry strength improvement and their use for retention purposes has been of a secondary nature. However, the recent use of cationic starch in a common microparticulate system has increased the importance of this material in the overall retention picture.

c. Synthetic organic polymers

Introduction - Thanks to the wide range of possibilities afforded by modern organic polymer synthesis technology, an array of water soluble, polymer flocculating agents are available to the papermaker today. **Table 15-5** lists some of the most important characteristics used to differentiate among the available materials.

Each backbone structure will be discussed below. Polymer molecular weight is a very important characteristic and a wide range of products are available. Since polymer synthesis techniques produce a molecular weight distribution in the final product, it is usual to refer to the average molecular weight of a particular material. Due to the proprietary nature of the business, polymer suppliers do not reveal quantitative molecular weight information about their products, but rather rely upon qualitative terms like *low, medium,* and *high.* **Table 15-5** lists approximate quantitative average molecular weight ranges for these terms, as suggested by Halverson.[10]

Nonionic, cationic, anionic, and amphoteric products having varying degrees of charge density are available. Again, industry practice is to use qualitative terms to designate the charge density associated with a specific product.

Paper industry application practice has the high molecular weight, low charge density flocculants used most often for the retention of fine particles under higher shear and turbulence conditions. The low molecular weight, high charge density (cationic) products are used for their charge modifying capabilities and for retention purposes in low shear papermaking systems.

Table 15-5. Synthetic polymer characteristics

- Backbone structure
  - Polyacrylamides
  - Polyamines
  - Polyethyleneimine
  - Polyamidoamines
  - Polyethylene oxide
- Average molecular weight
  - Low: 1,000-100,000
  - Medium: 100,000-1 million
  - High: 1-5 million
  - Very high: >5 million
- Charge type
  - Nonionic
  - Cationic
  - Anionic
  - Amphoteric
- Charge density
  - Low: 1-10%
  - Medium: 10-40%
  - High: 40-80%
  - Very high: 80-100%
- Delivered form
  - Emulsions
  - Aqueous solutions
  - Dry powders or granules

Polymer products are delivered in the three different forms. In general, the high molecular weight materials come as water-in-oil emulsions containing approximately 30% solids. The low and medium molecular weight products are usually shipped as aqueous solutions. Synthetic polymers are not often shipped as dry powders or granules. On the other hand, natural organic polymers, such as cationic starch and guar, are usually delivered in this form.

Polyacrylamides - Nonionic polyacrylamide polymers are synthesized by a free-radical chain reaction with acrylamide as the starting monomer (**Fig. 15-3**).

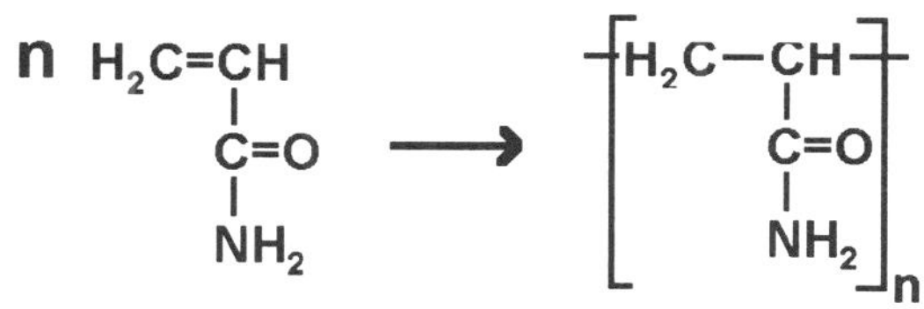

Figure 15-3. Synthesis of polyacrylamide by a chain reaction of acrylamide monomer

Since the polyacrylamide polymer is nonionic, it has little affinity for fibers and is not a very good retention aid. In order to be useful, it is necessary to incorporate ionic groups into the structure. There are two primary ways to accomplish this:

(1) Copolymerization of acrylamide and other anionic or cationic vinyl monomers;

(2) Modification of nonionic polyacrylamide.

An example of a cationic vinyl monomer commonly used in copolymerization, trimethylaminomethacrylate, is shown in **Fig. 15-4**. A quaternary cationic polyacrylamide is produced that has a positive charge at all pH values.

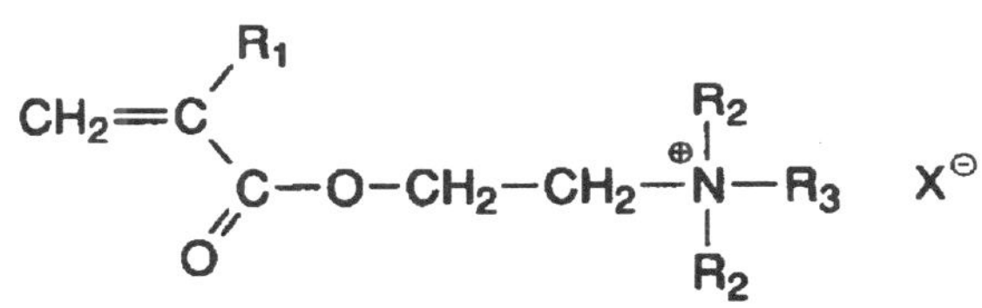

Figure 15-4. Trimethylaminomethacrylate. A cationic vinyl monomer that is copolymerized with acrylamide to produce cationic polyacrylamides.

The incorporation of cationic groups by modification of nonionic polyacrylamide is carried out often by the Mannich Reaction. **Fig. 15-5** illustrates one such reaction.

Cationic polyacrylamides used in the paper industry contain from 20-70% cationic groups, with the lower values most often accompanying the highest molecular weights.

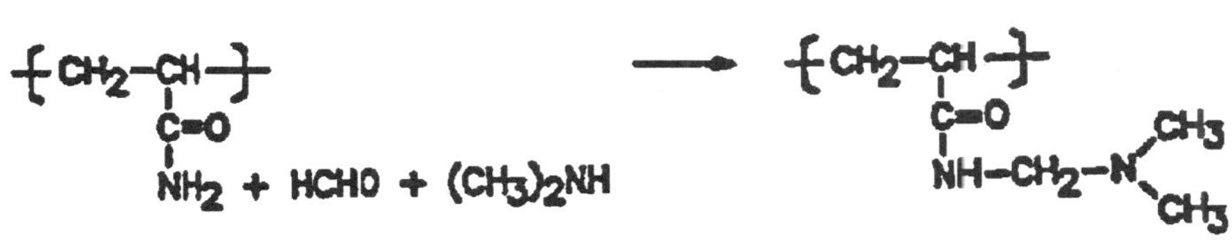

Figure 15-5. Synthesis of cationic polyacrylamides via the Mannich Reaction. Quaternary products are formed by reacting the Mannich reaction product with dimethyl sulfate or methyl chloride.

Anionic polyacrylamides can be produced either by copolymerization of acrylamide and acrylic acid, or by hydrolysis of some of the amide groups of polyacrylamide. The resulting material contains a mixture of acrylamide and acrylic acid groups. Since the acrylic acid carboxyl group is a weak acid, it carries less charge under acid pH conditions than under alkaline pH conditions. It is 100% protonated (zero charge state) at pH 2.8.

Polyacrylamides are commonly supplied as water-in-oil emulsions (ca. 30% solids), or as powders. The usual practice is to follow a two-step procedure when making down the chemical. Initially, the emulsion or powder is diluted to 1-2% solids under high shear with fresh water. The dissolution process is slow for solid products. Once the polymer particles are completely dissolved, the solutions are then diluted down to the very low concentrations employed for stock addition. The solutions should be used within 24 hours of preparation to minimize polymer degradation. The polymers also break down when exposed to high shear agitation.

Polyamines - Polyamines can be formed by the reaction between dimethylamine and epichlorohydrin (**Fig. 15-6**). The molecule has its charged nitrogen in the molecular backbone chain rather than residing on a pendant group.

The products are low to medium molecular weight and have high charge densities. They are not usually employed as single component retention systems, but are often used in conjunction with anionic polyacrylamide in dual polymer systems. Polyamines are usually supplied as 40-50% water solutions.

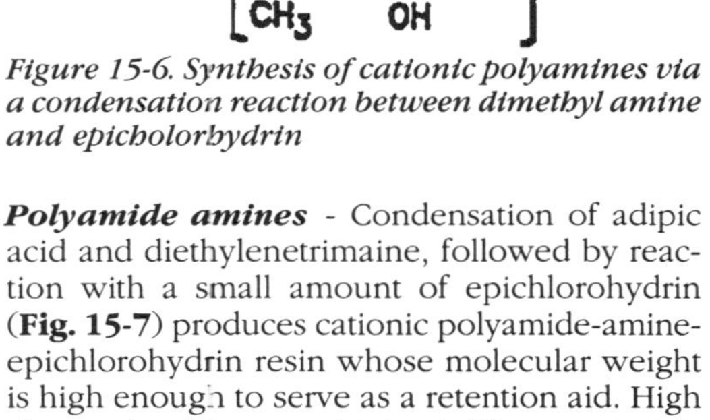

Figure 15-6. Synthesis of cationic polyamines via a condensation reaction between dimethyl amine and epicholorhydrin

Polyamide amines - Condensation of adipic acid and diethylenetrimaine, followed by reaction with a small amount of epichlorohydrin (**Fig. 15-7**) produces cationic polyamide-amine-epichlorohydrin resin whose molecular weight is high enough to serve as a retention aid. High charge densities are obtained. These materials also find wide use as wet strength resins. (See chapter 9.) They are supplied as 20-30% solids aqueous solutions.

Poly-DADMAC - Poly-DADMAC is a quaternary cationic polyelectrolyte of medium-high molecular weight (100,000-1 million) synthesized by polymerization of diallyl dimethyl ammonium chloride (DADMAC) according to the reaction shown in **Fig. 15-8**. High charge densities are achieved. The polymer is often used as a standard cationic titrant in the colloid titration method. (See Chapter 6.)

The polymer is supplied as 30-50% aqueous solutions. It is used primarily as a charge neutralizer to react with interfering substances and as the cationic component of dual polymer retention systems.

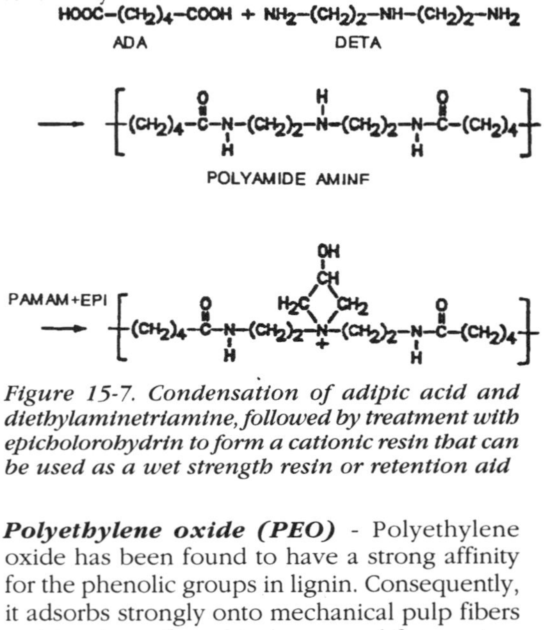

Figure 15-7. Condensation of adipic acid and diethylaminetriamine, followed by treatment with epicholorohydrin to form a cationic resin that can be used as a wet strength resin or retention aid

Polyethylene oxide (PEO) - Polyethylene oxide has been found to have a strong affinity for the phenolic groups in lignin. Consequently, it adsorbs strongly onto mechanical pulp fibers and has been found to be a good fines retention for mechanical pulps. It is synthesized by polymerizing ethylene oxide (**Fig. 15-9**) in the presence of a catalyst. PEO retention aids have molecular weights in the 4 million range. They have a very low anionic charge.

Polyethyleneimine (PEI) - Polyethyleneimine (PEI) is synthesized by a ring opening, acid-catalyzed polyaddition reaction of ethyleneimine (**Fig. 15-10**). A branched structure results that contains primary, secondary, and tertiary amine groups. The charge density is a function of pH. The polymer has a very high charge density under acid conditions. A wide range of molecular weights are available. PEI products are supplied as alkaline solutions having 15-20% solids. They are used as retention aids and charge neutralizing agents.

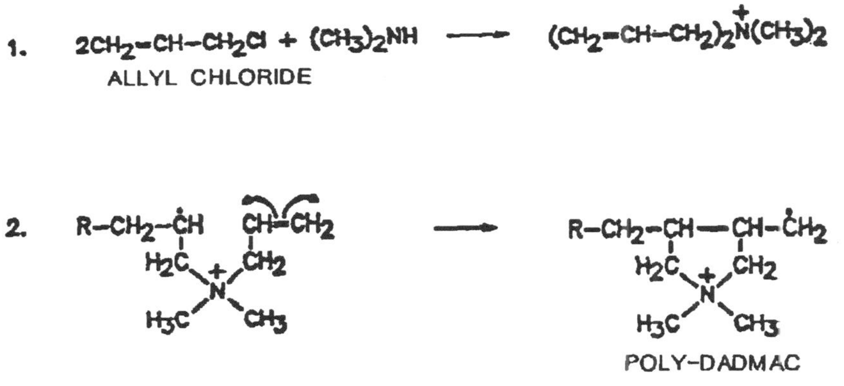

Figure 15-8. Synthesis of poly-DADMAC.

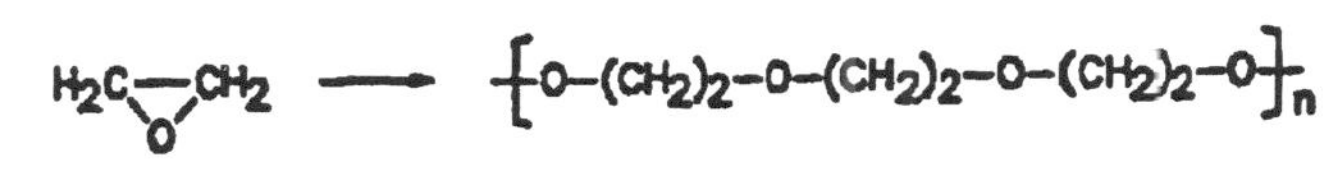

Figure 15-9. Synthesis of polyethylene oxide (PEO)

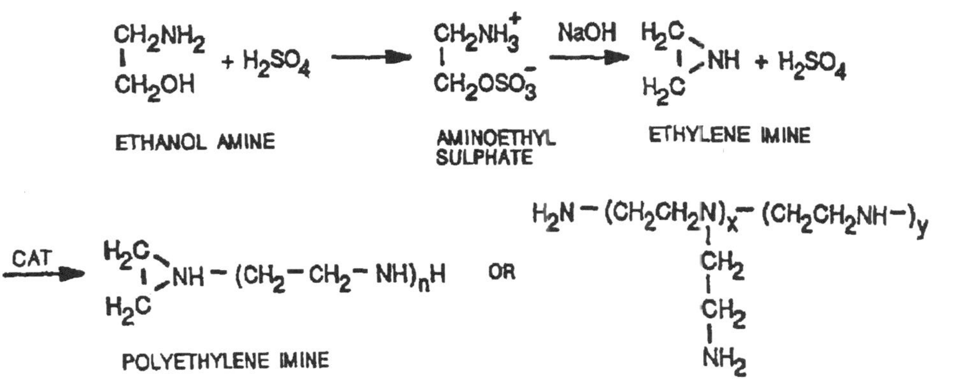

Figure 15-10. Synthesis of polyethyleneimine

d. Closing comments about chemical additives that promote the colloidal interactions of fines

It is clear that there are many different products available to the papermaker and that they have a wide range of charge densities, molecular weights, and backbone structures. Having introduced these materials, we now turn to a review of the fundamental mechanisms by which they operate to enhance fines aggregation and retention.

7. Overview of interactions between fibers and polyelectrolytes

In Chapter 5, it was stated that cationic polymers produce Langmuir-type equilibrium adsorption isotherms (**Fig. 5-6**). While equilibrium adsoption behavior is important in its own right, papermaking timeframes place greater emphasis on polumer-fiber interaction kinetics. Research has shown that when a polymer is

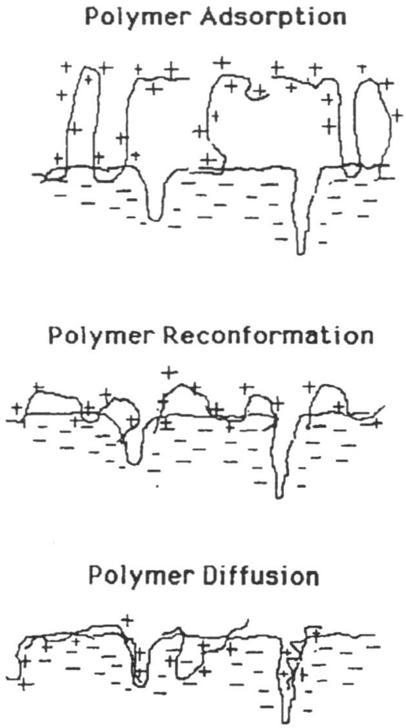

Figure 15-11. Three stages of polymer:fiber interaction [13]

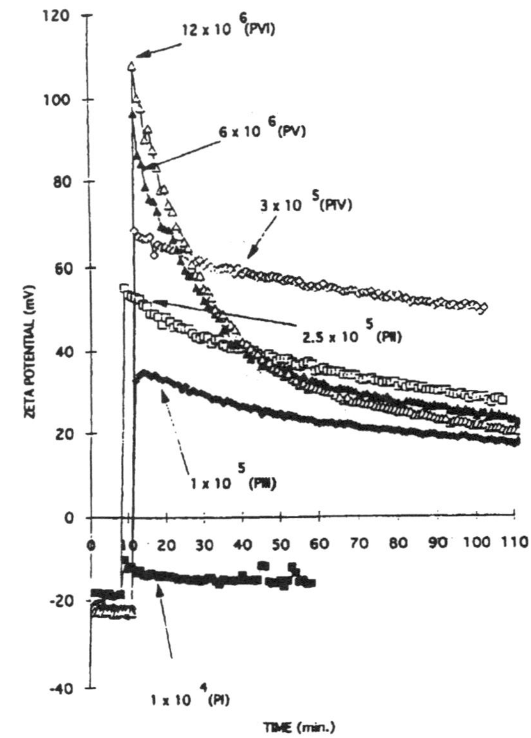

Figure 15-12. Charge decay exhibited by various synthetic cationic polyelectrolytes adsorbed onto 50:50 softwood:hardwood bleached kraft pulps [13]

introduced to a fiber slurry, a three-stage process happens. First, *adsorption* occurs extremely rapidly, on the order of a few seconds or less in many instances. Immediately following the adsorption of the polymer by the fibers, which often involves contact by only small portions of the polymer chain, *reconformation* begins. The reconformation process involves movement of previously extended parts of the polymer onto the particle surface so that the polymer molecule takes on a flatter adsorbed conformation that exists closer to the surface. The third stage is *diffusion* of the adsorbed and reconformed polymer chains into particle pores so that they no longer contribute to surface charge phenomena.[12] **Fig. 15-11** illustrates the three stages of polymer:fiber interactions.[13]

The adsorption, reconformation and diffusion processes can be followed by zeta potential measurements. **Fig. 15-12** illustrates this for the case of mostly linear, cationic, synthetic polymers interacting with bleached kraft softwood fibers. The initial adsorption produces a rapid increase in zeta potential to a maximum value.The highest molecular weight polymers produced the highest initial zeta potential, regardless of polymer charge density. Next, the zeta potential value is monitored with time and they are observed to decrease with time (charge decay) for all polymers, albeit at different rates. The charge decay curves have different initial slopes — which are inversely related to the polymer molecular weights. With time, all the decay curves transition to essentially the same slope. The following interpretation of these results has been suggested.[13]

The initial part of the charge decay curve, which varies from polymer to polymer, is believed to be associated with the reconformation process. It is logical that the highest molecular weight polymers would exhibit the greatest change in zeta potential since they have the greatest initial zeta potential and have chains that are extended the farthest away from the fiber surface that can reconform onto the fiber surface rapidly and produce a large zeta potential change. The magnitude of the anionic fiber surface charge also influences the rate of charge decay — greater fiber surface charge produces faster charge decay.

The second portion of the charge decay curve has essentially the same slope for all curves in **Fig. 15-12**. It is believed that this point is associated with diffusion of polymer chains into fiber pores and away from the surface. The diffusion process does not appear to be strongly influenced by molecular weight.

High amylopectin content cationic starch does not behave like linear synthetic polymers with respect to charge decay.[14] In this case, charge decay is a "product function" of charge density and addition level.[14] If the polymer charge density or addition rate is above a particular level, charge decay does not occur. It appears that the highly branched amylopectin structure does not lend itself to reconformation nor diffuse under certain charge density/dose level conditions. The specific effects of these two variables will depend on the starch molecular weight and charge density, the fiber pore size distribution, and the magnitude of the fiber surface charge.

High amylose starches tend to act like the synthetic polymers, rather than amylopectin, in that charge decay is observed at all levels of charge density and dose level.[14] This is undoubtedly a reflection of the lower molecular weight of amylose relative to amylopectin and the linear structure of the polymer chain.

8. Fines aggregation mechanisms

a. Introduction

A considerable body of research has been done concerning the fundamental mechanisms of fines aggregation by colloidal interactions. This work has produced a number of conceptual models upon which the papermaker can develop trouble-shooting and optimization strategies. Several of these models were introduced in Chapter 5 in the general introduction to papermaking colloid science. It will be useful to revisit them here. **Table 15-6** lists some of the relevant terminology.

Table 15-6. Fines aggregation terminology

Coagulation
- Charge neutralization
- Heterocoagulation
- Patching

Flocculation
- Bridging flocculation
- Network flocculation

b. Charge neutralization

Charge neutralization, or coagulation, was the first aggregation mechanism introduced in Chapter 5. This mechanism is the simplest in that the addition of an electrolyte salt or very low molecular weight polyelectrolyte compresses the electrical double layer enough that repulsion between particles is diminished and Van der Waals attractive forces can induce coagulation whenever two particles collide. **Fig. 15-13** illustrates that an electrolyte's coagulation effectiveness is dependent on the valence of the ion whose charge is opposite that of the aggregating particles. Since practically all papermaking particles are anionically charged, cations are the effective charge neutralizing ions and the higher the valence of the added ion, the greater its aggregating efficiency. Increased machine speeds and greater headbox and forming section shear forces have reduced the effectiveness of charge neutralization as a stand-alone retention mechanism.

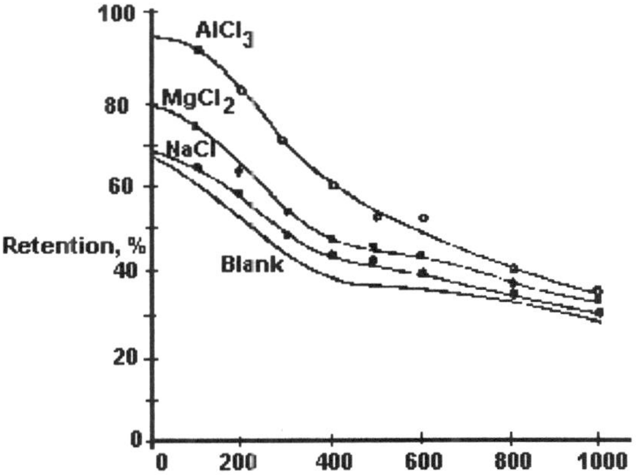

Figure 15-13. Relationship between a cation's fines retention capability and its valence

When polyelectrolytes are employed to promote charge neutralization, it is important that the molecular weight be sufficiently high that molecular diffusion into fiber pores occurs slowly. This usually means that the molecular weight should be greater than 100,000. High charge density is necessary. Polyamines, PEI, and polyamideamine epichlorohydrin polymers are preferred. Cationic starch is not a good charge neutralizing material due to its low charge density and high molecular weight.

c. Heterocoagulation

Heterocoagulation refers to the aggregation of oppositely charge particles. Examples include the precipitation of cationically charged size particles onto anionically charged fiber surfaces. The same mechanism applies to deposition of cationic precipitated calcium carbonate on anionic fiber surfaces.

d. Patching

Patch-type aggregation is illustrated schematically in **Fig. 15-14**. This mechanism depends upon the formation of cationic sites, or "patches", that have a high charge density and that are larger than the thickness of the fiber surface electrical double layer. This process requires the use of high charge density, low-to - medium molecular weight cationic polyelectrolytes that can form patches, such as modified polyethyleneimine, polyamines, and polyamideamine epichlorohydrin resins.[7]

The cationic polymer adsorbs strongly on the negatively charged fiber surface and charge reversal occurs at the adsorption sites. Aggregation then occurs by electrostatic attraction between positive patches and negative sites on adjacent particles. The extent of attraction depends on the charge density of the polymer and amount of surface coverage. The degree of surface coverage for maximum aggregation is 0.5 (i.e. 50% of surface area) for low molecular weight polymers.[7] Since patching depends solely upon electrostatic forces, it is classified as a coagulation mechanism.

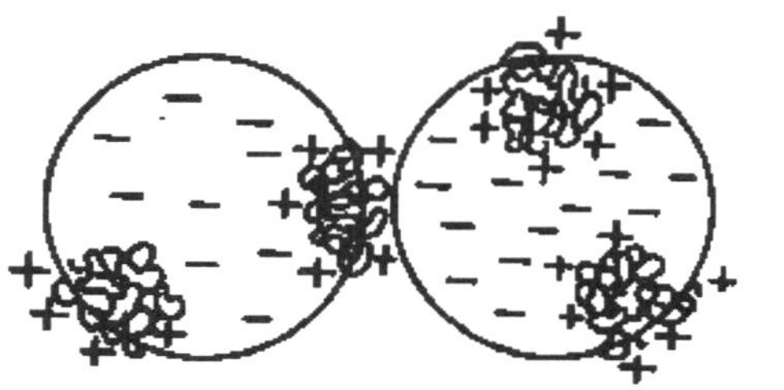

Figure 15-14. Patch-type aggregation mechanism

In order for sites having different charges to exist on a surface, it is necessary that the patch extend through the electrical double layer. In low conductivity systems, where the double layer can be quite thick, it may be difficult to have the patch extend beyond the double layer and aggregation can only occur by charge neutralization in this case. If electrolyte is added to collapse the double layer, then patch formation can occur.

When selecting a polymer, care must be taken that the molecular weight and charge density are high enough that the polymer will not rapidly diffuse into the fiber pores and cause charge decay to occur. It also helps if the polymer is branched, which retards reconformation and diffusion.

e. Bridging flocculation (Fig. 15-15)

Up to now, we have focused on coagulation aggregation mechanisms that utilize simple salts or low-to-medium molecular weight, high charge density polymers that adsorb onto fibers and fines in relatively flat conformations. We now turn to mechanisms that pertain to high molecular weight polymers that have low or no charge density.

When high molecular weight polyelectrolytes are employed to aggregate fines and fibers, a four-step mechanism occurs:

(1) A polymer molecule is attracted to a particle surface and collides with it, initiating adsorp tion.

(2) The polymer attaches to the particle in a fash ion so that a portion of it (the polymer) is attached and the rest of the molecule ex tends out into the surrounding water as a series of loops and trains.

(3) A second particle collides with these ex tended polymer loops and trains.

(4) The second particle adsorbs the extended polymer loops and trains, forming a bridge between the two particles and producing ag gregation.

Aggregation brought about by bridging is designated as flocculation.

f. Factors important to bridging flocculation

It is clear that polymer molecular weight is a critical variable in the flocculation mechanism. Other factors, such as polymer charge density, particle surface charge density, and the ionic concentration in the system are also very important. These factors, in turn, influence the rate of polymer adsorption, the strength of polymer attachment to particle surfaces, the rate of polymer reconformation, the thickness of particle electrical double layers, the inherent polymer extension potential, and polymer solution conformation. The following relationships among these variables have been identified.

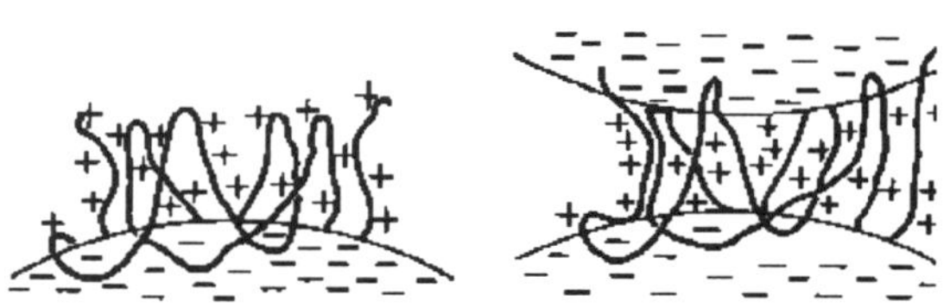

Figure 15-15. Bridging model. When a high molecular weight polyelectrolyte is added to the system, there is an initial adsorption of retention aid onto particle surfaces. The unattached parts of the polymer molecule form extended loops and trains that project out into the suspending medium. These loops and trains then collide with another particle, become attached, and form a floc.

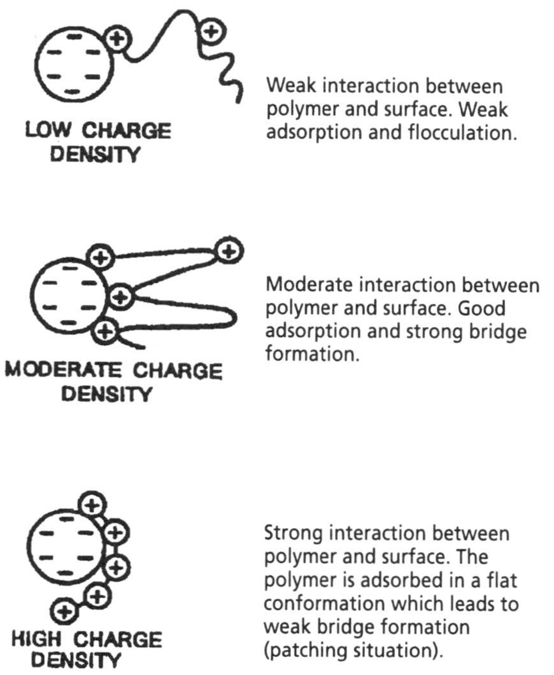

Figure 15-16. Effect of polymer charge density on the interactions of polymers with fine particle surfaces [7]

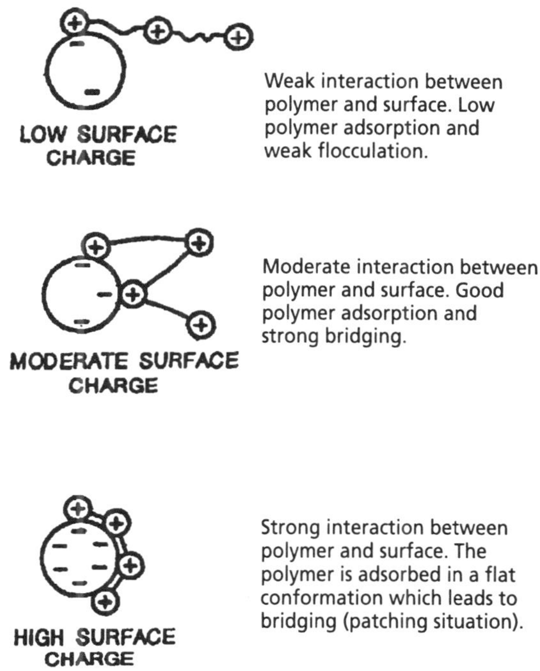

Figure 15-17. Effect of fine particle surface charge density on the interactions of fines with polymer molecules [7]

g. Molecular weight

Many studies have confirmed the importance of molecular weight to the flocculation effectiveness of polyelectrolytes. The reason for this is that the length of a polymer chain is directly related to its molecular weight. The longer the chain, the greater its ability to bridge between particles and form flocs.

h. Polymer charge density

Polymer charge density affects the flocculation process in the following ways. First, the rate of diffusion of a dissolved polymer molecule to a particle surface is affected by the electrostatic charge gradient that exists between the polymer and the particle surface. The larger the gradient, the faster the diffusion. Higher polymer charge densities produce larger gradients.

Second, greater polymer charge densities provide more points for the polymer to electrostatically bond to the fine particle surface and result in stronger polymer:particle attachment.

Third, greater polymer charge density produces more electrostatic repulsion between charged groups on a dissolved polyelectrolyte chain and causes the molecule to adopt a more expanded conformation, which promotes the extension of loops and trains away from a fine particle surface.

The above three polymer charge density effects all enhance the flocculation process. The fourth and fifth effects of greater polymer charge density — namely, increased polymer reconformation rate and flatter adsorbed conformations — negatively impact flocculation. **Fig. 15-16** summarizes some of the effects of polymer charge density.[7]

i. Fine particle surface charge

The positive effects of fine particle surface charge on flocculation are similar to polymer charge in that greater surface charge promotes increased adsorption rate and a stronger attachment of polymer chains to particle surfaces. On the adverse side, greater particle surface charge increases the polymer reconformation rate and causes the polymer molecule to adopt a flatter, less extended, conformation on the particle surface (**Fig. 15-17**).

j. Ionic concentration

The concentration of simple salts has several effects on flocculation. First, increased concentration of divalent and trivalent cations reduces the electrical double layer thickness and thereby enables two fines particles to approach one another more closely which, in turn, decreases the bridging distance a polymer must span in order to induce flocculation. Second, whenever divalent and trivalent cations undergo ion exchange reactions with fiber and fines surface carboxyl groups, the attraction between the particles and cationic polyelectrolytes is decreased, which adversely impacts polymer adsorption and the attachment of polymers to fiber and fines surfaces. Third, increased ionic concentration shields the repulsion between charged groups on a polymer chain and enables the molecule to coil up into a tighter, less extended conformation. This reduces the bridging capability of the polymer.

The above discussion should make it clear that the papermaker is faced with a very difficult task when selecting flocculating polymers for a given furnish and set of papermaking conditions. With the exception of molecular weight, which always has a positive impact on bridging, the other factors have both positive and negative impacts. Therefore, it is necessary to determine the molecular weights, charge densities, and dosage levels required for the best aggregation performance of a particular polymer aggregation system for a given furnish and paper machine environment. The complexity of the assignment means that this task usually must be undertaken on an empirical basis.

k. Network flocculation

The bridging flocculation systems described above rely upon polymer:particle interactions to initiate the process. Network flocculation systems involved solution reactions between two or more polymers as the first step. This reaction produces a network product that entraps the particles mechanically or interacts specifically with sites on particle surfaces. Electrostatic interactions between polymers and particles play less important roles in network flocculation than in bridging flocculation.

9. Different retention systems used in papermaking

We have now reviewed the individual types of polyelectrolytes used for colloid aggregation purposes and surveyed the various fundamental mechanisms by which they operate. In practice, these polymers and mechanisms are combined in a variety of ways to produce the different retention systems applied in papermaking. Common examples include single cationic polymers, single anionic polymers coupled with aluminum ions, dual polymers systems, microparticulate systems, and network systems.

a. Single cationic polymer systems

Single cationic polymers represent one of the most common retention systems employed by papermakers. High molecular weight, low charge density polyacrylamides are often used for this purpose. Addition levels typically range between 0.25-3.0 lbs/ton. All of the flocculation fundamentals discussion presented above applies directly to these retention systems.

b. Dual component systems: anionic polymer plus aluminum

It is often possible to gain synergistic effects by combining several retention agents that interact electrostatically with each other on fiber and fines particle surfaces. One system of this type is the alum/anionic polymer system. Alum is added first and adsorbs onto fiber and fines surfaces, imparting a positive character to sites. Then, a high molecular weight anionic polyacrylamide is added. This negatively charged polymer is attracted to the alum sites on the fibers and fines and interacts with them to form

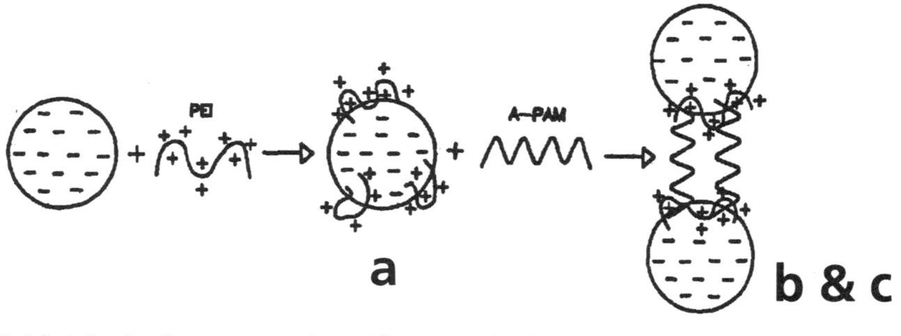

Figure 15-18. Model for dual polymer retention aid system. (a) Formation of a cationic patch by a low molecular weight, high charge density cationic polyelectrolyte (PEI in figure). (b) Adsorption of high molecular weight, low charge density anionic polyelectrolyte (A-PAM in the figure) to the positive patch on the particle to form a strong anchoring point. The surrounding negative surface on the particle causes the anionic polymer to extend out into the surrounding water. (c) Attachment of the extended anionic polymer to a positive patch on a second particle forming a strongly anchored bridge between the two particles.

strong polymer attachments to the fiber or fines particle. Loops and trains of anionic polymer extend out from the particle surface and attach themselves to other particles, forming a bridge.

It is important that the cationic aluminum has time to adsorb onto the fiber and fines surfaces and that the concentration of residual dissolved alum be low. If the dissolved alum concentration is not low, then it can react with the anionic polymer in solution and neutralize it, thereby rendering it ineffective.

c. Dual component systems: cationic polymer plus anionic polymer (Fig. 15-18)

Another popular multiple component system involves two polyelectrolytes — one a low molecular weight, high charge density cationic polymer (branched, if possible) and one a high molecular weight, low charge density anionic polymer. The two materials must react electrostatically with another on the fiber surface in order to be of use.

In practice, sequential addition of two oppositely charged polyelectrolytes has been found to produce very tenacious flocculation. The sequence of cationic polymer followed by anionic polymer has been found to be most useful. The original work on fine paper systems utilized the addition of an intermediate molecular weight cationic polyelectrolyte followed by treatment with a high molecular weight anionic polyelectrolyte. The mechanism for producing such flocculation involves the laying down of a cationic patch of cationic polyelectrolyte which provides an anchor point for the anionic polyelectrolyte. The high molecular weight anionic polymer bonds to the cationic site provided by the patch, but the rest of the chain is repelled by the negative sites surrounding the positive patch. This forces the anionic polymer to extend into the surrounding water, facilitating its adsorption onto a cationic patch site on another surface where it acts as a bridge, holding the two surfaces together.

It is important that little cationic polyelectrolyte remain in solution when the anionic polymer is added. Otherwise, a liquid phase reaction will occur to form a network. This network can serve to retain material by an occlusion process.

d. Microparticulate systems

Recent developments in retention aid technology involve systems where a microparticle is used in conjunction with a natural or synthetic polymer. Although such systems have been available since 1980, they have received the most attention in the past few years. In practice, a high molecular weight cationic polyacrylamide or cationic starch is added to the papermaking system prior to last point of high shear. After the shear point, an anionic microparticle (colloidal silica bentonite clay) is added.

The mechanism by which these systems operate is complex. Large flocs are formed by bridging when the polymer component is first added. These flocs are then degraded at shear points in the system. Addition of anionic microparticles provides many negative sites that can interact with the positively charged flocs to form a highly coagulated system of dense, small flocs that dewater readily. It is possible to achieve high retention, good formation, and good drainage with microparticulate systems.

e. Network forming flocculating systems

The various retention systems described above do not work well in groundwood containing furnishes or in the presence of high concentrations of charged interfering substances. One type of system that often works with such furnishes is a network forming system. In these systems, the flocculants react in solution to form transient, unstable networks that capture fines particles by an occlusion process. Two examples of such systems are polyethylene oxide (PEO)/ phenolic resin and anionic polyacrylamide/ montmorillonite clay.

10. The strength of papermaking fiber and fines aggregates: soft flocs and hard flocs

a. Introduction

The retention models and supporting information presented above find practical expression in the first-pass retention performance of a paper machine and in the nature of the aggregates (flocs) formed. Flocs are often characterized by their size, density, and shear resistance. Qualitatively speaking, aggregation by charge neutralizing and patching mechanisms produce small flocs that dewater well. As mentioned above, microparticulate systems produce small, dense flocs that are easy to dewater. On the other hand, bridging flocculants produce larger, open flocs that dewater less well.

With respect to shear resistance, the aggregates formed in papermaking can be characterized as either soft (reversible) flocs or hard (irreversible) flocs.

b. Soft flocs

A floc that exhibits good fines retention at low levels of turbulence and shear, whose retention decreases substantially upon exposure to greater turbulence; but, when allowed to reflocculate, will return to its original retention levels is called a soft floc.

Soft flocs form when simple salts or low molecular weight polyelectrolytes are used to induce aggregation. When the system is exposed to shear, the flocs break up but the ions or polyelectrolytes bound to the surface remain in place. When the turbulence subsides, the same degree of aggregation can occur and the original levels of retention are obtained. The data in **Table 15-7** illustrate this phenomenon.

Table 15-7. Reflocculation phenomena in charge neutralization and agglomeration systems.[a] [15]

System	Turbulence (rpm) 0	400	800
NaC1			
-A	67	43	34
-B	80	41	36
$A1C1_3$			
-A	96	58	40
-B	92	57	38
Low M.W. PEI			
-A	93	54	37
-B	91	50	34

[a] Retentions determined on the Britt Dynamic Drainage/Retention Jar per TAPPI Method T 261

A = Retention of original stock and additives

B = Retention after exposure to 1500 rpm for 1 min followed by a reflocculation period

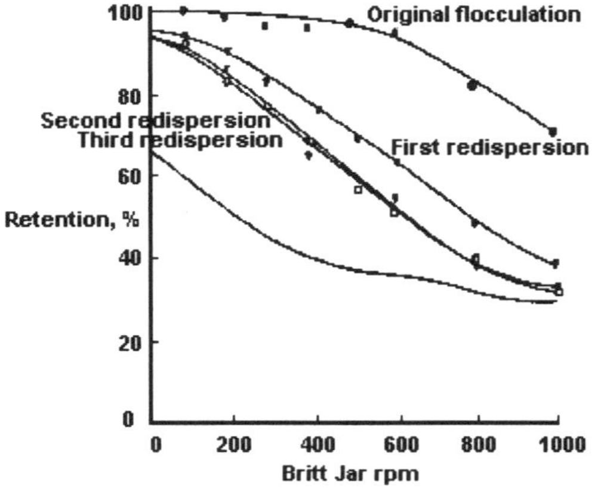

Figure 15-19. Flocculation-redispersion behavior of hard-floccing polyelectrolyte systems. The system produces a relatively shear-resistant floc that withstands high stirring speeds in a Britt Jar. Once the strong flocs have been disrupted by stirring and reformed in the absence of stirring, they do not exhibit the same tenacity. Subsequent exposure to increased stirring rates results in behavior characteristic of reversible, soft floccing systems. (The solid line curve in the figure represents the behavior of the system without polyelectrolyte addition.)

c. Hard floc

A hard floc retention system shows good fines retention over a wide range of turbulence and shear for brief periods of exposure. Subsequent exposure will cause a decrease in retention that cannot be re-established after the system is allowed to reflocculate. Upon reflocculation, the system reverts to the behavior of a soft floc system. This effect can be seen in the subsequent flocculation/redispersion/flocculation behavior shown in **Fig. 15-19**.

Fig. 15-20 illustrates the mechanism proposed for the transition to soft floc behavior. It is believed that bridging polymer chains are broken when the floc is subjected to shear forces. The polymer chain ends remain attached to individual particles, however, and then reconform onto the particle surface in a flat conformation — effectively forming a cationic patch. When the shear force is removed, the particles then re-aggregate by a patching mechanism that produces soft, irreversible flocs.

11. Effect of interfering substances on the effectiveness of retention systems

a. Nature and source of interfering substances

The notion of interfering substances was introduced in Chapter 4 and it would be a good idea to review that material at this time. There, it was indicated that interfering substances consist of a wide variety of materials that are dissolved from wood pulp (most significant in high yield pulps), carried over from the pulp mill and bleach plant, and introduced with recycled fiber and broke. They can also accompany virgin additives or be added as chemical processing aids. Examples include lignin, hemicelluloses, anionic starch, dispersants, pulping reaction products, bleaching reaction products, deinking chemicals, and coating adhesives.

The terms "anionic trash" or "accidental additives" have also been applied to these materials.

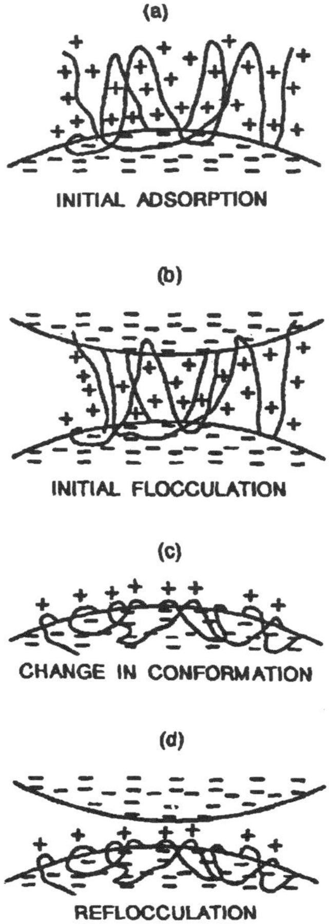

Figure 15-20. Proposed mechanism for the reversion of hard flocs to soft floc aggregation behavior when exposed to high shearing forces. Extended exposure to turbulence breaks the polymer bridge between the two surfaces. Separation occurs with the polyelectrolyte remaining attached to one surface or with chain rupture occurring and parts of the polymer molecule remaining on each surface (Fig. 15-20c). Because the molecule on the surface is undergoing motion due to hydrodynamic forces, and there are adjacent regions of opposite charge on the fines or fiber surface, the molecule can undergo a reconformation (Fig. 15-20c). This makes the polymer act more like the patch model and reaggregation occurs by a soft floc mechanism (Fig. 15-20d).

We will avoid these terms here because they imply that interfering substances are accidental in nature or extraneous substances that show up in the furnish due to bad practice. In fact, additives that are added to the furnish for a specific purpose, such as filler and pitch dispersants, can act as interfering substances and should not be overlooked.

b. Effect of interfering substances on retention

The effect of interfering substances was introduced in Chapter 6 and **Fig. 15-21** below is taken from that section. In the figure, the offset along the X-axis in going from a "clean" system to a "dirty" system is due to interfering substances — labeled "cationic demand." The added polymer reacts with the interfering substances first and until this reaction is complete — i.e. the interfering substances are "neutralized" — particle retention remains low. This is the typical response observed.

The term "cationic demand" was defined in Chapter 6 and refers to the method used to determine the concentration of interfering substances in a stock sample.

Interfering substances have always been present in papermaking systems, but only in the past 20 years have they become a significant papermaking problem. This development is the result of trends, such as increased closure of white water systems and the greater use of mechanical pulps and recycled fibers.

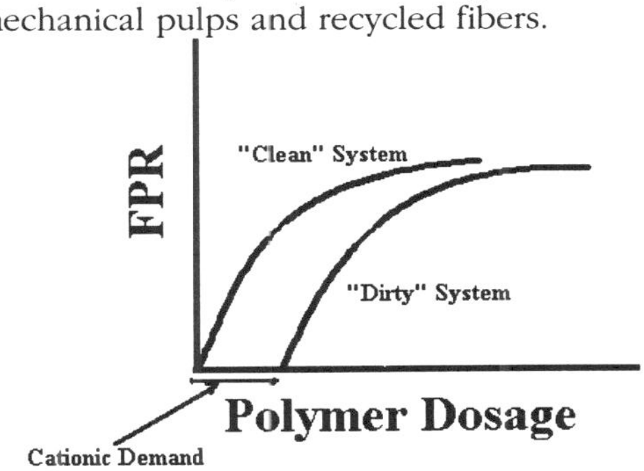

Figure 15-21. A typical effect of interfering substances (cationic demand) on the effectiveness of a retention aid polymer ("FPR" represents "first-pass retention").

c. Strategies for coping with interfering substances

Whenever possible, the first approach to reducing the impact of interfering substances should be to eliminate them. For example, in situations where pulp mill, bleach plant, or deinking plant carryover is the primary problem source, then every effort should be made to improve operations in that part of the mill. When operational solutions are unavailable, then the papermaker usually resorts to a chemical approach involving charge neutralizing additive. Included are alum, PAC, and low molecular weight, high charge density cationic polyelectrolytes. PEI is a particularly good polymer for this purpose due to its very high charge density. The strategy is to neutralize the interfering substances, usually in the thick stock, prior to the addition of retention agents. This topic will be addressed in more detail in Chapter 16.

12. Effect of fines management on retention

In earlier chapters, the theme that the relatively great specific surface area of fines causes them to adsorb large amounts of additives was introduced. **Table 15-8** , taken from Chapter 3, illustrates this point.

Table 15-8. Adsorption of wet end additives by fines

Additive	Relative Adsorption Intensity		
	Fibers	Filler Clay	Fines
Cationic Starch	1	4	5
Rosin Soap Size	1	4	16
Dispersed Rosin Size	1	16	20
Alum	1	2	3

Retention agents behave like other adsorbing additives and partition themselves onto furnish particles according to their available surface area. As a result, increasing the fines level, at fixed additive addition rates, reduces the efficiency of the retention aid because it gets spread over a greater area, i.e. it gets "diluted." This is a well-known phenomenon that has significant implications for retention process control.[16] It is important that the papermaker control closely the rate and quantity of fines input to the furnish, keeping it as low as possible and as uniform as possible.

Several factors contribute to the fines content of the headbox. On the materials side, filler levels and the type of fibers selected are important. Mechanical pulps have greater fines levels than chemical pulps. Hardwood pulps have greater fines levels than softwood pulps. Poor first-pass retention leads to increased headbox fines levels, as does the increased use of broke — particularly coated broke — and secondary fibers. It is critical that the papermaker maintain close control over these fines sources. Failure to do so will result in erratic retention performance.

13. Effect of first-pass fines retention on fines distribution in paper

The manner in which fines are distributed in the z-direction of paper depends upon a number of mechanical and colloidal considerations. The principal mechanism at work on the paper machine appears to be self-filtration which can lead to decreased filler and fiber fines content in each hypothetical layer from the felt (top) side to the wire side.

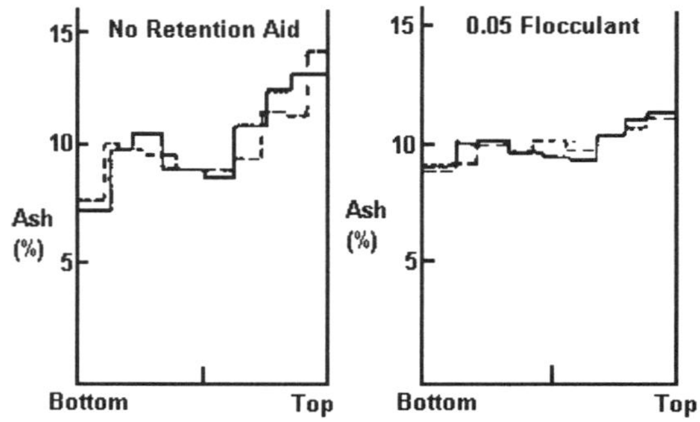

Figure 15-22. Influence of first-pass filler retention levels on the z-direction distribution of filler in handsheets [17]

The forming medium is the first structure to filter the stock components. Because of its relatively coarse nature, few fines are trapped and they pass through the medium into the white water. As the sheet consolidates and more of the network is built up, increasing amounts of fines are entrapped in the interstices of the fibrous mat. The fines in the last portion of the stock drained have a greater chance of being filtered. Consequently, there are relatively more fines in the top side of the sheet.

The second mechanism controlling the z-direction distribution of fillers is that of colloidal interaction. Tanaka and coworkers reported that the higher the first-pass retention of filler, the more uniform the distribution of filler in the z-direction (**Fig. 15-22**).[17]

14. Effect of first-pass fines retention on optical properties and sheet formation

Retention aids can greatly increase fines retention and produce a subsequent decrease in white water solids. These positive aspects do not exist without a negative counterpart, a decrease in filler efficiency that often accompanies high retention. [18,19] When filler retention is increased by the use of chemical additives or polyelectrolytes, a corresponding reduction in the light scattering coefficient of the filler fraction can occur due to the flocculation of filler particles. The maximum light scattering power of a material occurs when its particle size is in the range of one-half the wavelength of the light being scattered. As particles agglomerate and act collectively as one large particle, the number of filler-to-air interface surfaces decrease and their ability to scatter light diminishes rapidly. [20] To this end, care must be taken to guard against overflocculation of filler particles in cases where optical properties are important.

Overflocculation of fines can produce very large flocs that will appear as poor formation when the sheet is viewed by transmittal light. Consequently, excessive fines flocculation is to also be avoided.

15. Fiber flocculation

a. Introduction

There are two principal aspects to fiber flocculation. In the first instance, fibers become physically entangled with one another as they flow through pipes and are stored in chests and tanks. The primary approaches taken to minimize this type of aggregation are dilution and the judicious application of shear and turbulence forces. The second type of fiber flocculation behavior derives from the gathering together of fibers through electrostatic attraction and other attractive forces (e.g. hydrogen bonding).

Much work has been done on the effect of fiber length on formation.[21–24] ("Formation" is used here to denote the sheet property of the sheet and the physical measurement of this phenomenon.) Formation is often measured by monitoring the attenuation of beta radiation as the sheet moves past a radioactive source and recording the resulting power spectrum of the transmitted beta particles. Analysis of this power spectrum gives a measure of the size and distribution of flocs. It has been generally found that the longer the fibers, the greater the tendency to form large "cottonball" flocs. Longer fiber lengths are also very resistant to mechanical deflocculation. In many cases, substitution of short hardwood fibers for softwood fibers, maximum possible dilution, and the application of turbulence and cross-direction shear immediately prior to sheet formation on the paper machine have been used to improve formation. Synthetic polyelectrolytes, starches, gums, and other chemical additives can have both beneficial and adverse effects. Polymer additives that are utilized specifically for improving formation are called formation aids.

b. Formation aids

The ability of high molecular weight, viscosity-enhancing materials like the extracts of Hibiscus manikot root to facilitate the formation of long-fibered papers has been known and used for centuries in Japan. [25] Naturally occurring products such as guar gum, locust bean gum, and deacetylated Karaya gum have also been used for this purpose commercially.[26]

In recent years, synthetic polymers have gained prominence in this area. It has been found that high molecular weight is a primary factor and that anionic polymers are more effective than nonionics. In addition, the active functional groups (carboxyl) must be ionized to be effective. [27,28] Any condition that causes the polymer to expand and increase the viscosity of the medium improves its effectiveness. The mechanism of action in all cases appears to be that the expanded polymer (estimated end-to-end distance = 1 µ) prevents close contact of fiber surfaces so that flocculation cannot occur. Additionally, viscosity increases restrain the motion of the fibers in the suspending water form. There is also a certain amount of fiber-fiber repulsion generated by the anionic charges on the polymers. Along with these desired effects, adverse side-effects can occur, such as sticking on press roll surfaces if excess anionic polymeric material is used. There can also be a large decrease in drainage rate associated with the use of large quantities of formation aids.

c. Fiber flocculation by retention aids

Retention aids can promote flocculation of the fiber fraction of a furnish. Different materials affect fibers in different ways. Alum and low molecular weight polyelectrolytes seem to form small, weak, easily dispersed flocs (termed "soft flocs") that readily reform after exposure to shear and turbulence forces. These normally do not produce undesirable sheet properties. On the other hand, high molecular weight cationic polyelectrolytes and dual polymer systems can produce tenacious fiber flocs. After exposure to prolonged high levels of shear and turbulence, these flocs (termed "hard flocs") tend to break down and do not readily reform after the shear and turbulence subside.

In practice, it is impossible to prevent the occurrence of fiber flocculation when using the high molecular weight polyelectrolyte retention aids required on modern, high-speed paper machines. Excessive fiber flocculation can cause many problems with sheet properties. For example, the fiber flocs result in large differences in mass distribution throughout the sheet. Because of the weak spots that develop between these flocs in the sheet, wet web strength is reduced, crushing can occur in the press section, and the overall strength properties of the final sheet suffer. Appearance properties are also adversely affected.

The papermaker can overcome the potentially adverse effects of fiber flocculation by the use of mechanical (hydrodynamic) action in the headbox and on the forming table, by running as low a headbox consistency as possible within the drainage constraints of the machine, and by taking care that the levels of retention aid utilized are not excessive. Unfortunately, all of these approaches can have adverse effects on fines retention.

16. Drainage and dewatering [29]

Once the various components have been assembled into the papermaking furnish and sent to the paper machine headbox, the actual papermaking process begins when the stock passes through the headbox slice and onto the forming section. Two major functions occur on the machine. First, the sheet is formed, consolidated, bonded, and many of its surface characteristics are established. Second, the suspending water associated with the headbox furnish is removed so that the solids content increases from approximately 0.5% to 95%. It is this latter function, dewatering, that we will address now.

Water removal processes are very important because they impact paper quality and paper machine productivity. Consequently, the subject has received much attention from researchers, papermakers, and machinery and chemical suppliers. We will focus our discussion on those aspects of dewatering that are affected by wet end chemistry.

17. Fiber-water relationships

Water associates very strongly with paper webs. Two primary mechanisms are involved: capillary inclusion and hydrogen bonding. The following three types of fiber-water relationships have been described.[30]

Free water — Free water is water held within the interstices and capillaries of the paper web. It can flow relatively unimpeded from the sheet when gravitational and hydraulic pressure forces are applied to a paper web.

Imbibed (sorbed) water — Imbibed water is held on fiber surfaces by hydrogen bonding interactions between water and cellulose, hemicellulose, and other web components. It is generally considered to include water held within the fiber walls, but not within fiber lumens. It can be determined by the water retention value (WRV) test.[31]

Water of constitution — Water of constitution is bound into the crystalline packing structure of fibers and is essentially inaccessible to papermaking dewatering processes.

18. Water removal zones

Water removal occurs in the following four distinct zones on the paper machine.

Forming (free drainage) zone — The forming zone extends from the point that the stock jet impinges on the forming fabric to a point just before the first vacuum suction box. The web consistency typically increases from 0.5% to 2-4% in this zone.

Vacuum zone — The vacuum zone begins with the first vacuum box and extends to where the sheet leaves the suction couch. The web consistency typically increases from 2-4% to 20-22% in this zone.

Pressing zone — The pressing zone includes all the press nips and associated equipment included in the press section. The web consistency typically increases from 20-22% to 35-45% in the pressing zone.

Dryer zone — The dryer zone includes all heated units after the press section (including the calender stack if it is heated) up to the reel. The web consistency typically increases to 95% or greater in the dryer zone.

In general, much of the free water in a paper web can be removed in the drainage, vacuum and pressing zones. The free water held in the smallest capillaries and interstices and imbibed water is accessible to the dryer section.

Experience has shown that wet end chemistry has its greatest effects in the forming and vacuum zones. It has less effect on pressing and very little effect on the dryer zone. Consequently, our discussion will concentrate primarily on the first two areas.

19. The forming zone

As the stock jet issues from the headbox slice and impinges on the forming board, the fibrous mat begins to form and dewatering commences. Most fiber losses occur at this point where fibers can be driven vertically down through the forming fabric (depending on the jet impingement angle) before the mat has achieved significant thickness. Fines that are not attached to retained fibers can also pass through the fabric along with the draining water.

The initial water removal occurs mainly by gravity and jet impingement forces. Water continues to drain as the sheet moves down the forming section and passes over various drainage elements — mainly foils, nowadays — and the web is subjected to alternating pressure-vacuum pulses. As the web approaches the leading edge of an element, it experiences a pressure pulse that disrupts the sheet structure and promotes dewatering. As it passes over the trailing edge of the element, the web is subjected to a vacuum pulse and water is drawn from the sheet. A point — usually 2.5-5% consistency, depending on the furnish — is reached where the combination of gravity, pressure, and hydraulic drainage forces will not remove any more water and vacuum must be applied to effect further dewatering.

The pressure-vacuum forces produced by foils, while benefical to drainage, can be detrimental to fines retention when they are too severe and disrupt the fiber-fines aggregates in the web.

20. Filtration and thickening [32]

Two different dewatering processes occur in the free drainage zone — filtration and thickening. They are defined as follows:

Filtration — Filtration occurs when fibers and fines are able to move freely in a suspension and a sharp demarcation exists between the suspended particles and forming mat. The consistency in the suspension remains essentially constant throughout the filtration process.

Thickening — Thickening occurs when the particles in a suspension are in contact with one another and are entangled or immobilized in a network. The structure collapses as drainage proceeds. There is no sharp boundary between the formed mat and the suspension and water is removed from all parts of the network simultaneously and at equal rates.

It has been deduced from studies of paper structure that filtration is the predominant dewatering mechanism that occurs in the free drainage zone.

21. Specific filtration resistance

One of the most important factors pertaining to drainage rate is the *specific filtration resistance* of the forming web — a term derived from the Kozeny-Carman analysis of paper drainage. [33] Many variables contribute to this property, such as the pressure drop across the forming web and the flexibility of the fibers in the furnish. Urick and Fisher have listed eight wet end chemistry factors that are likely to affect how much resistance a forming web will present to drainage. [34] These are presented in **Table 15-9**.

Table 15-9. Wet end chemistry factors that affect drainage resistance

- Freeness
- Stock temperature
- Surfactants
- Entrained air
- Filler content
- Fiber fines content
- Fiber surface chemistry
- Degree and type of aggregation of furnish particles

Highly refined fibers drain slowly. The papermaker often increases the stock temperature to counteract this — presumably by decreasing the viscosity of the suspending water.

Surfactants increase dewatering rate, but have a number of adverse side effects, such as reduced interfiber bonding and sizing. They also promote foaming.

Entrained air exists as tiny bubbles attached to fibers and other furnish particles. They fill the interstices between particles in the forming web and impede the flow of draining water.

Fillers can have two, competing, effects. First, fillers do not interact with water as strongly as fibers and this promotes drainage. On the other hand, fillers do become lodged in fiber mat pores where they can retard the free flow of water through the network.

Fiber fines interact strongly with water and, as a consequence, have large amounts of associated water. They also swell as a result of this interaction. Both effects reduce drainage rate.

The last two entries in **Table 15-9** are related, in that they depend on the presence of charge neutralizing and/or flocculating agents. Salts and low molecular weight polyelectrolytes can reduce the zeta potential of fiber surfaces and thereby cause fibrils and the internal surfaces to collapse, reducing the surface area accessible to water. This increases the drainage rate. Flocculants can have both positive and negative impacts on drainage rate. This very important subject will be addressed in the next section.

22. Drainage aids

a. Introduction

As defined in the paper industry, *drainage aids* are chemical additives that will improve the dewatering of the paper web, at any point in the process. This means that a material might not affect free drainage, but have a significant effect on vacuum drainage or pressing response.

Several potential benefits accrue from the use of drainage aids. Examples include increased production rate, improved formation (the property), and lower dryer steam consumption. Of these, increasing the production rate on drainage-limited machines is probably the most common reason for using drainage aids. Sheet formation can be improved through the use of drainage aids by taking advantage of the improved drainage to decrease headbox consistency. This, in turn, decreases fiber-fiber flocculation and improves the distribution of solids in the sheet.

b. Materials used as drainage aids

In general, all agents that serve as retention aids and charge neutralizers can also be used as drainage aids. Thus, alum, PAC, calcium salts, cationic starch, synthetic polyelectrolytes, and microparticulate retention systems are all applied for this purpose.

Alum, PAC, and calcium salts — enhance drainage by two mechanisms: charge neutralization and coagulation. In the first instance, the materials reduce the zeta potential of all fiber surfaces, internal and external, causing them to collapse onto themselves, thereby reducing the total area and volume available to associate with water.

The second mechanism is also charge related, but involves the coagulation of fines onto fiber surfaces where they are no longer able to block web pores and impede water removal.

Cationic starch — Cationic starch, when used alone, probably increases drainage primarily by aggregating fines onto fiber surfaces. The flocs formed by cationic starch tend to be of a small to medium size and do not hold water strongly. Cationic starch is also an important component of some of the microparticulate systems discussed below.

Synthetic polyelectrolytes — It is clear from our earlier discussion that a wide variety of synthetic polyelectrolyte-based systems are available. They all operate by the same general mechanism — fines aggregation — but their specific effects can differ significantly. For example, the low-to-medium molecular weight, high charge density polymers produce small, tight flocs that do not hold water strongly and, hence, provide good dewatering performance. However, the flocs are not highly shear resistant.

On the other hand, high molecular weight, low cationic charge polyacrylamides form large flocs that occlude water and do not promote drainage significantly. Because of this, when polyacrylamides are used as drainage aids, it is best to select in the lower molecular weight, higher charge density range of available products.

Microparticulate systems — These systems, which usually combine a cationic polymer with a very small, anionic pigment are designed primarily for drainage and formation enhancement, with improved retention as a secondary benefit. The mechanism by which these systems operate is to produce very small, tight flocs that do not hold water strongly and produce a uniform distribution of material throughout the sheet.

c. Closing comments about free drainage

This concludes our brief discussion of free drainage. Papermakers often tend to think solely in terms of free drainage when dealing with drainage aids. This is a mistake. Experience has shown that these materials can have significant, often adverse, effects on the succeeding vacuum drainage and pressing water removal operations, which will serve as our next topic of discussion.

23. Vacuum water release

As the stock suspension approaches the flat boxes, it reaches consistencies of 4-5%. When the stock passes over the suction box holes or slots, three water removal mechanisms operate.[35] First, the pressure differential causes the fiber network to compact, squeezing water from the sheet. Second, water in capillaries is displaced by air. Third, as the sheet passes the trailing edge of suction box openings, water that has been sucked out of the sheet is scraped off the bottom of the forming media. This decreases rewet after the vacuum zone has been passed.[36]

Table 15-10. Forming and vacuum zone dewatering: the effect of flocculants

Run No.	Additive	HB Consistency, %	Wire Consistency, %	Consistency After the Couch, %
1	None	0.46	4.4	16.3
2	PEI(0.8%)	0.35	2.1	17.6
3	PEI(0.8%) + An. PAM (0.1%)	0.31	8.8	15.5
4	Same as Run 3 + dispersion	0.33	10.35	20.9

The effect of polyelectrolytes in this zone has been studied to a lesser degree than the drainage zone, but some interesting observations have been made. In mill trials and pilot machine runs, it has been observed that products designated as drainage aids often improve water removal over table rolls and foils, but result in a wetter sheet at the couch. A number of laboratory studies have confirmed this finding. [35,37–39] **Table 15-10**, illustrates the phenomenon.[39]

The data were collected on a 200 ft/min pilot paper machine. In Run 3, a strong flocculating dual polymer system that gave a dramatic freeness increase was added to the papermaking stock. This system produced an increased fines retention, which subsequently lowered the headbox consistency because the recycled tray white water had a lower fines level. The wire consistency, measured just before the first vacuum box, doubled, indicating good drainage. However, the solids after the couch decreased almost 1%, compared to control run 1. Run 4 illustrates that subjecting the strongly flocculated stock to mechanical dispersion prior to the forming section produced retained the improved drainage but also enhanced the vacuum response.

The mechanism that causes the reduction in couch dryness is probably due to a change in sheet porosity. [35,37] By tying up the fines through flocculation by polyelectrolytes, water and air are able to flow more freely through the sheet. On exposure to vacuum, water is quickly pulled from the capillaries and displaced by air. The air flow then satisfies some of the vacuum, reducing the pressure drop and decreasing the compaction (squeezing) of the sheet.

If the suspension is overdosed so that macroflocs (large fiber flocs) are formed, this condition is aggravated. Water will be removed from the areas between the flocs and air can easily satisfy the vacuum. Water in the flocs is not removed. Therefore, the sheet is wetter after the couch than before the polyelectrolyte was added. This phenomenon can be avoided if the macroflocs are broken down and reformed as smaller, tighter structure, as was done in Run 4, **Table 15-10**. This mechanism also operates in microparticulate systems where a multitude of tiny flocs form around anionic microparticles. These flocs do not retain water strong, thereby enhancing free drainage, and they provide a very uniform distribution of material throughout the web, which enhances vacuum response.

The notion that a balance must be found between having the sheet open enough to allow water to flow freely, but compressible enough and tight enough to optimize vacuum water removal is not a new idea. Well before drainage aids were utilized by papermakers, Campbell and Lodge [40] recognized this paradox when they stated:

> *.... A slow stock is slow because it forms a compact sheet with high resistance to flow. Such stock will drain slowly on the wire. However, this same compact sheet will contain a small volume of space between the fibers, and it is this volume which determine the amount of water retained by the sheet leaving the couch. It will come off dryer than stock which drains more freely on the wire though it may seem a bit slow when passing over the first flat boxes.*

This is exactly the phenomenon often observed with drainage aids.

It should be clear from the above discussion that the effective application of drainage aids is not a simple matter. One must take care to evaluate their effect on dewatering along the entire forming section and not just in the free drainage zone. As we will see below, this statement applies to the pressing zone as well.

24. The effect of drainage additives on press zone dewatering

The effect of drainage aids on press dewatering has not been investigated much, but the few studies reported suggest that both beneficial and adverse effects occur.[41–43] For example, Klungness[41] reported that cationic starch increased both the solids after the couch and the press section. Stratton found that this was not true in all cases.[42] He reported that some polymers improved pressing, while others hindered it. He concluded that the way in which fines were retained and the resulting sheet macrostructure had a great deal to do with the way a sheet responded to pressing. If a flocculant gave fine flocs, where fines were immobilized and produced good formation on the macro scale, the pressing response would be enhanced. On the other hand, if a flocculant retained fines but promoted the macroflocculation of fibers, it tended to reduce the pressing response of the sheet. It was hypothesized that in this instance, pressing pressures would be borne by high spots in the sheet associated with fiber flocs and the water pressed out of them would flow to adjacent, lower basis weight areas and not out of the sheet into the press felt as desired. On the outgoing side of the press nip, when the pressure is reduced, the flocs then expand, pulling water back from adjacent areas and thereby reducing the overall efficiency of the pressing operation.

Our state of understanding of the effect of retention and drainage additives on pressing is limited and does not warrant further discussion here. However, it is important that you recognize that these materials can have effects, both positive and negative, over the entire forming and pressing sections of a paper machine and make the appropriate measurements when running trials. Otherwise, you could easily draw incorrect conclusions regarding the overall effectiveness of a drainage aid if your data is limited to a single dewatering zone.

25. References

[1] Marton, J. "Effects of machine speed on formation, drainage, and retention." 1989 TAPPI Retention and Drainage Short Course Notes, Washington, DC, 19-25(April 12-14, 1989).

[2] Han, S. T. and Chang, N. L. "Retention of fines in fiber mats." Tappi J. 52(4):688(1969).

[3] TAPPI Test Method T 261. "Percent fines by Briott Jar screening." TAPPI PRESS, Atlanta.

[4] Tarvin, R. F. and Westwood, R. L. "Use of modified Britt Jar for drainage testing." Proceedings of 1982 TAPPI Papermakers Conference, 243(1982).

[5] Davison, R. W. "Vacuum pulsation drainage procedure for determination of fine particle retention." Proceedings of 1989 TAPPI Papermakers Conference, Washington, DC, 75(1989).

[6] Unbehend, J. E. "The dynamic retention/drainage jar — increasing the credibility of retention measurements." Tappi J. 60(7): 120(1977).

[7] Lindstrom, T. and Eklund, D. "Retention and dewatering." Chapter VII in Paper Chemistry: An Introduction, DT Paper Science Publications, Grankulla, Finland, 145-191(1991).

[8] Au, C. O., Johansson, K. and Thorn, I. "The use of retention and drainage aids in the wet-end." Chapter 2 in Applications of Wet-End Paper Chemistry, Au, C. O. and Thorn, I., Editors, Blackie Academic and Professional, Chapman and Hall, 13-26(1995).

[9] Horn, D. and Linhart, F. "Retention Aids." Chapter 4 in Paper Chemistry, J. C. Roberts, Editor, Chapman and Hall, 44-62(1991).

[10] Halverson, F. "Retention aids, drainage aids, and flocculants." Chapter 8 in Chemical Processing Aids in Papermaking: A Practical Guide, K. Hipolit, Editor, TAPPI PRESS, 104-127(1992).

[11] Phillip, B. "Polyelectrolytes: formation, characterization, and application." Hanser/Gardner, Cincinnati, 343 pages, 1994.

[12] Odberg, L., Swerin, A., Tanaka, H. "Some kinetic aspect of wet end chemistry." Proceedings of 1995 TAPPI Papermakers Conference, 65-70(1995); Wagberg, L., Odberg, L., Lindstrom, T., and Aksberg, R. J. Colloid Interface Science 123:287(1988).

[13] Scott, W. E. and Koethe, J. "Polyelectrolyte interactions with papermaking fibers: the mechanism of charge decay." Tappi Journal 76(12):123-133 (1993).

[14] Gupta, B. and Scott, W. E. "Interactions between cationic starches and papermaking fibers: the effect of fiber surface charge and addition level on starch retention." Proceedings of 1995 TAPPI Papermakers Conference, 85-95(1995).

[15] Unbehend, J. E. "Mechanisms of soft and hard floc formation in dynamic retention measurements." Tappi Journal 59(10): 74(1976).

[16] Scott, W. E. "Fines management and control in wet end chemistry." Tappi Journal 19(11): 31-34(1986).]

[17] Tanaka, H., Luner, P. L., and Cote, W. "How retention aids change the distribution of filler in paper." Tappi Journal 65(4): 95(1982).

[18] Dobbins, R. J. "Overflocculation reduces opacity and affects the strength of paper." Paper Trade Journal 162(20): 42(1978).

[19] Brill, H. C. and Hecklau, F. L., "Titanium dioxide retention." Tappi Journal 43(4): 229A(1960).

[20] Gill, R. and Scott, W. E. "The relative effects of different calcium carbonate filler pigments on optical properties." Tappi Journal70(1): 93(1987).

[21] Hubley, C. E., Mason, S., and Robertson, A. "Flocculation in suspensions of large particles." Canadian Journal of Research. 28(8): 770(1950).

[22] Balostis, Y., McKenzie, A. W., and Harrington, K. J., Consolidation of Paper Webs. Volume 2. Editor, F. Bolam. British Paper and Board Manufacturers Association, 659(1966).

[23] Wahren, D. "On the three-dimensional fiber network." Svensk Papperstidning. 13:536(1964).

[24] Wrist, P. E. Formation and Structure of Paper, Volume 2. Editor, F. Bolam. British Paper and Board Manufacturers Association, 839(1962).

[25] Hofreiter, B. T. "Natural products for wet end addition." In Pulp and Paper: Chemistry and Chemical Technology, Volume III. Editor, J. P. Casey. Wiley-Interscience, 1475(1981).

[26] Yanc, C., "Effect of freeness, consistency, and additives on formation." Tappi Journal 46(10): 169A(1963).

[27] Wasser, R. B. "Formation aids for paper: An evaluation of chemical additives for dispersing long fibered pulps." Tappi Journal 61(11): 115(1978).

[28] Wilharm, M. T., Novak, R. W. and Savino, C. A., Proceedings of 1987 TAPPI Nonwovens Conference. 101(1987).

[29] Unbehend, J. E. "Wet-end chemistry of retention, drainage and formation aids." Chapter VII in Pulp and Paper Manufacture Volume 6, Stock Preparation. Third Edition. Hagemeyer, R. W. and Manson, D. W. Editors. Joint Textbook Committee of the Paper Industry, 112-157(1992).

[30] Christensen, P. K. and Giertz, H. W. Consolidation of the Paper Web, Volume 1. F. P. Bolam, Editor. British Paper and Board Manufacturers Association, London. (1966).

[31] Jayme, G., Tappi Journal, 41(11): 180a-182a(1958).

[32] Parker, J. D. The Sheet Forming Process. STAP No. 9. TAPPI PRESS, Atlanta (1972).)

[33] Parker, Mason, S. G. and Robertson, A. A. Pulp and Paper Magazine of Canada. 50(12): 103(1949).

[34] Urick, J. M. and Fisher, B. D. Tappi Journal 59(10): 78(1976).

[35] Gavelin, G., Pulp and Paper Mag. Canada, 55(C)(1954).

[36] Attwood, B. W. "A study of vacuum box operation." Paper Technology, 3(5): 446(1962).

[37] Unbehend, J. and Britt, K. "Retention, drainage, and sheet consolidation." Industrial and Engineering Chemistry product Res. and Dev. 21(2): 150-153(1982)].

[38] Abson, D., Bailey, R. M., Lenderman, C. D., Nelson, J. A. and Simons, P. B. "Predicting the performance of shear-sensitive additives." Tappi Journal 63(6):55(1980).

[39] Britt, K. W. and Unbehend, J. E., Tappi Journal 63(4):67(1980).

[40] Campbell, W. B. and Lodge, W. C., Pulp Paper Magazine Canada, 38(2): 189(1937).

[41] Klungness, J. H. Paper Trade. J. 162(5):46 (1978).

[42] Stratton, R. A. Proceedings of the 1982 TAPPI Papermakers Conference, TAPPI PRESS, Atlanta, p.179(1982).

[43] Springer, A. M., Nabors, L. and Bhatia, O. "The influence of fiber, sheet structural properties, and chemical additives on wet pressing." Tappi Journal 74(4):221 (1991).

Chapter 16

Overview of Wet End Chemistry Process Control

1. Introduction

In the previous fifteen chapters we have focused on the concepts, properties, and interactions associated with the various components of a papermaking furnish. While interactions among components were mentioned, the primary objective was to learn about each specific component or component system. In practice, the papermaker cannot focus solely on a single wet end chemistry component to the exclusion of the rest of the furnish because they all interact and contribute to the final paper properties and machine performance achieved. Consequently, it is important that we also synthesize the bits and pieces of knowledge presented in earlier chapters into a more comprehensive picture. We shall do this under the heading of "process control."

This subject is huge and it would not be appropriate to attempt a comprehensive treatment in an introductory work. Rather, we shall strive to provide a broad overview that builds on the concepts and interactions introduced in the preceding chapters. You are encouraged to consult the technical literature of the past ten years, where a number of excellent discussions of specific topics can be found. As for other topics, the author has found the TAPPI Papermakers Conference Proceedings to be particularly useful in this regard.

2. Definition of wet end chemistry process control

The term "process control" means different things to different people. We shall adopt the following simple operating definition for wet end chemistry process control:

> *Wet end chemistry process control refers to what is done or should be done in the mill by operators and engineers to keep the papermaking chemistry of a machine within some desired set of conditions and reduce process variability.*

Note that the definition is independent of the approach taken by a specific mill toward the implementation of wet end chemistry control — whether it be based strictly on infrequent, manual measurements and operator adjustments or closed-loop, on-line sensor, computer-based control systems.

The chapter is organized in three parts. First, an overview of the subject is presented that brings together many of the concepts covered in previous chapters. Next, the measurements required for wet end chemistry control are reviewed. Finally, the very important topic of retention control is discussed.

3. Trends making wet end chemistry process control important

During the past 20-30 years, many papermaking trends have appeared that have increased the need for improved wet end chemistry process control. **Table 16-1** lists some of the more important ones.

Table 16-1. Papermaking trends that require improved wet end chemistry process control

- Increasing machine speeds
- Increasing twin wire and top former utilization
- Decreasing basis weights
- Conversion from acid to alkaline papermaking
- Increasing white water reuse
- Requirement of reduced wastewater treatment and landfill loading
- Increasing wet end additive costs
- Demand for higher paper quality

The first three trends all have made it more difficult to achieve high first pass fines retention. On the other hand, a successful conversion from acid papermaking to alkaline papermaking (trend four), requires increased first-pass fines retention. The build-up of interfering substances in white water systems associated with increased white water reuse (trends five and six) impacts a wide range of wet end chemistry phenomena. Trends seven and eight have obvious implications for the cost-effective and efficient use of wet end additives.

4. Benefits of improved wet end chemistry control

While the trends listed in **Table 16-1** provide sufficient incentives for improved wet end chemistry control, there are many other reasons why it is to the papermakers advantage to achieve this end. Some typical examples are listed in **Table 16-2**.

Table 16-2. Advantages to be gained from improved wet end chemistry control

- Improved sheet strength
- Better machine runnability
- Decreased two-sidedness
- Reduced MD variation of basis weight
- Reduced MD variation of ash content
- Improved machine cleanliness (reduced deposits)
- Improved sizing, wet strength, dry strength uniformity
- Improved uniformity of appearance properties
- Improved uniformity of sheet formation

5. Prerequisites to the design of comprehensive wet end chemistry control for a given system[1]

According to Brewster,[1] a prerequisite for the design of comprehensive wet end chemistry control is an approximate, quantitative understanding of the relationships that exist among the parameters involved, such as chemical addition, chemical environment variables, retention, formation, sizing, wet strength, and dry strength. As we indicated in previous chapters, many of these are non-linear, interacting relationships.

The subsequent implementation of a comprehensive control scheme requires cost-effective, maintainable sensors and the design and implementation of mult-variable control functions, including optimization, to take care of the inevitable trade-offs that arise.

The effects of additive rates on the chemical environment and, in turn, on wet end chemistry phenomena such as retention and formation, also depend on the specific physical environment of the system (e.g. turbulence and shear). These relationships must be also determined (or confirmed) on-line, which points to the availability of adequate sensors as being key to development of the control scheme.

6. Models of comprehensive wet end chemistry control strategies

In the past, various models of comprehensive wet end chemistry control strategies have been proposed. [2,3] **Fig. 16-1** illustrates one such approach.

It does not take much reflection to understand why no comprehensive wet end chemistry process control scheme like the one illustrated has ever been implemented. In the first place, the required process models simply do not exist, and it is highly unlikely that they ever will. Some reasons for this are presented in **Table 16-3**.

7. Conclusion: the comprehensive approach will not work

The above introduction was intended to make it clear that one must not think in terms of a comprehensive approach to wet end chemistry control. There are too many variables and interactions to make this possible. If this is the situation, then what should one do?

It is the conclusion of this author that a more realistic approach to wet end chemistry control is to adopt a "reductionist" strategy. In other words, divide the overall process into subsystems and then strive to reduce the variability in each of them — always keeping in mind the larger implementations of what you are doing. In other words, to use an environmentalist slogan: *Act locally, think globally.*

This approach is in tune with modern quality management thinking where the emphasis is on reducing product (process output) variability through the reduction of process input variability. In other words, if you reduce the variability in early process inputs, then it becomes easier to do the same with later process inputs, and the end result is a more uniform product. This is a perfect model for wet end chemistry, which consists of a series of input processes that interact to eventually effect the final paper properties.

Table 16-3. Obstacles to the development of comprehensive wet end chemistry process models.

- The paper machine systems of interest have many controlled and uncontrolled primary inputs. Some are desired. Others are not. In addition, there are inputs whose effects are unknown.
- Many feedback loops exist that introduce disturbances into the process.
- Many wet end chemistry phenomena are irreversible in nature.
- Mechanical factors influence wet end chemistry interactions.
- Many wet end chemistry variables interact with one another.
- Wet end chemistry phenomena are dynamic, i.e. they are not usually at equilibrium in the system.
- Competing chemistries (dispersion vs. aggregation) are employed.

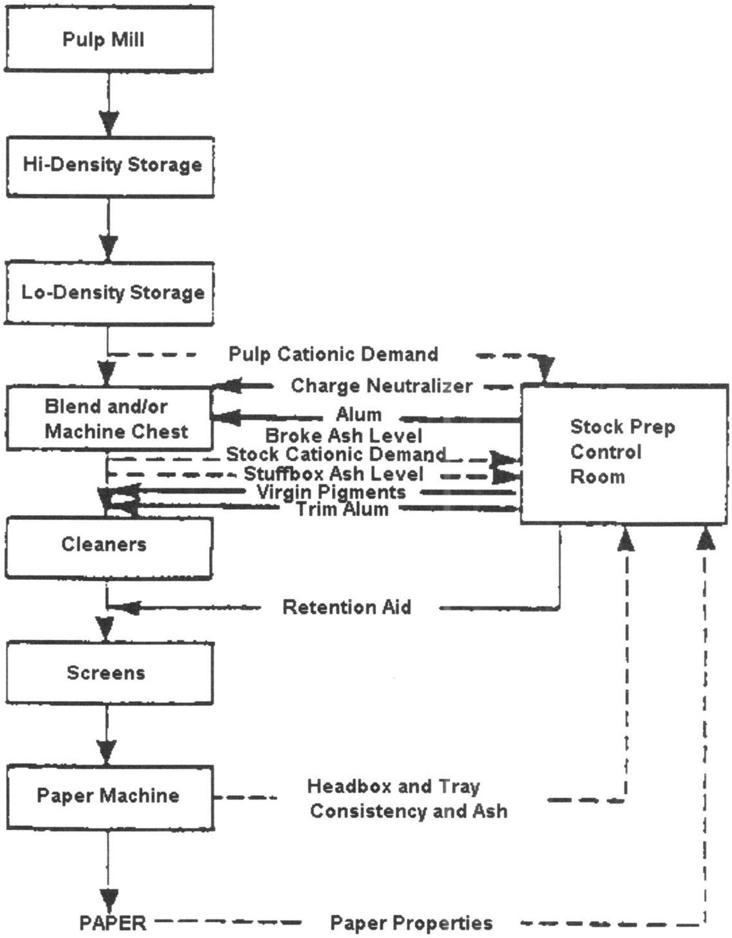

Figure 16-1. One model of a comprehensive retention control scheme. In order to implement this strategy, quantitative models for the interactions that occur among fiber fines, interfering substance, pigments, retention aids, and machine parameters must be known and sensors must be in place to measure fines and pigment retention, interfering substances, fiber flow rates, and the consistency of various process streams. [1]

8. A conceptual model upon which to base wet end chemistry control strategies

In the introduction to this book, we emphasized that models would play a very important role in our discussion. This is particularly true in the case of wet end chemistry process control. **Fig. 16-2** illustrates the model that will guide our thinking in this instance.

The model illustrated in the figure can be thought of as having three "layers" or "levels." The first layer, represents the chemical environment within which we carry out wet end chemistry. **Table 16-4** lists some of the most important chemical environment variables that we have discussed previously. Note that these variables are mainly water quality parameters.

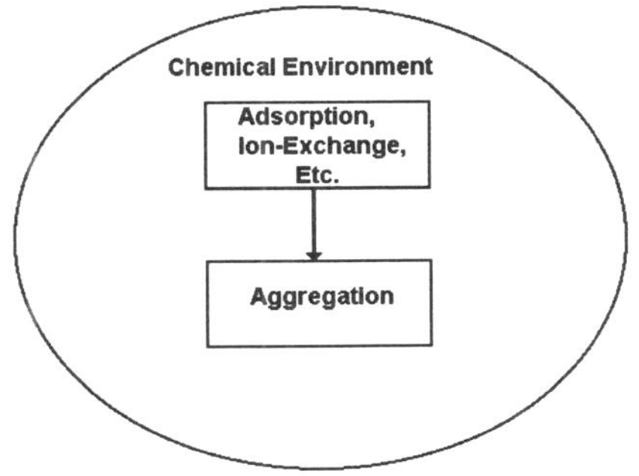

Figure 16-2. A model of the overall wet end chemistry process

Table 16-4. Chemical environment variables

- pH
- Acidity, alkalinity
- Temperature
- Conductivity
- Hardness
- Interfering substances

The second layer or level in **Fig. 16-2** contains the adsorption and ion exchange phenomena important to wet end chemistry. These phenomena affect paper properties directly (e.g. dyes, wet strength resins) or indirectly (e.g. retention aids). **Table 16-5** lists some of the materials that participate in these phenomena.

It is understood that adsorption phenomena require adsorbing surfaces. **Table 16-5** also lists some of the predominant adsorbing surfaces in a papermaking furnish.

Table 16-5. Materials that participate in adsorption, ion-exchange, polyelectrolyte interaction phenomena important in wet end chemistry

Adsorbing Molecules

- Dyes
- Retention aids
- Dry strength additives
- Wet strength resins
- Alum

Adsorbing Surfaces

- Fibers
- Fiber fines
- Fillers
- Sizing agents

The third layer in our model (**Fig. 16-2**) represents the various aggregation phenomena that occur in papermaking. **Table 16-6** lists the most important ones.

*Table 16-6. Important aggregation phenomena that occur in papermaking**

- Fines retention
- Filler retention
- Size retention

**Note that fiber retention is not an aggregation phenomenon, but rather involves mostly filtration.*

Table 16-7 contains examples of phenomena that occur when wet end chemistry gets out of control.

Finally, it will be useful at this point to remind ourselves once again of exactly why we are doing wet end chemistry. In other words, let's list the various paper properties and machine runnability considerations that are impacted.

Table 16-7. Phenomena that occur when wet end chemistry gets out of control

- Foaming
- Deposits
- Scaling

Table 16-8. Properties and machine factors impacted by wet end chemistry control

- Sheet formation
- Drainage
- Machine speed
- Machine cleanliness
- Basis weight
- Moisture content
- Sizing
- Wet strength
- Dry strength
- Opacity
- Brightness
- Color

We will now proceed to discuss the general factors that are important to control in the adsorption/ion-exchange and aggregation layers of the model. Control of the chemical environment will be dealt with as part of this discussion. Later in the chapter we will consider the specific measurements needed to do control.

9. Controlling adsorption phenomena

As they are applied in the paper industry, adsorption processes are relatively simple in concept. **Fig. 16-3** illustrates such a process. The system consists of an input stock stream, an input additive stream, one output stream, and a mixing point. Adsorption occurs at the mixing point and in the output stream.

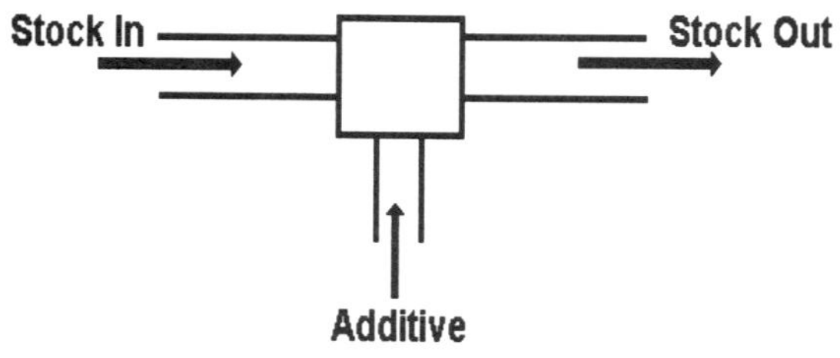

Figure 16-3. Conceptual illustration of a typical additive adsorption process employed in the paper industry. The system consists of an input stock stream, an input additive stream, one output stream, and a mixing point. Adsorption occurs at the mixing point and in the output stream.

Table 16-5 listed several important papermaking additives that rely upon adsorption as their chief retention mechanism. In order for these materials to be effective, it is imperative that the papermaking system promote rapid and complete adsorption of the additive and that the material be adsorbed in a form such that the desired effect (i.e. paper property, retention, charge modification) is achieved. If we consider this to be the "goal" of the adsorption process, then the conditions listed in **Table 16-9** must be met in order achieve this goal.

Table 16-9. Objectives and conditions facilitating accomplishment of the adsorption process goal

1. One hundred percent retention of additive, within a given period, all the time
   - Uniformly high charge difference between additive and particle surface
   - Uniformly high additive concentration
   - Uniformly high incoming stock consistency
   - Uniformly benign chemical environment in all input streams
2. Uniform distribution of additive over particle surfaces at desired addition level
   - Good mixing and turbulence at the addition point
   - Uniformly low additive concentrations in additive input stream
   - Uniformly low input stock consistency
   - Good distribution of additive throughout the stock at the addition point
   - Uniform solids level in additive stream
   - Uniform additive system pumping rate
   - Uniform input stock flow rate
3. Constant "surface concentration" of additive on particle surfaces

   All of the above, plus:
   - Uniform overall stock consistency
   - Uniform stock fiber fines concentration
   - Uniform stock filler concentration

Consideration of the items in the table immediately reveals that practically all of them refer to input stream conditions. It is also clear that the table contains some internal inconsistencies. For example, rapid retention is promoted by high stock consistency and high additive concentrations, but uniform distribution is favored by low stock consistency and low additive concentrations. Since both conditions cannot be met for a given process, the best compromise must be found by experimentation.

A serious shortcoming of most papermaking adsorption processes is that no immediate feedback measurement is made that informs the papermaker about the effectiveness of the adsorption process. The feedback information usually available — paper lab tests — comes much too late to be useful in adsorption process control (**Fig. 16-4**).

The inescapable conclusion to be drawn, then, is that the key to successful adsorption process control lies in minimizing input stream disturbances (**Fig. 16-5**). **Table 16-9** provides a "checklist" of the parameters to be controlled.

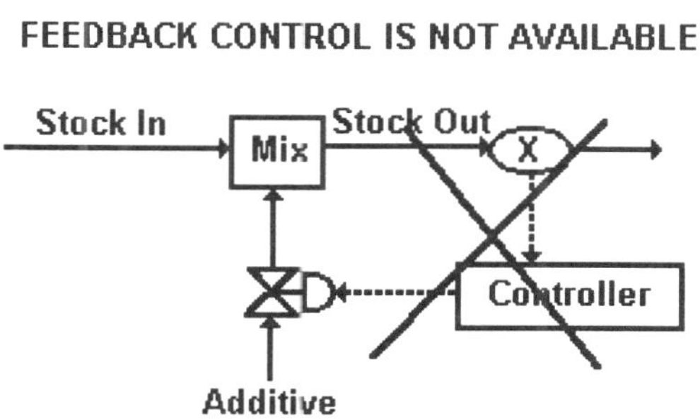

Figure 16-4. Immediate feedback control measurements are not usually available for paper industry adsorption processes

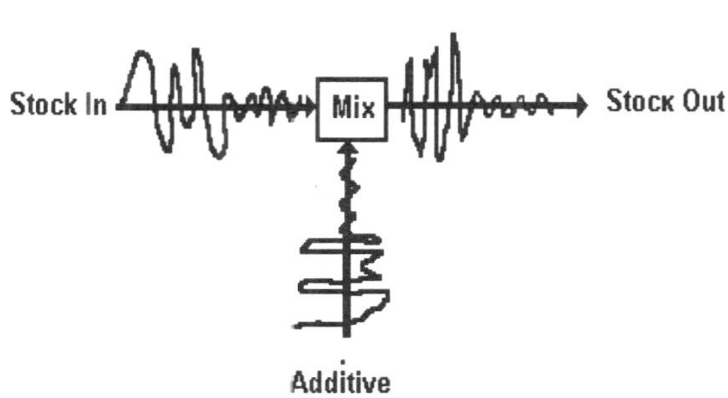

Figure 16-5. Schematic illustration of fact that effective control of adsorption processes requires the reduction of disturbances in incoming stock and additive streams

Examples of controllable disturbances in adsorption process input streams - While each additive addition point will have its own set of unique conditions, **Table 16-10** lists some typical examples of common input stream disturbances that can have significant impacts on adsorption processes.

Table 16-10. Common adsorption process input stream disturbances

<u>Input Stock Stream</u>

- pH
- Interfering substances
- Residual cationic & anionic additives
- Consistency
- Flow rate
- Fiber fines level
- Filler level

<u>Additive Input Stream</u>

- Additive concentration
- Flow rate

Later in the chapter we will review some common approaches adopted by the paper industry to minimize these disturbances. For the moment, we will continue our conceptual treatment of the subject and move on to the control of aggregation processes.

10. Controlling aggregation phenomena

Similar to adsorption processes, aggregation processes have goals, such as those listed in **Table 16-11.**

Table 16-11. Ultimate "goals" of aggregation processes

- Improved machine runnability
- Decreased two-sidedness
- Reduced MD variation of basis weight
- Reduced MD variation of ash content
- Improved machine cleanliness (Reduced deposits)
- Improved sizing, wet strength, dry strength uniformity
- Improved uniformity of appearance properties
- Improved sheet formation

The accomplishment of these goals requires the attainment of uniformly high first-pass retention with uniform distribution of materials throughout the web. When single polymer retention systems or microparticle retention systems are employed, primary control of aggregation occurs within the thin stock loop (**Fig. 16-6**). The thick and thin stock areas are both involved when dual polymer systems are used.

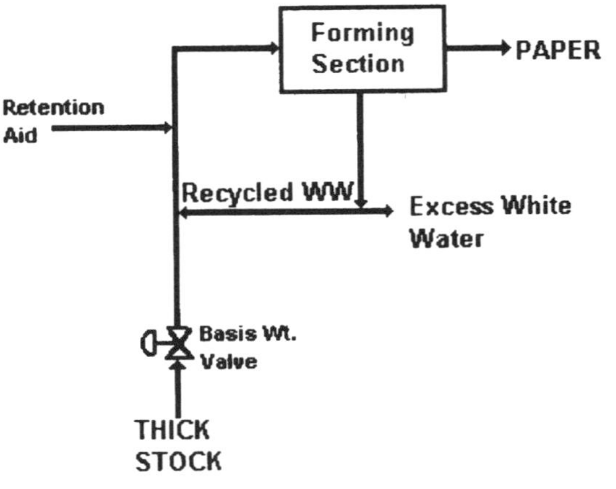

Figure 16-6. First-pass retention control loop when single polymer retention systems are used

Thin stock loop measurements related to aggregation phenomena are available and feedback control is possible.

Since aggregation is accomplished through the use of adsorbing polyelectrolytes, it is necessary that the conditions discussed previously for adsorbing processes be met. In addition, the following conditions unique to aggregation processes must be met (**Table 16-12**).

It turns out that the two conditions interact. The more highly variable the thick stock, the more highly variable the first-pass retention and hence, the more variable the recycled white water consistency. Therefore, the papermaker's strategy has been to try to minimize the variation in the recycled white water consistency. This topic will be further explored below.

Table 16-12. Conditions that facilitate accomplishment of the aggregation goals

1. Maximize particle-particle collisions under low energy conditions
   - Uniform mixing
   - Minimize excessive shear and turbulence
   - Maximize contact time

(These conditions are set by the physical attributes of the system and are more-or-less designed into the process.)

2. Reduce inter particle repulsion forces and/or increase inter particle attractive forces.
   - Add alum to reduce zeta potential
   - Add polyelectrolytes to produce cationic patches on particles
   - Add high molecular weight retention aid to induce inter particle bridging

(This represents an adsorption process or ion exchange process. Consequently, the conditions described above under adsorption processes apply.)

3. Maintain uniform concentration of fibers, fines, fillers, size particles in stock flow to the headbox.
   - Control recycled white water consistency and ash content
   - Minimize variation in thick stock flow to thin stock system

(The first of these two conditions can be met by undertaking on-line feedback control of the recycled white water consistency. (See below.) The second condition is largely a function of how well the thick stock wet end chemistry processes are gotten under control and by the uniformity of regulation of the basis weight valve. Significant machine direction basis weight variation will lead to a more highly variable basis weight valve opening.)

11. Basic measurements made for wet end chemistry process control

a. Introduction

A very wide range of measurements are employed by the paper industry to control wet end chemistry. It will be useful to review them briefly before examining some typical examples of adsorption and aggregation control schemes. We will do a quick overview here. More extensive reviews can be found in References [2] and [4].

The majority of the experimental procedures and methods used in paper machine papermaking chemistry work can be grouped under one of the categories listed below in **Table 16-13**. Several parameters can be derived from data collected by the primary measurements listed in the table. Foremost are the first-pass retention (single-pass retention) levels of wet end components. Depending on the number of wet end streams investigated, it will be possible to carry out various material balances. The extreme being a complete balance that includes the stock preparation, white water recycle, and broke reclamation areas. If the data points are collected sequentially with time, it will be possible to determine the steady-state or nonsteady-state performance of the machine. The determination of first-pass retention, steady-state and nonsteady-state character, and the calculation of material balances will be discussed later. Before doing so, we will consider some of the details of the measurements listed in **Table 16-13**.

Table 16-13. Categories of papermaking chemistry experimental measurements and procedures used in process control

1. Component levels
2. Fiber qualities
3. Slurry drainage characteristics
4. Electrokinetic quantities
5. Flow rates (liquid, mass)
6. Paper machine operating parameters.

b. Experimental methods and measurements

Laboratory methods for studying component levels, fiber qualities, electrokinetic quantities, and slurry drainage characteristics include the particular determinations that fall under each of these four categories (**Table 16-13**). It would be beyond the scope of this discussion to describe each of the determinations listed in **Table 16-14**. They cover a wide variety of analytical procedures and methods ranging from rather standard water analysis techniques to very specific measurements unique to papermaking chemistry. Many of the details of the procedures can be found in References [5-30].

Table 16-14. Specific papermaking chemistry determinations

Component Levels

A. Suspended solids
   1) Percent consistency
   2) Percent ash
   3) Filler components
   4) Percent fines

B. Dissolved solids
   1) Dissolved inorganic salts
      a) Specific conductance
      b) Water hardness
      c) pH
      d) Acidity and alkalinity
      e) Cations and anions
   2) Dissolved organic materials
      a) Starch
      b) Retention aids and other polyelectrolyte additives

Fiber qualities

A. Fiber composition
   1) Hardwood: Softwood percentages
   2) Groundwood: Chemical fiber percentages
   3) Broke: Virgin fiber ratio

B. Fiber length distribution

C. Pulp viscosity

D. Degree of refining
   1) Microscopic inspection
   2) Freeness
   3) Fines classification

Electrokinetic characteristics

A. Zeta potential

B. Electrophoretic mobility

C. Cationic demand

D. Colloid demand titration

Slurry drainage characteristics

A. Freeness

B. Handsheet mold drainage time

C. Dynamic drainage jar drainage time.

c. On-line measurements

Most of the parameters listed in **Table 16-14** are carried out by manual measurement methods. However, a few of them can be determined by on-line instruments. **Table 16-15** lists the parameters that can currently be measured on-line.

Table 16-15. Papermaking chemistry parameters for which on-line measurements are available

Percent consistency

A. Thick stock

B. Thin stock

Thin stock filler levels

Dissolved solids

A. Specific conductance

B. pH

C. Certain inorganic cations and anions

Electrokinetic characteristics

A. Zeta potential

B. Cationic demand

Slurry drainage characteristics

A. Freeness

On-line thick stock consistency measurements have been carried out routinely in the pulp and paper mill for many years. Thin stock consistency sensors have been available for several years but have only recently begun to be routinely utilized. These new sensors are capable of measuring both the total solids consistency and the filler concentrations in the slurries. First-pass furnish and filler retentions can be determined from these measurements.

A survey conducted by TAPPI's Retention and Drainage Subcommittee revealed that pH is by far the most frequent on-line dissolved solids measurement made in the paper industry. [2] Other on-line determinations of dissolved materials are done very infrequently.

The TAPPI survey also revealed that an on-line measurement of cationic demand or some other comparable furnish charge measurement was the "most needed" measurement as indicated by papermakers and chemical suppliers. Presently, zeta potential (streaming potential) and dissolved charge can be measured on-line.

On-line freeness measuring instruments have been available to the industry for many years. Most of these devices have been aimed at the mechanical pulping industry but could be used for measuring the freeness of papermaking stock. Many mills have installed these devices on the wet ends of paper machines.

It is strongly suggested that on-line papermaking chemistry data should be used whenever possible. It is very important, however, that the accuracy and reliability of all on-line data be verified. This means that all sensors must be calibrated regularly and the precision of their output known. Reference [31] provides a good overview of the factors involved in on-line measurement.

d. Liquid and mass flow rates

Process stream liquid and mass flow rates are necessary for carrying out material and energy balance calculations on a wet-end system. They are also needed for determining material addition rates and materials loss rates. There are a variety of flow measurement methods available to the engineer — several of which are listed in **Table 16-16**.

Table 16-16. Flow measurement methods

1. Magnetic flow meters and other on-line devices
2. "Bucket and stopwatch" methods
3. Chemical dilution methods
4. Ultrasonic flow meters
5. Flow balance calculations

Several of the methods listed in **Table 16-16** are important enough to deserve very extensive discussion themselves. Indeed, flow rates are the most common measurement made in the paper industry. Here, we will provide only a cursory treatment that should serve to guide you if you wish to pursue any of the specific methods in detail.

Magnetic flow meters and other on-line devices are found in great numbers throughout the paper mill. They provide basic information required by operators. Process engineers rely upon these sensors for carrying out a variety of trouble-shooting, process monitoring, and process optimization activities. As with all such on-line instruments, it is absolutely vital that they be maintained and kept in good calibration. This means that a mill should have a routine calibrating schedule for each meter and the routine should be adhered to closely. If this is done and there are sufficient meters in the system, then the collection of flow data will be greatly simplified for the process engineer. Unfortunately, it is a rare mill where the engineer can rely solely on on-line instruments for flow data. Because of this, he or she must be familiar with other flow measuring techniques.

"Bucket-and-stopwatch" methods refer to any method where either the length of time required to collect a given quantity of material is determined, or the quantity of material collected in a given length of time is measured. The time-measuring device is usually a stopwatch, while the volume-measuring device can range anywhere from a beaker or graduated cylinder to a barrel or large tank, depending on the magnitude of the flow rate.

This method is usually of limited usefulness around the wet end of a paper machine because it requires collecting the entire flow from a given process stream during a given period and wet end process stream flow rates are far too great to be amenable to such an approach. The method can be very useful for measuring additive flow rates, however.

One method of getting around the high wet-end flow rates is to use a chemical dilution method for flow determination. In such an approach, the flow can be determined by introducing a salt solution of known concentration into a stream. By determining the dilution of the salt by the stream, the flow may be calculated with reasonable accuracy.

The development of portable ultrasonic flow meters has proven to be a significant advance for the papermaking chemist.[32] These meters provide a convenient, non-intrusive method for measuring flow rates in a wide variety of wet end process streams. Ultrasonic flow meters come in two complementary versions — the Doppler reflective type, and the clamp-on transmission type. The Doppler reflective type of meter is used in streams having suspended solids or air bubbles, while clamp-on transmission meters are applied to streams of clear liquids. Both types are available as portable meters that have proven to be very valuable troubleshooting tools.

Material balance calculations can be used for determining the flow rates of streams that are inaccessible to the types of flow measurements discussed above. The primary requirement is that sufficient data be available to enable the calculation of a material balance that includes the process stream of interest. If this condition is met, then the flow rate in the stream of interest can be computed without making any direct measurements in that stream. This is a powerful technique that can be extensively useful in many instances.

e. Paper machine operating data

Even though we are dealing primarily with papermaking chemistry topics in this chapter, it should be understood that the mode of operation of the stock preparation system and wet end of a paper machine can significantly affect fines retention, sizing, and other papermaking chemistry phenomena. Consequently, it is always advisable to monitor wet-end system conditions when carrying out papermaking chemistry work. For example, **Table 16-17** lists some of the paper machine operating data that should be routinely monitored.

Table 16-17. Stock preparation and paper machine conditions that can impact wet end chemistry performance

1. Couch vacuum
2. Suction box vacuums
3. Machine speed
4. Jet-to-wire speed ratio
5. Dryer section steam use
6. Press loading
7. Refiner settings

In addition to the parameters listed in the table, the configuration of the drainage elements on the forming table and the stock jet impingement angle will strongly influence retention and drainage performance.

12. Mill process control practices: adsorption processes

We will now resume our earlier discussion of controlling adsorption process input stream disturbances. **Table 16-18** (a copy of **Table 16-10**) lists some common sources of variability.

Table 16-18. Common adsorption process input stream disturbances

Input Stock Stream

pH
Interfering substances
Residual cationic & anionic additives
Consistency
Flow rate
Fiber fines level
Filler level
State of fiber, fines, and filler surface charge

Additive Input Stream

Additive concentration
Flow rate

The pH, interfering substances, and previously added additives all pertain to the chemical environment in the stock stream. Some typical control schemes employed with these parameters are the following:

pH - pH is a commonly controlled parameter in the paper industry. Traditional practice employed alum to achieve acid pH levels and soda ash or some other alkali to gain alkaline pH levels. The many documented adverse effects of alum have led to its replacement by acid in recent years. Stock streams containing calcium carbonate fillers often do not need pH control due to the buffering effect of the filler.

Interfering substances - The increased levels of interfering substances found in papermaking systems these days requires the papermaker to add charge neutralizing chemicals to many stock streams in order to ensure the effectiveness of chemicals added downstream. Alum PAC and low molecular weight, high charge density (e.g. PEI) are the preferred control agents. Addition close to the source of the interfering substance is preferred. Dissolved charge measurements can be employed both to determine the addition levels of charge neutralizer required and to evaluate the effectiveness of the treatment.

Most interfering substance neutralization is done manually, however, on-line control technology is available now. For example, the on-line, closed loop control of interfering substances associated with coated broke slurries has been demonstrated.[33] **Fig. 16-7** illustrates the control scheme employed for this purpose. Both the flow rate and consistency of the coated broke stream was monitored and the information fed forward to the coagulant (charge neutralizer) setpoint controller. This information was critical to good control.

The effectiveness of the neutralization process was monitored by measuring the streaming potential or turbidity of the output stream. This information was fed back to the coagulant controller.

A similar control loop could be used to control the neutralization of other interfering substances by substituting one of the new on-line dissolved charge sensors for the streaming potential or turbidity sensor shown in **Fig. 16-7** .

Residual cationic and anionic additives in the input stock stream - Information about residual cationic and anionic additives can be gotten from two sources. First, the information can be fed forward from an upstream point of addition. Second, dissolved and total charge titrations can be done on the input stock stream to determine the charge state of the solid particle surfaces and the amount of unadsorbed additives remaining in solution (which can act as interfering substances).

The next four input stream parameters in **Table 16-18** (consistency, flow rate, fiber fines level, and filler level) refer to the quantity of adsorbing surface in the input stock stream — i.e. the specific surface area.

Consistency and flow rate - Consistencies and flow rates are among the most commonly controlled parameters in a paper mill. They are also two of the most critical factors that determine the effectiveness of adsorption processes. In practice, it often not possible to control the consistency or flow rate to satisfy the need to a specific additive addition point due to the demands of other parts of the papermaking process. If this is the case, then it is important that consistency and flow rate information be passed

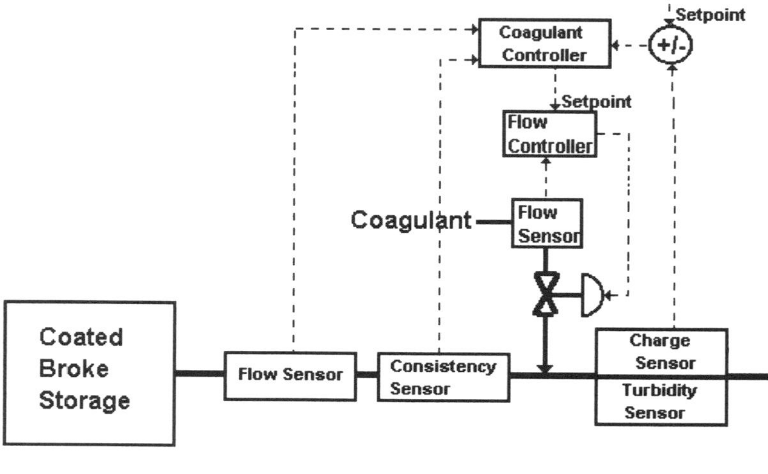

Figure 16-7. On-line system for the closed loop control of interfering substances in a coated broke process stream. Flow, consistency, and particle charge (or stream turbidity) are monitored continuously and the information used to control the addition of a charge neutralizing agent. [33]

to the additive flow rate controller so that the appropriate adjustments can be made in order to maintain a constant additive-to-stock ratio. This practice has become more common in recent years.

Fiber fines and filler levels - The reason for maintaining uniform fines and filler levels in input stock streams should be obvious from the many times we have discussed their ability to adsorb relatively large quantities of additives compared to fibers. Thus, variations in the flow of fines and filler can introduce large disturbances into adsorption processes and drastically affect their effectiveness. This is one of the reasons that the careful management of coated broke addition rates is so important.

At this time, there are no on-line fines or filler sensors available for use under thick stock consistency conditions. This means that manual measurements must be employed, which limits measurement frequency and the frequency with which control actions can be taken. Nevertheless, any fines and filler data that can be gathered will be a valuable contribution to the control scheme and should be included. The primary goal, of course, is to have constant fines and filler level in the input stock stream. In most cases this will not be possible, however, so the information should be fed to the additive flow rate controller along with the stock flow rate and stock consistency information.

State of fiber, fines, and filler surface charge - The attractive force between an additive molecule and the surface of a furnish particle is usually established by arranging to have the two substances be oppositely charged. The greater the charge difference, the greater will be the rate at which additive molecules diffuse to the particle surface and become adsorbed. If the additive is cationic, as is often the case, then the fiber or filler surface must be anionic in order for strong attraction to occur between the two. If both surfaces have the same charge, then the adsorption rate will be low and the process will be ineffective.

The surface charge on input stock particles will depend on many factors, such as the surface chemical composition of the stock particles, the stock pH, the concentration and nature of interfering substances, and the presence of previously adsorbed additives. It can be determined by zeta potential or total charge measurements. This subject was discussed in detail in Chapter 5 and you should review that material if this topic is of particular interest to you.

The usual practice is to determine empirically the input stock charge and system conditions that correlate with the best additive performance and then try to control at these conditions.

Solids level and flow rate of additive input stream - The control of disturbances in additive input streams depends upon the uniformity with which chemicals are prepared and delivered which, in turn, is primarily a function of the technology employed in the mill chemical preparation area. It is impossible to achieve uniform delivery of additives without installing reliable, up-to-date equipment and maintaining a rigorous calibration program.

Modern practice is to feed forward stock flow and consistency information to the additive ad-

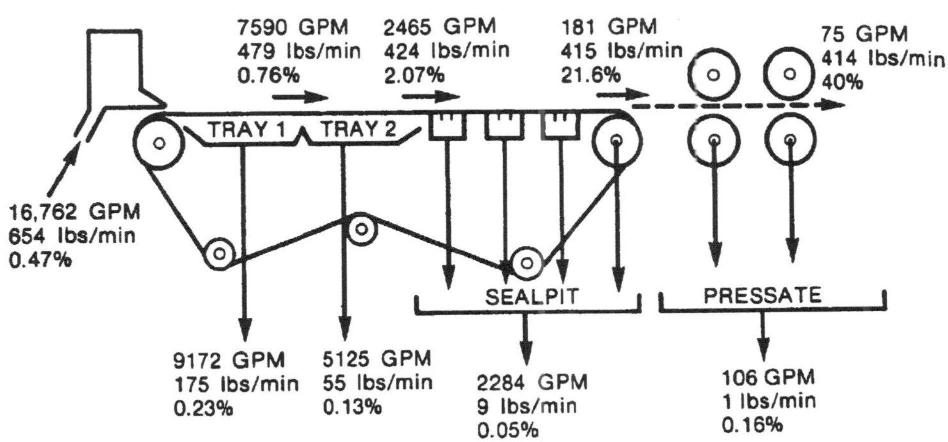

Figure 16-8. Sample calculation of water and solids material balance around the wet end of a paper machine

dition control system, so that variations in stock flow and consistency can be compensated for continuously. (See **Fig. 16-7**.)

13. Mill process control practices: aggregation control

a. Introduction

Of the various aggregation processes that occur in the paper mill, those associated with first-pass retention control receive by the far the most attention in day-to-day operation. For that reason, we will focus our attention on this subject.

In Chapter 15, we defined first-pass retention to be: *the ratio of the amount of material that leaves the headbox slice compared to the amount of material that is contained in the paper web leaving the couch roll.*

In order to practice retention control, it is necessary to convert this definition into a mathematical relationship. This can be done in the following way. First-pass retention values are usually derived from the steady-state material balance conditions existing around a forming section. Since first-pass retention is a steady-state parameter, it is not strictly correct to refer to it when a paper machine is not operating under transient conditions. In such instances, it is often more useful simply to monitor the key parameters involved, like headbox consistency and tray consistency.

In this section, we will first present the results of an example calculation of a material balance around the forming section of a paper machine and then show how the balance can be used to derive information about the retention performance of the machine. Next, we will relate the retention parameters calculated from the material balance to those calculated from headbox and tray consistencies and demonstrate the assumptions that underlie the latter calculation.

b. Wet end material balances and first-pass retention

Fig. 16-8 contains the results of a typical material balance calculation that can be computed from the flow and consistency data collected during a paper machine drainage performance evaluation. Water and forming table solids are the variables of interest. The most common practice is to measure the headbox flow rate, the consistencies of various white water streams flowing through the forming fabric, and the consistency of the paper on top of the forming fabric. The information contained in **Fig. 16-9** can be used in several ways to describe the performance of the paper machine wet end. We will concentrate on what it tells us about first-pass retention behavior.

If we apply the definition of first-pass retention given above, it can be expressed by **Equation (1)**:

First-pass total solids retention, %

$$= 100\% \times \frac{\text{total solids flow in paper}}{\text{total solids flow from headbox}} \quad (1)$$

Equation (1) is a general expression that applies to the entire paper machine or any part of the machine that is related to the headbox by a continuous process flow. The distinguishing feature is where you measure the total solids of the paper at the reel, after the press section, at the couch, etc. For this discussion we will concentrate on the first-pass total solids retention on the fourdrinier section and, thus, the total solids in the paper of interest refers to the solids in the paper after the couch roll.

From **Fig. 16-8** we see that the total solids coming onto the wire from the headbox is 654 lb/min (297 kg/min). The total solids in the paper leaving the couch roll is 415 lb/min (188 kg/min). Applying **Equation (1)** results in a first-pass total solids retention on the fourdrinier of 63%.

The papermaker does not usually measure the total solids coming onto or leaving the forming fabric in order to determine the first-pass retention performance of a machine. Rather, the papermaker most often uses **Equation (2)**, which relates first-pass total solids retention to headbox and tray consistencies.

First-pass total solids retention, %

$$= 100\% \times \frac{C_H - C_T}{C_H} \quad (2)$$

where C_H and C_T refer to the headbox consistency and tray consistency, respectively.

Utilizing data from **Fig. 16-8** in **Equation (2)**, the calculated first-pass total solids retention is 51% and not the 63% we found earlier. Why the difference? This question can best be answered by considering how **Equation (2)** is derived. In deriving **Equation (2)** from the material balance, two assumptions must be made. First, it is assumed that the headbox solids are all divided between the sampled tray and the paper. In other words, the total headbox solids equals the total paper solids plus the total sampled tray solids. The second assumption is that all the water flowing from the headbox flows into the sampled tray, i.e. none goes with the paper.

In practice, the paper leaving the couch roll contains only 1-2% of the total water coming from the headbox. Therefore, the second assumption would appear to be reasonable. However, the situation is not quite so straightforward with respect to the first assumption. The assumption that all headbox solids flow to the sampled tray can be met in at least three ways on a given paper machine. First, if a paper machine has only one white water tray, then most of the solids that flow from the headbox that are not retained in the paper will flow into the tray with the white water and **Equation (2)** should be applicable. Second, if the machine is configured so that all the white water from various trays is collected and mixed at some point and a representative sample of the mixed white water can be gotten, then **Equation (2)** will also hold. (Note that if all white water is not mixed, then it is important that the mixing be done in the ratio of the relative flow rates of the white water to the various trays.) Finally, if the white water flowing to all trays has the same consistency, then **Equation (2)** will provide a reasonable estimate of first-pass retention.

With these considerations in mind, it is easy to understand why **Equation (2)** did not give a correct value for the first-pass solids retention of the paper machine in **Fig. 16-8**. Referring to the figure, it can be seen that only 73% of the headbox solids and 55% of the headbox water flow into the sampled tray. If it were necessary to obtain an accurate measure of the first-pass solids retention on this machine, it is clear that one would have to measure the white water consistencies of both trays and also have a good estimate of the flow rates to the two trays. With this approach you could account for approximately 95% of the total solids and the calculated first-pass retention would be 65%, a reasonably close estimate to the true value of 63%. In many mills the flow rates are not easy to obtain. Headbox side bleeds can introduce error into the way headbox flows are used and can even enrich white water consistency.

It is not always necessary to collect the flow rates and consistencies required to compute an accurate first-pass retention value. It all depends on the purpose of the work. For example, if the objective is simply to monitor the first-pass retention performance of a paper machine over time, then the primary information required is related to the changes in first-pass retention that occur. Thus, the absolute value of the parameter is not so important and headbox and tray consistency data may be all that is needed.

On the other hand, if one is following the results of a retention aid trial, then it is advisable to carry out a more exact retention analysis because the material under investigation can affect retention performance at a number of points on the paper machine — including the vacuum drainage section, the couch roll, and the press section.

First-pass ash retention is often calculated from the formula presented in **Equation (3)**, which is closely related to **Equation (2)**.

First pass ash retention, %

$$= 100\% \times \frac{C_H A_H - C_T A_T}{C_H A_H} \qquad (3)$$

where A_T and A_H refer to the tray and headbox ash levels, respectively.

The derivation of **Equation (3)** involves exactly the same assumptions as those discussed above for **Equation (2)** and it should therefore be used with the same caution.

Another popular method of calculating first-pass ash retention is to compute the ratio of the ash content of the paper leaving the couch roll to the ash content of the furnish in the headbox. This relationship provides a reasonable value for first-pass ash retention only when the first-pass total solids retention is very high. The lower the first-pass total solids retention, the lower the accuracy with which the couch-to-headbox ash ratio approximates the true first-pass ash retention.

c. Importance of precise consistency measurements

Regardless of which formula is used to calculate first-pass total solids retention or first-pass ash retention, it is absolutely critical that the consistencies and ash levels be determined with good precision. The reason for this is that the retention equations involve ratios of the experimental variables and errors in the measured consistency data can be greatly amplified in the final calculated retention values.

Putting it another way, small errors in the headbox and tray consistency and ash data can produce large variations in the calculated first-pass retention values. For example, it can be shown that a 0.02% deviation in the absolute headbox or tray consistencies can result in a 6% change in the calculated absolute first-pass total solids retention. To avoid these problems, it is absolutely necessary that all weighing associated with first pass retention consistency and ash content data be made to the nearest 0.001 measurement unit.

d. Using the Britt Jar in first-pass retention process control

The use of the Britt Jar and other devices for the laboratory measurement of first-pass retention was described in Chapter 15. In many mills, Britt Jar measurements are employed as part of the retention control program. In these instances, the practice usually involves determining the fines level in headbox and tray samples. First-pass fines retention is then calculated with **Equation (2)**.

e. On-line, closed loop control of first-pass retention

Earlier in the chapter (**Table 16-12**), the importance to aggregation control of maintaining uniform concentration of fibers, fines, fillers, and size particles in the stock flow to the headbox was introduced. There, it was pointed out that the on-line feedback control of recycled white water consistency was a good way to achieve this objective.

It was also pointed out there that thick stock wet end chemistry processes and machine direction basis weight variation play key roles in this regard. The more highly variable the thick stock, the more highly variable the first-pass retention and, hence, the more variable the recycled white water consistency. Therefore, the papermaker's control strategy should be to minimize the variation in the recycled white water consistency.

f. Feedback control of recycled white water consistency and ash content as a means of reducing disturbances in the white water loop [34,35]

In recent years, there has been increased interest in the development of closed loop retention control systems. Reference [34] describes one such project. In this work, total first-pass retention, first-pass ash retention, and total recycled white water consistency were evaluated to determine which parameter would give the most rapid response to a significant process disturbance. The control system configuration is illustrated in **Fig. 16-9**. The scheme utilized two consistency sensors (headbox and white tray) and a computational module. The sensors were capable of determining both the total consistency and the ash consistency of the sampled streams. The following table contains the experimental results.

Table 16-19. Comparison of total retention, ash retention, and white water consistency as first-pass retention control variables [34]

	Minutes to Dampen Change*		
Variables Studied	**Total First-pass Retention**	**First-pass Ash Retention**	**Total Recycled White Water Consistency**
Total Retention	18	13	10
Total White Water Consistency	19	13	8

**The change was a 20% dilution of retention aid.*

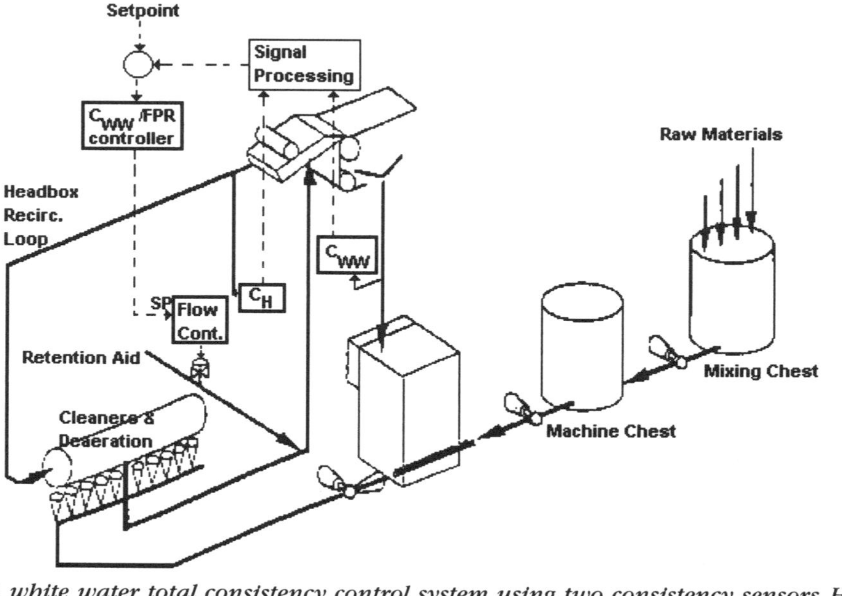

Figure 16-9. A white water total consistency control system using two consistency sensors. Headbox and tray consistency and ash contents were determined on-line and the information was used to calculate total first-pass retention and first-pass ash retention. [34]

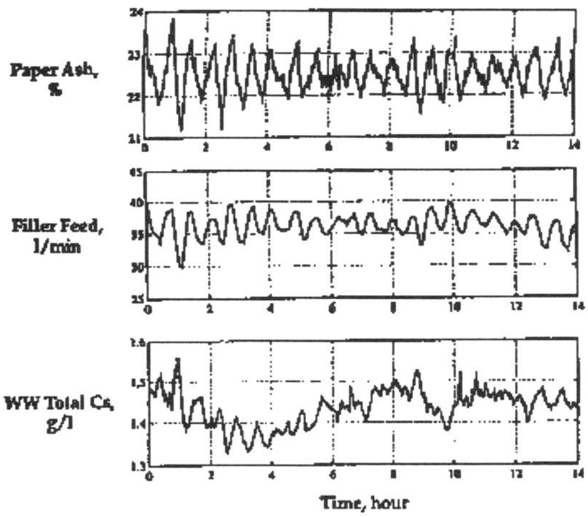

Figure 16-10. Illustration of improvement in paper ash uniformity with white water consistency control. Situation before tuning of paper ash control.[34]

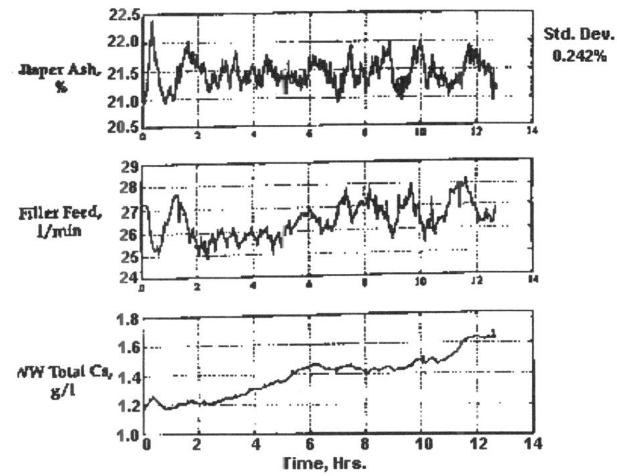

Figure 16-11. Illustration of improvement in paper ash uniformity with white water consistency control. Situation after tuning of paper ash control.[34]

It was concluded that the total recycled white water consistency measurement provided the quickest response to a process disturbance and would therefore be the best control variable.

A summary of the results collected from a six month study of the system are presented in **Table 16-20** and **Figs. 16-10** and **16-11.** There, significant reductions in several important paper quality and operating variables are documented. These improvements are examples of what one would expect in moving from manual, infrequent control to on-line, real-time control.

Table 16-20. Long term (six months) experience with white water consistency control: effect on process variables[34]

Process Variable	Average Change in Standard Deviation, %
White Water Total Consistency	-81.4
White Water Ash Consistency	-68.8
Headbox Stock Total Consistency	-22.3
Headbox Stock Ash Consistency	-47.6
First-Pass Furnish Retention	-59.8
First-Pass Ash Retention	-23.5
Retention Chemical Consumption	No Change
Paper Basis Weight	-7.7
Paper Ash Content	-27.1

The above example employed two separate consistency sensors in the control system. Another approach to the task is to utilize a single sensor to alternately measure the consistency of several different streams. A manifold sampling system is required in order to accomplish this. Both single sensor and multiple sensor system are in use today.

g. Limitations of thin stock loop retention control systems

It should be obvious that the types of retention control systems described above have limitations in terms of the magnitude of disturbances they can handle. They should really be considered to be "trimming" control systems. The really large disturbances that affect retention should be dealt with in the thick stock area before the furnish reaches the basis weight valve. Included in this category are variations in fines and ash levels, interfering substance concentration, and dissolved and surface charge. Again, the guiding philosophy should be to distribute the control back into the stock preparation area. By so doing, considerable progress is possible through application of existing measurement technology and knowledge.

14. Summary

We will close this introduction to wet end chemistry process control at this point. By now, you should recognize that we have barely touched the surface of this complex and important topic. Until recently, the papermaker was limited primarily to manual measurements and operator control for most of the parameters of interest. Fortunately, that situation is changing. The recent introduction of new consistency, ash, streaming potential, and dissolved charge sensors have provided the industry with powerful tools around which to devise control strategies. Recognition that papermakers should concentrate on reducing the variability of process subsystems also provides a clear path forward. It is expected that the near future will see significant advances in this arena.

15. References

[1] Brewster, D. B. "Papermaking area." Chapter XII in Pulp and Paper Manufacture, Vol. 10, Joint Textbook Committee of the Paper Industry, 191-192(1993).

[2] Scott, W. E. Tappi J. 67(11): 72(1984).

[3] Springer, A. M., Noe, J. S., Wegner, T. H., "Fundamental strategy for control of retention and drainage on a modern paper machine." Tappi J. 70(1): 43(1987).

[4] Onabe, F. "Measurement and Control." Chapter 11 in Paper Chemistry, edited by J. C. Roberts, Blackie: Chapman and Hall, London, pp. 198-216 (1991).

[5] Unbehend, J. E., "The dynamic retention/drainage jar — increasing the credibility of retnetino measurements." Tappi J. 60(7): 110(1977).

[6] TAPPI Test Method T240 (Percent Consistency), TAPPI PRESS, Atlanta.

[7] TAPPI Test Method T 413 (Percent Ash), TAPPI PRESS, Atlanta.

[8] TAPPI Test Method T 233 (Percent Fines and Fiber Length by Classification), TAPPI PRESS, Atlanta.

[9] TAPPI Test Method T 261 (Percent Fines by Britt Jar Screening), TAPPI PRESS, Atlanta.

[10] Tappi J. Useful Method 508 (Analysis of TiO2 in Process Streams and Paper), TAPPI PRESS, Atlanta.

[11] TAPPI Test Method T 627 (Analysis of Ti02 in Process Streams and Paper), TAPPI PRESS, Atlanta.

[12] "Methods for Chemical Analysis of Water and Wastes," U.S. Envir. Prot. Agency, EPA-625/6-674003) (1974)(Sample Preparation Methods for Atomic Adsorption Analysis of Process Streams).

[13] TAPPI. Useful Method 214 (Measurement of pH in Stock and White Water Samples), TAPPI PRESS, Atlanta.

[14] Avery, L.P. "Evaluation of Retention Aids: The Quantitative Alum Analysis of a Papermaking Furnish and the Effect of Alum on Retention." Tappi J. 62(2):43 (1979) (Analysis of Alum in Process Streams).

[15] Aldrich, L.C. "A Volumetric Method for the Determination of Sulfates." Tappi J. 57(7):122 (1974) (Sulfate Analysis Methods).

[15] TAPPI Test Method T 419 (Starch in Paper), TAPPI PRESS, Atlanta.

[16] Allen, L.H., Trans. Tech. Sect., CPPA 3(2):32 (1977) (Pitch Determination Method).

[17] Browning, B.L. "Methods of Wood Chemistry," Vol. II, Interscience Publishers, p. 812-823 (1967)(Methods of Lignin Analysis).

[18] Strazdins, E. "Factors Affecting the Electrokinetic Properties of Cellulose Fibers". Tappi J. 55:1691 (1972) (Use of Fines in Zeta Potential Measurement).

[19] Halibisky, D.D. "Wet-end Control for the Effective Use of Cationic Starch." Tappi J. 60(12):125 (1977) (Colloid Titration Method).

[20] Hubbe, M.A. "Modified Reporting Procedure for Polyelectrolyte Titrations." Tappi J. 62(8):120 (1979) (Refinement of Colloid Titration Method).

[21] TAPPI Test Method T B (Identification of Wood and Fibers from Conifers), TAPPI PRESS, Atlanta.

[22] Parham, R.A. and Gray, R.L. The Practical Identification of Wood Pulp Fibers," TAPPI PRESS (1982) (SEM Microscopy of Wood Pulp Fibers).

[23] TAPPI Test Method T 401 (Fiber Analysis of Paper and Paperboard), TAPPI PRESS, Atlanta.

[24] Clark, J.d'A. Pulp Technology and Treatment for Paper. Miller Freeman Publications, (1978) (Discussion of Handsheet Devices and Other Pulp Treatment Devices).

[25] Hergert, R.E., and Waech, T.G. "Newsprint Furnish Solids Retention. Field and Experimental Machine Studies." Tappi J. 62(11):79 (1979) (Description of Britt Jar Survey of Paper Machine First-Pass Retention).

[26] Pelton, R., Allen, L.H., and Nugent, H. "Measuring Fines Retention of Newsprint Pulps." Pulp Paper Can. 80(12):125 (1979) (Experimental Variables Associated with the Britt Jar).

[27] Werdouschegg, F.M.K. "Interrelating a Practical Retention Testing Procedure with Wet-end Retention Results -Theory and Practice." Tappi J. 60(7):105 (1977) (Use of Minidrinier Retention Tester).

[28] Kerekes, J.R. and Harvey, D.M. "Drainage Tester Survey". Tappi J. 63(5):89 (1980) (Survey of Drainage Testing Practices).

[29] Tarvin, R.F. and Westwood, R.L., Proc. 1982 TAPPI Papermakers Conf., p. 243 (1982) (Use of Modified Britt Jar for Drainage Testing), TAPPI PRESS, Atlanta.

[30] LeBel, R.G., Nobleza, G.C., and Pacquet, R. "Water Retention Value Indicates Machine Runnability of Pulp." Pulp Paper Can. 80(5):64 (1979) (Use of Water Retention Value in Papermaking).

[31] Thomas, K. V. "Retention — On-line determination and systems factors." In 1989 TAPPI Retention and Drainage Short Course Notes, Washington, DC, TAPPI PRESS, 37(1989).

[32] DiNapoli, L. D. and Weber, H., "Paper stock consistency and pipe size effects on L&N Doppler flowmeter performance." In Proceedings of the 1981 TAPPI Engineering Conference, TAPPI PRESS, 559(1981).

[33] R. Gossage, "On-Line Treatment Control of Coated Broke Slurries" in Proceedings of 1993 TAPPI Papermakers Conference. Atlanta, GA 551-554(April 18-21, 1993).

[34] Nolelainen, J., Piirainen, R. and White, C. "Practical experiences of white water consistency control of a paper machine wet end." In Proceedings of 1993 TAPPI Papermakers Conference. Atlanta, GA. pp. 227-238, (April 18-21, 1993]

[35] Sopenlehto, T. "On-line monitoring and control in the wet end." Chapter 10 in Applications of Wet End Paper Chemistry. Au, C. O, and Thorn, I., Editors, Blackie Academic and Professional, 172-182(1995).

Glossary

Term	Definition
Abietic Acid	One of the resin acids derived from conifers that is used to produce rosin size materials. Chemically, it is a member of the triterpenoid family and contains three six-membered carbon rings that impart hydrophobicity to the structure. A carboxyl group is attached to the structure and provides the bonding point with fibers.
AC Streaming Current	In this measurement, a piston is used to force liquid back and forth through a small clearance between the piston and enclosing cylinder. A sample of paper making slurry is placed in the cylinder. The pressure created by the piston motion produces an AC current that can be detected by electrodes imbedded in the sides of the cylinder. The sign and magnitude of the streaming current is related to the sign and magnitude of the zeta potential of the particles in the suspension under test. Experience has shown that streaming current is most reliable as an indicator of the sign of the furnish charge (cationic or anionic) and as a zero point indicator. it does not correlate well with other electrokinetic methods at nonzero zeta potential values. AC streaming current is commonly used as an endpoint indicator in colloid titration determinations and to control flocculant addition rate in water treatment systems.
Accessibility	Degree to which chemical reagents and solvents are accessible for reaction with cellulose, hemicellulose, lignin, and other polymeric materials. Accessibility is primarily controlled by the closeness with which polymeric molecules are packed together in the material of interest. In wood fibers, accessibility and degree of crystallinity are inversely related.
Acetate	Ester formed by reacting acetic acid with an alcohol. It contains the acetyl, group Ch_3CO. Hemicelluloses contain acetate groups that get hydrolyzed during the kraft pulping process and produce acetic acid which consumes alkali in acid-base neutralization reactions.
Acid Dye	A group of dyes that contain the salts of sulfonic acids as functional groups. They are not naturally attracted to cellulose fibers and must be retained through the use of alum or some other cationic dye promoter. Acid dyes are not widely used by the paper industry.
Acid Hydrolysis	See Hydrolysis.
Acid Size	Rosin size which consists primarily of unsaponified resin acids in their free acid form.
Acid-Base Neutralization	Chemical reaction between acid and base such that the resultant solution contains neither an excess of hydrogen ions nor hydroxyl ions.
Acidity	Acidity is the total capacity of a substance to contribute hydrogen ions. Acidity represents the ability of a water solution or suspension to react with bases. It is often expressed in terms of the amount of base, of a given concentration, that is required to raise the pH of a sample to a given value, such as 8.0. Alum contributes acidity, as do many pulp bleaching products.
Acids and acid groups	Inorganic and organic compounds that contain hydrogen atoms which dissociate from the compound to form hydrogen ions(or ions). Acids contribute to an excess of hydrogen ions in water solutions and produce an acid pH. Acids undergo acid-base neutralization reactions to form salts and water. In cellulose, and hemicelluloses and their derivatives, the most common acid groups are usually carboxyl groups. In lignin, they are phenolic OH groups. In CTMP pulps, sulfonic acid groups also occur.
Additive Systems	A term that applies to the equipment and materials employed in the delivery of chemical additives in the papermaking process.

Adsorption Physical or chemical phenomenon, consisting of the adhesion of molecules or colloids to solid surfaces with which they are in contact. In papermaking, adsorption usually refers to the adhesion of polymeric additives and dyes to fiber surfaces during papermaking. Adsorption is the primary mechanism for retention of many papermaking additives. The bonds that hold molecules to solid surfaces include hydrogen bonding, dispersion forces, and electrostatic attraction.

Air Content Percentage by volume of air in papermaking stock, consisting of large gas bubbles and tiny bubbles either trapped in the fiber lumens or firmly attached to the fiber surfaces. The former form of air produces foam, while the latter is associated with entrained air.

Air Entrainment Trapping and holding of free air in stock or white water, mainly by means of agitation and cascading. Entrained air exists as tiny bubbles that attach to fibers, fines, and other surfaces in the papermaking system. Many adverse effects are associated with entrained air in papermaking.

Alcohol Organic compounds containing one or more hydroxyl (OH) groups. Cellulose and hemicellulose contain many alcoholic hydroxyl groups that undergo the same chemical reactions as common alcohols.

Aldehyde Organic compounds containing CHO groups. They are oxidized to acids and reduced to alcohols. Sugars can be aldehydes when they exist in a straight chain form. Cellulose and hemicellulose chains contain aldehyde groups on the *reducing end* of the molecule.

Aldehyde Starch Starch to which aldehyde groups have been added by oxidation.. Aldehyde starches are used as wet strength agents. They impart temporary wet strength and papers containing them can be readily repulped.

Alkali Compounds containing a reactive OH (hydroxyl) group that is removed when the compound is dissolved in water to form a hydroxide anion. Alkalis react with acids to form salts and water. Alkalis are also called bases.

Alkaline Papermaking Papermaking systems run at pH levels above 7.

Alkalinity Alkalinity is the total capacity of a substance to contribute hydroxide ions. Alkalinity represents the ability of a water solution or suspension to react with acids. It is often expressed in terms of the amount of acid, of a given concentration, that is required to reduce the pH of a sample to a given value, such as 4.0. Calcium carbonate, kraft liquor carryover, and incoming process water contribute alkalinity to papermaking systems.

Alkenyl Succinic Anhydride (ASA) Synthetic sizing agent added to the stock in the form of a cationic starch-stabilized emulsion. Since this agent has limited stability in water, ASA must be emulsified in the mill just prior to use to prevent formation of the undesirable hydrolyzate which contributes to runnability problems. ASA reacts relatively rapidly and develops sizing quickly on the paper machine.

Alkyl Ketene Dimers (AKD) Internal synthetic size derived from fatty acids sizing agents. Applied in the form of cationic starch-stabilized emulsions. AKD emulsions are stable and can be stored prior to use on the paper machine. AKD emulsions are less reactive than ASA sizes and do not develop a high degree of sizing immediately on the paper machine.

Alum Term applied by papermakers to aluminum sulfate, $Al_2(SO_4)_3$, which is widely used in papermaking systems for pH control, as an acidic source of alumina to precipitate rosin size, as a retention aid, and as a mordant for dyes. It is also used in small amounts in alkaline papermaking to control interfering substances. Alum produces an acidic solution when dissolved in water.

Aluminates Sources of aluminum for papermaking systems which do not lower the stock pH. For example, sodium aluminate ($NaAlO_2$) is alkaline rather than acidic. Small quantities of these materials have found application in systems that are sensitive to pH or sulfate ion concentration.

Aluminum Resinate Non-stoichiometric precipitate formed by the reaction of alum (pH 4.0-4.5) with rosin soap size. The positively charged precipitate has great affinity for cellulose fibers.

Amorphous Cellulose Regions of the fiber wall where adjacent cellulose molecules are not regularly packed together to form a highly ordered structure, but exist in a more disordered array.

Amphipathic Molecule A molecule having both hydrophilic and hydrophobic chemical groups. The foremost example in papermaking systems is rosin, which contains a hydrophobic hydrocarbon portion and a hydrophilic carboxyl group.

Amphoteric Starches Amphoteric starches contain both cationic and anionic (derived from carboxyl or phosphate groups) functional groups. They are useful in situations where the use of a cationic starch would lead to overcationization. They also function well in high hardness environments because their anionic groups complex with Ca^{+2} and Mg^{+2} ions.

Amylopectin In natural starches, the form of starch composed of alpha-1,4-linked glucan chains connected together in a "tree-like" structure by alpha-1,6-links. It has very high molecular weights. Amylopectin forms clear solutions and does not retrograde (gel) upon standing. The cationic starches used in papermaking contain a very high percentage of amylopectin.

Amylose The fraction of starch composed of α-1,4 glucan chains. Amylose molecules form solutions in water that tend to gel upon standing. The amylose portion of papermaking cationic starches is very small in most cases.

Anatase One form of titanium dioxide filler.

Anion Negatively charge ion.

Anionic Demand The specific capacity of a system to neutralize anionic polymers. Expressed *as lb polymer per ton of OD solids* in the titrated sample or *as meq charge per volume of sample*.

Anionic Starch Modified starches which carry a negative charge in solution. Starch phosphate is the principal anionic starch used in papermaking. Anionic starches are not used as a wet end additive because they have no natural affinity for fibers.

Anionic Trash Anionic oligomer and polymers that accumulate in the paper machine white water system and are detrimental to the papermaking process. The accumulation depends on the degree to which these materials are adsorbed onto the fibers, fines, and fillers, by their solubility and the degree of closure of the white water system. In practice, the terms "anionic trash," "extraneous substances," "interfering substances," and "cationic demand" are used interchangeably.

Antichlor A chemical, such as sodium sulfite, which is added to react with residual chlorine in a papermaking system and render it ineffective. Residual chlorine can be the result of bleach plant carryover or the carryover of chemical from a wet strength repulping operation. Residual chlorine adversely affects some retention aids and interferes with PAE wet strength resins.

Arabinose Five carbon sugar which is liberated by hydrolysis of xylan and mannan wood hemicelluloses.

Ash Inorganic residue from burning a paper sample. Since wood fibers have a very low natural ash level, the weight of ash can be used to determine the percentage of filler pigment in a sample.

Ash Retention Ratio of the amount of ash in the sheet to the total amount of ash in the papermaking system, expressed as a percent ("overall ash retention"). The papermaker is also interested in "first pass ash retention", the ratio of ash retained in the sheet on the forming section to the amount of ash in the headbox furnish. Ash retention is used as a measure of filler retention.

Association Colloids Colloid systems formed by aggregating small molecules to form aggregates which are much larger than the individual molecules themselves and fall into the colloidal size range. Most surface active substances used as wetting agents, detergents, and emulsifying agents behave like association colloids. Association colloids fall into the hydrophilic class of colloids. The molecules of association colloids are amphipathic. In other words, they have a dual character and exhibit both hydrophobic and hydrophilic traits.

Average Fiber Length Fundamental property or characteristic of papermaking pulps. The weight average fiber length is the traditional measurement of this property, as distinct from the number average fiber length. The average fiber length can be measured or estimated by fiber classification methods, by microscopic visual analysis, by fiber length grid methods, or by automated instrumental techniques.

Base See Alkali

Basic Dye Any dye which is the salt of a colored organic base combined with a colorless acid, such as hydrochloric or sulfuric. Basic dyes as a class are important for coloring paper, combining the advantages of low cost, high tinctorial strength and great brilliance. Syn. Cationic Dye.

Basis Weight The weight, in pounds, of a predetermined number of sheets having a specific size, known as the *basic size*. The number of sheets is known as the *ream size.* The most common ream size is 500 sheets. Common basic sizes include 25 inches by 38 inches, 17 inches by 22 inches, and 24 inches by 36 inches.

Batch Coloring Any method involving the addition of dye or pigment in a batch pulper or beater.

Beating Equipment for mechanically treating pulp fibers, consisting of a heavy roll revolving against a bedplate set into a partitioned tub around which the stock is pumped. When the fibers pass between the rotating beater roll and the stationary bedplate, they are subjected to alternating pressure pulses, brushing, and cutting actions. Taken together, these cause the fibers to take up water, swell, and become more pliable and conformable. As a result, fiber-fiber bonding is promoted and paper properties are developed. The cutting action of beaters improves the distribution of fibers during sheet formation. Beaters are traditional pieces of papermaking equipment, but have now been largely superseded by refiners.

Bentonite A naturally occurring clay mineral containing smectite, a hydrous aluminum silicate. Because of its fine particle size and high sorptivity , smectite is used in deinking processes to adsorb ink particles and in papermaking systems to absorb pitch. Bentonite is also used in a microparticulate retention system in conjunction with high molecular weight polymers.

Biocide Class of chemical utilized in the papermaking system to control slime formation by killing microorganisms. A bacteriocide or germicide is an agent that kills bacteria (but not necessarily bacterial spores), while a fungicide or myocide is a chemical agent that kills fungi or molds. Syn. Slimicide, Microbiocide.

Biostat Class of chemical utilized in the papermaking system to control slime formation by inhibiting the growth of microorganisms, thereby preventing them from producing adverse effects. A bacteriostat is a chemical agent that inhibits bacteriological growth, while a fungistat or mycostat is a chemical agent that inhibits the growth of fungi or molds.

Bonding Strength The magnitude of the forces associated with the bonds that occur at fiber-fiber crossings in a sheet of paper. Bond strength is enhanced by refining and by the use of dry strength additives, such as cationic starch.

Bridging Aggregation See Flocculation.

Bridging Flocculation See Flocculation.

Brightness	Brightness, or papermaker's brightness as it is commonly referred to, is the reflectance by white or near white papers of light having a wavelength of 457 nm (a blue light) under standard conditions of illumination and viewing. It is primarily a measure of freedom from pulp yellowness associated with the presence of lignin and other impurities left by incomplete bleaching.
Britt Dynamic Drainage Jar	A device used for the laboratory evaluation of retention and drainage. Named after the inventor, K. W. Britt.
Buffer	Chemical solution that resists change in pH when acids or alkalis are added. Or the action of any substance to resist a change in pH. For example, papers containing calcium carbonate filler pigments will buffer the web against pH changes -- particularly by acids.
Bursting Strength	A measure of the ability of a specimen to resist rupture when pressure is applied to one of its sides while being held between annular clamps by a specified instrument, under specified conditions . It is largely determined by the tensile strength and extensibility of the paper or paperboard.
Calcined Kaolin	Kaolin clay which has been thermally treated (at 1000-1050 °C) to increase its brightness and light-scattering properties. (The treatment also increases its abrasiveness.) Syn. Calcined Clay.
Calcium Carbonate	White pigment used as a filler in alkaline papermaking furnishes. Although available in both ground and precipitated forms, the trend is toward precipitated calcium carbonate, which is produced on-site at a paper mill in high purity, high brightness and controlled particle size and shape. The preferred particle form is a scalenohedral calcite in which the filler particles exist as rosettes of average size 0.5-3.0 microns.
Caliper of Paper	The perpendicular distance between the two principal surfaces of paper and paperboard under specified conditions. The customary unit for thickness in the US has been inches, with the terms "point" and "mil" being used sometimes to refer to a thousandth of an inch. The millimeter is the standard thickness unit elsewhere. In the SI system of units all thicknesses are in micrometers.
Capillary System	Network of small, connected cavities or capillaries. Most woods have a capillary system containing 50-75% of void spaces, consisting mainly of the lumens of fibers, tracheids and vessels. The capillaries are filled with air and/or water.
Carbohydrate	Large group of compounds synthesized by plants, containing carbon, hydrogen, and oxygen. The hydrogen:oxygen ratio is usually two-to-one, like in water. Cellulose, sugars and starches are all carbohydrates. The chemistry of carbohydrates is controlled by the properties of their hydroxyl groups, their aldehyde groups. In the case of the polymer carbohydrates (polysaccharides), the chemical properties of the acetal linkages between sugar units is also important.
Carbonyl Group	Organic functional group C=O, which is characteristic of aldehydes, ketones and carboxylic acids. When carbohydrates exist in their straight chain form they have a terminal aldehyde carbonyl group.
Carboxyl Group	Organic acid functional group COOH. Carboxyl groups are formed on cellulose by oxidation during pulping and bleaching. In wood fibers, most carboxyl groups occur naturally on the xylan hemicelluloses. They contribute to the bonding between fibers and to the retention of cationic additives. They also represent the ion exchange capacity of a pulp. When ionized, they produce the negative surface charge characteristic of wood pulp fibers.
Cationic Demand	The specific capacity of a system to neutralize cationic polymer titrants. Expressed as lb titrant/ton of OD solids in the titrated sample or as meq charge per volume of sample..

Cationic Direct Dyes — Newest class of dyes which behave as direct dyes, but are cationic rather than anionic in nature. Many of these dyes have excellent affinity for cellulose fibers.

Cationic Starch — Any of a group of chemically modified starches which ionize with a positive charge. The electrochemical affinity of cationic starch with negatively charged cellulose fibers results in excellent wet end retention. Cationic starches are primarily used as dry strength agents, but can also be used as a fines retention aid.

Cellulose — Material that forms the structure of plant cell walls -- which means that cellulose forms the structure of papermaking fibers. It is the most abundant organic compound in nature. It is a straight chain (linear) polysaccharide composed of repeating glucose residues bonded together by β-1,4 linkages. Its DP can be very high -- as great as 15,000. Cellulose adopts a flat, ribbon-like conformation which is ideal for structure functions. It is the principal participant in the interfiber hydrogen bonding between pulp fibers in paper.

Cellulose Fiber — Another term for a wood fiber which is derived from the pulping of wood. Since the fiber structure consists primarily of cellulose, wood fibers are often referred to as cellulose fibers.

Cellulose Viscosity — Property of a pulp or other cellulosic material which is expressed by the viscosity of a dilute solution of the material in a suitable solvent under specified conditions. Viscosity is related to the molecular DP and thus to the strength and other properties of fibers.

Centrifugal Cleaners — Device which removes particles from fiber suspensions by a combination of centrifugal forces and fluid shear. The power source is the stock pump which generates a free vortex in the cleaner at the expense of a pressure drop. The centrifugal action causes dense particles to migrate to the outside of the cleaner, while the lighter materials migrate to the vortex of the cleaner. Syn. Centricleaner, Vortex Cleaner, Liquid Cyclone, Hydrocyclone, Cyclone Cleaner.

Chalk — Natural, relatively pure form of calcium carbonate with a fine, soft structure consisting of fragments of tiny marine organisms.

Charge Decay — The reduction in zeta potential over time of a particle that has been reacted with a cationic or anionic polymer. It is thought that the polymer molecules reconform onto the particle surface and then diffuse into the fiber pores, where they no longer have any influence on the particle surface charge.

Charge Neutralization — (1) Aggregation of colloidal particles through the addition of salts or very low molecular weight, high charge density, cationic polyelectrolytes that cause the electrical double layer to shrink and enable particles to approach one another closely enough that their attractive forces can cause them to adhere to one another. Ions whose charges are opposite to that of the particle surfaces (counterions) are the effective agents. Trivalent counterions are more effective than divalent counterions, which are more effective than monovalent counterions. Charge neutralization is classified as a coagulation aggregation mechanism

(2) Neutralization reaction between a high charge density, low molecular weight cationic polymer and dissolved anionic, polymeric substances (extraneous substances) in papermaking systems.

China Clay — Fine grade of clay used in fillers and coatings.

Clay — Natural, fine-grained pigment with a plate-like morphology, composed primarily of silica and aluminum, widely used as a papermaking filler. A number of grades are available depending on particle size, and shape , as well as brightness. These characteristics are determined primarily by the degree of processing undergone by the material.

Coagulation The destabilization of a colloidal suspension with salts or very low molecular weight, high charge density polyelectrolytes. See also Charge Neutralization and Patching Coagulation

Coarseness Weight per length of wood fibers. Primarily determined by the thickness of the S2 layer in the fiber wall. Coarse fibers produce bulky, rough papers having high tearing resistance and lower tensile strength. Southern regions produce coarser fibers than northern regions due to the longer growing seasons in the south. The opposite of coarseness is "slenderness."

Cobb Size Test Common test for degree of sizing, or water resistance, where the amount of water absorbed by a given area of paper in a specified period of time is measured.

Colloid See Colloidal Particles.

Colloid Titration The titration of an unknown polyelectrolyte with a standard polyelectrolyte solution whose concentration and charge density are known. It is assumed that the two materials react in a one-to-one charge relationship. It is necessary to have a way to detect the charge neutralization endpoint. This can be accomplished with a dye, such as O-toluidine blue, that reacts with one of the standard polymers to cause a color change at the endpoint of the charge neutralization reaction. AC streaming current can also be used to detect the charge neutralization endpoint.

Colloid Titration Ratio (CTR) The ratio of the anionic demand of a sample to the cationic demand of the sample. The parameter has been found to be useful to practitioners who undertake charge surveys on papermaking systems.

Colloidal Particles Particles having at least one dimension in the 10 nm to 1 micron range. Such particles have a very high surface area-to-mass ratio and have many surface molecules that participate in secondary bonding interactions. Fiber fines, filler particles and sizing particles all fall into the colloidal size range.

Colloidal Retention The retention of fine particles in papermaking through colloidal aggregation mechanisms, such as coagulation and flocculation. Colloidal retention is the predominant fines retention mechanism in papermaking.

Colloidal System Intimate mixture of two substances, one of which (called the dispersed phase) is uniformly distributed in a finely divided state and has at least one dimension in the submicron range. The dispersed phase is distributed through the second substance (called the dispersing medium). Several types of colloidal systems are possible: Two examples of importance in papermaking are solid-in-liquid (called a "sol") and liquid-in-liquid (called an "emulsion"). Colloidal systems have properties distinct from true solutions because of the size of their particles and their high specific surface area, which places a large fraction of molecules in the surface of particles. Colloidal particles carry an electric charge on their surface, usually negative, which contributes to the stability of the dispersion.

Color of Paper The color, or shade, of paper is the property that relates to its ability to absorb specific wavelengths of white visible light so that only selected wavelengths are reflected to a viewer's eye. In this manner, the paper appears colored. Dyes and colored pigments are added to the papermaking furnish to accomplish this end.

Colored Pigments Grouping that includes minerals (e.g. iron oxides, ultramarine, umber, sienna, and chrome yellows), lampblack, and both organic and inorganic synthetic pigments. They are used in paper coatings most often, but do not find extensive use in paper coloring.

Conditioning Exposure of pulp or paper samples to a standard environment for a sufficient time for them to come to equilibrium with the environment. In North America, the standard conditioning environment is 50% relative humidity and 21°C.

Conductivity Conductivity, or specific conductance, is defined as the conductance of a material between opposite sides of a cube, 1 cm on a side. The traditional units for conductivity are ohm^{-1}, or mhos. SI units are millSiemens (mS), where 1 mS = 10 μmhos. The conductivities of paper mill slurries and white water streams range from a few hundred μmhos to 10,000 μmhos.

Conical Refiner Equipment used in the mechanical treatment of pulp fibers consisting of a rotating conical plug (rotor) and a stationary conical housing (stator), each fitted with metal bars oriented lengthwise. The fibers flow parallel to the bars. The relative position of the plug determines the clearance of the bars and controls the amount of work done on the fibers. Conical refiners can be differentiated into low-angle types (Jordans) and high-angle types (Claflins).

Consistency Mass or weight percentage of bone dry fiber and other solids in a stock suspension. Consistency is often described qualitatively as low, medium, or high without reference to a standard nomenclature. The following ranges are given as a general guide: very low consistency (0-1%); low consistency (1-8%); medium consistency (8-16%); high consistency (16-40%).

In the stock preparation area, operations utilizing consistencies in the 0.1-1% range are sometimes referred to as "thin stock" operations, while operations having consistencies of 2-5% are called "thick stock" operations.

Contact Angle When a drop of liquid is not strongly attracted to a surface, it forms a droplet having a finite internal angle, θ, between the surface and the tangent to the droplet at the point defining the surface/liquid/air interface. When the droplet does not wet the surface, $\theta > 90$ degrees. When the droplet wets the surface, $\theta < 90$ degrees. If the drop spontaneously spreads on the surface, no contact angle is formed. This behavior is referred to as *wetting*. The contact angle is a function of the interaction forces existing between the liquid and surface.

Continuous Coloring Metered addition of dye solution or pigment dispersion to the papermaking stock in the stock preparation area so that the dyed stock goes directly to the machine.

Contraries Term applied to all contaminant substances introduced into a secondary fiber system with wastepapers.

Cooker Starch make-down device where mechanical, thermal or thermal-chemical energy is used to produce low-viscosity solutions from dry starch starting materials.

Cross Direction of Paper See Directionality of Paper.

Cross Links Chemical bonds between adjacent molecules of polymers. In papermaking, wet strength resins form cross links that enable paper to maintain strength when saturated with water.

Crystalline Cellulose The portion of the fiber structure where cellulose chains are arranged in a regular and repeating manner having the three dimensional structure of a crystal lattice. Crystalline materials are characterized by the fact that they will diffract x-rays and have low accessibility to chemical reagents.

Crystallite Region in the fiber wall where the cellulose is arranged in a highly ordered, crystal lattice of parallel chains; these regions are of limited size and are separated by amorphous regions.

CUENE Abbreviation for cupriethylenediamine hydroxide, the cellulose solvent most often used in the viscosity test for measuring the chain length of cellulose. The test is either called "cuene viscosity" or "CED viscosity."

Cupriethylenediamine Hydroxide Solvent for cellulose obtained by mixing together cupric hydroxide with ethylenediamine. Note that the "ene" is used in the chemical formula as an abbreviation for the diamine complex.

DADMAC See Poly-DADMAC.

Deaeration — Removal of entrained air from papermaking stock, usually accomplished by spraying dilute stock (usually centrifugal cleaner accepts) into a vacuum compartment where the air is boiled off. Chemical deaerating agents can also be utilized for this purpose.

Debonding Agent — Chemical which can be added to kraft pulp to aid in a fluffing operation. Although a debonding agent significantly reduces the power requirement for fluffing, it also makes fibers more hydrophobic, and thus adversely affects their absorption properties.

Deciduous — Any tree shedding its leaves in the autumn, usually a broadleaf (i.e. hardwood) tree.

Deflocculants — Class of chemicals that can be added to the papermaking stock to reduce the flocculating tendency of fibers. Such materials are sometimes used as formation aids.

Deflocculation — Breaking down of fiber flocs in a papermaking stock, either by mechanical or chemical action.

Defoamer — Class of chemical that inhibits foam formation and/or reduces foam stability. Often the terms "defoamer" and "antifoam" are used interchangeably. Where applied differentially, an antifoam is used to prevent foam formation and a defoamer is used to destroy foam that has already formed.

Degradation — With respect to cellulose, refers to the excessive shortening of the long chain molecular structure either by chemical or physical action. In starch cooking, the amylose and amylopectin chains are purposely shortened in order to reduce the viscosity of starch solutions.

Degree of Crystallinity — Measure of the relative amount of crystalline cellulose in wood cellulose. It can be used to characterize the physical and chemical properties of pulp fibers.

Degree of Polymerization (DP) — In general, the number of monomer units in an average polymer molecule in a given sample of material. The weight average number of glucose units in a sample of cellulose is usually inferred by measuring the viscosity of a dilute cupriethylenediamine (cuene) solution of cellulose. Cuene viscosities are related to degree of polymerization through light scattering measurements on polymer solution standards.

Degree of Polymerization (DP) Distribution — Characterization of the distribution of molecular sizes of a sample of cellulose, hemicellulose, starch, or other polymer by fractional separation in increments of degree of polymerization (DP). Typically, the results of this laboratory procedure are shown graphically with increments of DP on the horizontal scale and mass fraction on the vertical axis. Native cellulose is thought to have a rather narrow range of DP values, while wood pulp cellulose and other cellulose samples that have received severe processing have broader ranges.

Degree of Refining — Degree of mechanical action on pulp fibers as measured by changes in pulp properties, most commonly by changes in pulp drainage or freeness.

Degree of Substitution (DS) — (1) Average number of hydroxyl groups that have been replaced per monomer unit of a polysaccharide. For cellulose, hemicellulose, and starch, the maximum DS is 3.0.

(2) In cationic starch, degree of substitution is usually defined in terms of the number of cationic groups per 100 anhydroglucose residues, expressed as a fraction. For example, 3 cationic groups per 100 anhydroglucose residues is a DS of 0.03.

Deinking — Removal of ink and other extraneous materials (contraries), fillers, and fiber fines, from reclaimed printed papers by mechanical disintegration and chemical treatment, with subsequent washing or flotation separation.

Delaminated Clay Grade of filler clay produced by splitting apart the larger particles of kaolin (called "books" or "stacks"), yielding particles called "platelets," which tend to be thinner and wider than normal clay particles. Delaminated clays have higher specific surface areas than undelaminated clays. They are often used in coatings and higher quality filled papers.

Deposit The precipitation or deposition of a solid on a surface by a natural process. Typical pulp and papermill deposits include scale, pitch, slime, and various combinations of materials with fillers. Deposits almost always arise from situations where chemical equilibria are upset or the wet end chemistry of the systems gets out of control.

Detergents Molecules that are used to remove dirt from solid surfaces by surface chemical means. A detergent has good wetting characteristics and facilitates the removal of dirt into the bulk liquid. It has the ability to solubilize or disperse removed dirt and prevent it from being redeposited onto the cleaned surface or from forming a scum. The removal of dirt involves a combination of mechanical and surface chemical mechanisms.

Detrimental Substances All dissolved and colloidal materials in paper stock that interfere with functional additives and chemical processing aids. Syn. Anionic Trash, Cationic Demand, Interfering Substances.

Dewatering Agent Polymer flocculant which, when added to stock, accelerates drainage. See Drainage Aid.

Diffusion Net movement of molecules or ions of a given substance from a region of higher concentration to a region of lower concentration.

Dimer Compound formed by the union of two molecules. For example, alkyl ketene dimer is a sizing agent.

Dipolar Bond Molecule in which the electrons forming the valency bond are not symmetrically arranged, thus giving a net positive charge on one end and a net negative charge on the other end. The foremost example of a polar molecule is water. Water interacts with cellulose and itself through the polar attraction of water molecules for each other and for the hydroxyl groups on cellulose, which are also polar.

Direct Dyes Direct dyes are the sodium salts of dye acids that are chemically similar to acid dyes but, unlike acid dyes, have a high affinity for cellulose fibers. The affinity of direct dyes for cellulose fibers can be attributed, in part, to the low solubility of the materials. In fact, some direct dyes are colloidal in nature and form gels upon standing.

Direct dyes can also be made cationic by incorporation of quaternary amine groups in their chemical structure. These substances have even greater affinity for fibers than anionic direct dyes. Because of these attributes, direct dyes are the most common dyes used in the paper industry. Syn. Substantive Dye

Directionality of Paper Directionality refers to the anisotropy of many paper properties, particularly the mechanical properties, with respect to the two principal directions in paper -- called the *machine direction* and *cross direction*. The machine direction of the web is the direction in which paper travels when it is manufactured. The cross direction lies in the principal plane of the paper and is perpendicular to the machine direction. The third direction in paper, the thin dimension, is sometimes referred to as the *z-direction*.

Disc Refiner Equipment utilizing two opposing discs for mechanical treatment of papermaking fibers. The discs have bars and grooves that treat the fibers with alternating pressure pulses that causes fiber fibrillation and swelling. The following configurations are used: Rotating disc opposite stationary disc; two opposing rotating discs; rotating double-sided disc between two stationary discs.

Dispersant Class of surfactant chemical utilized in papermaking systems to reduce deposits of pitch and slime, in deinking systems to disperse ink particles, and in coating formulations to keep pigment particles in suspension.

Dispersed Rosin Size High (90-95%) free rosin dispersions of resin acids. The dispersed size particles are stabilized by surfactants, starch, polymers, or proteins. Dispersed sizes can be anionic or cationic. They are widely used in papermaking today.

Dissolution Process of molecularly dispersing a material into a solvent, i.e., passing into solution. Syn. Dissolving.

Dissolved Charge Dissolved charge refers to electrostatic charges attached to dissolved polyelectrolytes in papermaking systems. Dissolved charge is measured by the colloid titration technique. It provides a method for determining the concentration of charged polymer species in the system. Anionic and cationic charges can be determined independently, depending upon the standard titrating polymers selected.

DLVO Theory A theory that combines the attractive and repulsive forces that exist between particles and describes the potential energy that exists between them.

Drainability Relative resistance of a forming fiber mat to the drainage or passage of water. Measurements of this property are referred to as freeness, slowness, wetness, or drain time, according to the instrument or method used.

Drainage Flow of water from a fiber suspension during the process of forming a web or sheet of paper on the paper machine.

Drainage Aid Chemical substance that is added to the papermaking furnish to increase water removal on the forming section. Both inorganic salts and polyelectrolytes can serve as drainage aids. The mechanism involved is thought to involve fines aggregation onto fibers so that the forming mat has an increased average pore size and thereby lower filtration resistance.

Dry Size A 100% neutralized, dry rosin size product that dissolves easily in water.

Dry Strength Additives Chemical additives that increase the dry strength of paper at a given level of refining or enable the papermaker to maintain a given level of dry strength with decreased refining. Cationic starch is the principal dry strength agent employed by the paper industry. Polyacrylamides and guar gum are also used for this purpose. Syn Dry Strength Adhesive. See also Dry Strength Resin.

Dry Strength Resins Type of dry strength additive; primarily water soluble acrylamide based polymers and copolymers which are effective at low concentration in developing fiber-to-fiber bonds. See also Dry Strength Additives.

Dual Polymer Systems Combination of a low molecular weight, high charge density cationic, patching polyelectrolyte and a high molecular weight, low charge density anionic polymer in a retention system. The cationic polymer is added first and forms cationic patches on particles. The anionic polymer molecules are then added, are attracted to the positive patches, and form flocculating bridges between particles.

Dye Colored soluble substance which is used to impart a more or less permanent color to another material.

Dynamic Drainage Jar Apparatus used for testing the flocculation or coagulation ability of a paper stock slurry under conditions of shear and turbulence induced by rapid stirring. No pad is formed during the use of the apparatus. It is often used to test retention aids and other retention-related materials. Syn. Britt Jar.

Electrical Double Layer Electrically charged surface of a colloidal particle together with a diffuse surrounding layer of counterions. This condition exists when the particle is suspended in water or some other solvent in which it is insoluble.

Electrokinetic Charge See Zeta Potential

Electrolyte Compound which conducts electrical current when dissolved in water or other suitable solvent.

Electrophoresis Migration of charged colloidal particles in a suspension under the influence of an external potential gradient.

Electrophoretic Mobility Relative ability of colloidal charged particles to migrate in a solution under the influence of an external potential gradient. Expressed as the ratio of the particle speed to the electric field strength. Electrophoretic mobility values are related to particle surface charge and, as such, provide an indication of the state of the particle surface with respect to adsorbing materials.

Electrostatic Attractions Actions or effects that charged particles exert on one another due to their forces of attraction and repulsion.

Emulsifier (1) Dispersing agent or emulsifying agent. (2) Mixing system for preparing a stable dispersion from two or more nonmiscible liquids, usually with addition of a chemical additive such as a detergent or other surfactant.

Emulsion Colloidal dispersion of a liquid in another liquid with which it is not normally miscible. The dispersion is stabilized by the addition of a surfactant (e.g. a soap, detergent) called an emulsifying agent, which forms an adsorbed film around the dispersed droplets and prevents their coagulation and coalescence. The most important physical property of an emulsion is its stability. When an emulsion breaks, it forms two physically distinct phases.

Emulsion Polymers High molecular weight polymers that are supplied as water-in-oil emulsions containing approximately 30% solids. They must be diluted under high shear conditions in order to avoid precipitation of the polymer.

End Groups Functional groups at either end of a cellulose molecule which are distinguished from those in the body of the chain, one by the presence of a reducing hemiacetal group, the other by the presence of an extra hydroxyl group.

End Point Final stage in a titration as indicated by a color change or potentiometric inflection point. Employed frequently in determining the completion of a colloid titration procedure.

Entrained Air Small bubbles of air, captured by cascading white water, that ultimately become attached to pulp fibers and filler particles in the papermaking system. A high level of entrained air can cause instability in the approach flow system, reduced drainage on the forming fabric, and blemishes in the sheet due to holes or spots.

Extended Size A 50% solids size product consisting of a 50-50 mixture of completely neutralized resin acids and urea.

Extender Inexpensive mineral filler which is normally used in conjunction with a more expensive filler, such as titanium dioxide, to space the particles and thus increase their overall light scattering efficiency.

Extractive Chemical wood components that can be dissolved out by a solvent that does not react chemically (neutral solvent) with them. Examples are starches, pectins, terpenes, resin acids, fatty acids, fatty acid esters, alcohols. Extractives play important roles of food storage and protection from microbial attack in trees. Resin acids are used as internal sizing agents in papermaking. Volatile terpenes also are used as starting materials in the production of turpentine and other solvents. Nonvolatile extractives present in small quantities in papermaking systems can be the source of pitch problems. Extractives are non-fiber wall constituents of wood.

Fiber Long, cylindrical cell, tapered and closed at both ends; includes fiber tracheids with bordered pits and libriform fibers with simple pits.

Fiber Analysis Determination and counting of fibers to determine the approximate percentages of species or type in a given sample of pulp or paper. Fiber analysis is often done manually by visually observing fibers suspended on a slide under a microscope. Hardwood:softwood fiber ratios can also be done instrumentally.

Fiber Classification Separation of pulp into fractions of different fiber length, by utilizing a specific apparatus, such as a multiple-screen classifier or an automated fiber analysis instrument.

Fiber Coarseness Weight of fiber wall material per unit length of fiber, usually measured in mg per 100 m. Fiber coarseness is s fundamental property of papermaking pulps and is used as a relative indicator of the fineness or slenderness of the fibers and, therefore, their flexibility. Coarse fibers produce rougher, bulkier papers having higher tearing resistance and lower tensile strength, while slender fibers produce smoother, dense papers having lower tearing resistance and higher tensile strength.

Fiber Floc Entanglement of fibers in a papermaking stock.

Fiber Flocculation Aggregation of fibers in suspension. Usually, papermakers wish to avoid fiber flocculation because it leads to poor paper formation. Microturbulence and shear forces are employed in the headbox and on the forming section to disrupt fiber flocs and improve the distribution of these materials throughout the web.

Fiber Saturation Point Stage in the drying or wetting of wood when the cell cavities are free of water, but the cell walls are saturated with moisture.

Fiber Ultrastructure Lamellar structure of wood fiber cell walls. Consists of the primary (P), secondary one (S1), secondary two (S2), and secondary three (S3) layers. The center of a fiber consists of a hole designated as the lumen.

Fibers (Hardwood) Wood fibers derived from deciduous trees. Fiber lengths of hardwood fibers lie between 1-2 mm. They produce the smoothest, best formed papers.

Fibers (Softwood) Wood fibers derived from coniferous trees. Fiber lengths of softwood fibers lie between 3-5 mm. They produce the strongest papers.

Fibril Threadlike component of cell walls visible under an optical microscope. Fibrils are composed of more elementary morphological units called microfibrils, which are only visible under an electron microscope. Without qualification, the term "fibril" is taken as equivalent to "macrofibril," a opposed to the ultrastructural "microfibril."

Fibrillation Refining action which disrupts lateral bonds between adjacent fibrillar layers of pulp fibers. "External fibrillation" refers to the loosening of fibrils and raising or particle detachment of fibrils on the surface of fibers. "Internal fibrillation" refers to the less visible disruption of bonds within the fiber.

Filler Inert, finely-divided mineral material added to a papermaking furnish to modify sheet properties by filling in the void spaces between fibers. Fillers enhance optical properties, such as brightness, opacity, and gloss; improve formation and smoothness; and sometimes effect cost savings.

Filtration Retention Retention of fine particles in the forming paper web by a sieving effect. One of the two main mechanisms of retention in papermaking -- the other being colloidal retention. Syn. Mechanical Entrapment.

Filtration Water Removal Filtration occurs when the fibers and fines are able to move freely in a suspension and a sharp demarcation exists between the suspended particles and the forming mat. The consistency in the suspension remains essentially constant throughout the filtration process.

Fines Very short fibers, fiber fragments, ray cells or debris from mechanical treatment. Typically, all material that passes through a 200 mesh screen. In a headbox furnish, filler particles get classified as fines. Fines play key roles in many wet end chemistry phenomena. Fiber fines fall into three categories: (1)Primary fines: Naturally occurring fines in a wood pulp sample. (2)Secondary fines: Fines generated by refining. (3)Tertiary fines: Fines generated during the flow of stock around the paper machine white water system as a result of interaction with pump impellers and tank agitators. Fines have high specific surface areas and retard drainage on the forming section. They also adsorb large amounts of chemical additives.

First-Pass Retention The ratio of the amount of material that leaves the headbox slice compared to the amount of material that is contained in the web leaving the couch roll, expressed as a percent. The first-pass retention of the entire furnish or individual furnish components (e.g. fines, filler, size) can be referred to. Typical values are 20%-90%. Syn. Single-Pass Retention.

Flocculation 1. Process of forming fiber flocs. 2. Process of aggregating colloidal particles though the action of high molecular weight, bridging polymers. See also Heterflocculation.

Fluorescent Dye Fluorescent dyes absorb light in the ultraviolet region of the spectrum and reemit it in the visible blue range (peaking at 435 nm), giving a fluorescent effect that produces a bright white in daylight, masking the inherent yellowness in bleached pulps. Syn. Optical Brightener, Fluorescent Whitener.

Fluorocarbon Size Any of a family of water soluble, synthetic fluorocarbon compounds used as internal or surface sizing agents. When used as an internal sizing agent, retention is dependent on the use of a strongly cationic retention aid which serves as a bonding agent between the anionic fiber and the anionic fluorocarbon. Used as wet-end additives in the manufacture of oil-proofing papers.

Foam Gas dispersed in a liquid in a ratio such that the bulk density of the mixture approaches that of a gas rather than a liquid. When foam collects at the air/liquid interface, it is called "surface foam." When foam is mixed into liquid and is only slightly visible on the surface, it is called "entrained air."

Foam Inhibition Ability of a defoamer to sustain foamless conditions for a sustained period.

Foam Stability Relative persistence of a foam as measured by time dependent changes in such foam properties as bulk density, rheology, and structure.

Foaming Agent Any substance which promotes foam formation in papermaking stocks. Surfactants, fines, starch, and other materials all can serve to promote foam formation though a variety of mechanisms. Syn. Frothing Agent, Frother.

Folding Endurance The number of folds under specified conditions in a standard instrument which a paper will withstand before failure. The property is important to papers that are subjected to repeated folding in use. The most common instruments employed in this test are the Schopper and the MIT fold testers.

Formation The uniformity with which fibers and other solid components are distributed in paper determines its *formation*. In practice, the term formation refers to the appearance of a sheet when viewed by transmitted light. A mottled, or streaked, appearance is said to be a *poor*, or *wild*, formation. When the appearance approaches the uniform texture of ground glass, it is said to be a *closed*, or *fine* formation.

Formation Aids High molecular weight anionic or nonionic polymers that enhance sheet formation when added to the furnish prior to the headbox. Both natural (root extracts, guar gum, locust bean gum) and synthetic polymers (anionic polyacrylamides) have been employed in this application -- mostly in long fibered stocks.

Fortified Rosin Chemically modified form of rosin (e.g. with maleic anhydride) used as internal sizing, either alone or in combination with unmodified rosin. Less fortified rosin is required for a given sizing effect as compared to normal rosin size. Practically all rosin sizing agents are prepared from fortified rosin.

Free Drainage Water removal over the early portion of the forming section where foils are the primary water removal elements and the web consistency goes from approximately 0.5% to 2-4%.

Free Rosin Unsaponified rosin used in colloidal suspension for internal sizing. Under some conditions, high free-rosin size is more efficient than other types of size.

Free Stock — Stock that freely releases water in the free drainage zone of a paper forming section. Produces a high Canadian Standard Freeness test.

Freeness — Measurement of pulp drainability, usually by means of the Canadian Standard Freeness Test (CSF). The CSF is defined as the number of ml of water collected from the side orifice of the standard tester when a dilute stock drains through a perforated plate under carefully controlled conditions. The higher the number, the freer the pulp. Freeness is strongly affected by the fines content of a stock.

Fugitive Sizing — The disappearance or significant reduction of water resistance in a paper that exhibits sizing immediately after it is manufactured.

Furnish — Water suspension of fibrous materials and additives being prepared for conversion into paper or paperboard.

Galactan — Polymer from which the sugar, galactose, is derived from hydrolysis.

Galactomannan — Polymer from which the sugars, galactose and mannose, are derived from hydrolysis.

Galactose — Hexose sugar which is the C-4 epimer of glucose. Found in softwood hemicellulose galactoglucomannans.

Gel — Structure formed in aqueous media by hydrophilic colloidal polymeric substances such as hemicellulose, pectin and starch. Gels absorb many times their own weight of water with consequent swelling and can set to a viscous or semisolid jelly (gel) when cool.

Gelatinization — Absorption of water with unlimited, irreversible swelling as occurs when starch granules are cooked in an aqueous suspension to form a viscous solution for papermaking applications. Syn. Solvation.

Gloss — Gloss is the geometrically selective reflectance by a surface responsible for its shiny or lustrous appearance. Gloss depends on the kind of illumination, the angles of its incidence and reflection, and the relative position of the paper and the observer. Different gloss tests have been devised for different kinds of paper and uses.

Glucan — Polysaccharide from which the sugar, glucose, is derived by hydrolysis. Cellulose is a glucan.

Glucomannan — Hardwood and softwood hemicellulose composed of glucose and mannose units.

Glucose — Hexose which is polymerized in wood and other vascular plants to form cellulose.

Glucuronic Acid — Glucose molecule that contains a carboxylic acid group (COOH) in place of the primary hydroxyl group on C6. Generally, carboxyl groups in the sugar units are referred to as "uronic acid". Glucuronic acids are present in xylan hemicelluloses and are responsible for the surface charge on fibers. They also are the sites for ion exchange reactions and adsorption of cationic additives.

Glycosidic Bond — Specifically, the C-O-C bond between sugar residues in a polysaccharide molecule. Even though it appears to be similar to an ether linkage, it is not an ether. It is an acetal linkage that is susceptible to hydrolysis by the action of acids.

Grammage — The weight, in grams, of an area of paper equal to one square meter.

Granule — Small discrete mass of a solid substance. For example, starch is naturally obtained in granules which range in size from 2 to 100 μm.

Ground Calcium Carbonate — Calcium carbonate fillers produced by grinding natural limestone. These materials are used both as paper fillers and as coating pigments.

Guar Gum Natural polysaccharide utilized as a dry strength agent in papermaking to a small extent. Chemically, guar is a galactoglucomannan which consists of a main chain of beta-D-glucose units linked through carbons one and four. Single sugar alpha-D-galactose side chains linked through carbon six occur frequently, also. In order to insure high retention, quaternary amine functional groups are substituted in the structure to impart a cationic charge to the molecule.

Gymnosperms Woody plant that produces naked seeds, i.e. not enclosed in an ovary. The conifers are the only common trees of this group. See Softwood.

Gypsum Naturally occurring hydrate of calcium sulfate used as a paper filler to a small extent is some specialty grades.

Handsheet Sheet made from a stock suspension on a laboratory sheet mold, followed by pressing and drying, all operations carried out under carefully controlled conditions. In wet end chemistry, handsheets are used to evaluate functional additives. Handsheets are not useful for studying retention, or other machine-related phenomena.

Hard Flocs Aggregates formed by bridging mechanisms. They are strongly resistant to high shear conditions. However, prolonged exposure to such conditions leads to the eventual breakdown of these flocs. When the shear is removed, the flocs reform. Subsequent re-exposure to high shear conditions reveals a diminished floc shear resistance. Multiple re-exposures eventually causes the aggregation behavior to become *reversible* in character. This means that they have very little shear resistance, yet reform as soon as the shear is removed.

Hard Sized Papers Very highly sized papers that have high water resistance. E.g. cup stock, milk carton raw stock.

Hardness Traditionally defined as the soap consuming capacity of a water sample. Evaluated by titrating a water sample with a standard liquid soap to a foam producing endpoint. More recently, hardness is defined as the concentration of calcium and magnesium ions in a water sample, expressed as ppm calcium carbonate. Hardness is often determined by titration with disodium ethylenediamine tetraacetate (EDTA) plus Calgamite indicator. Hardness is an indication of the scaling potential of process streams and also interferes with alum-rosin sizing and wet strength development at very high levels.

Hardwood Generally, wood produced by a broadleaf tree and containing pores. The term has no relation to the actual hardness of the wood. See also Angiosperm

Heartwood Dead central portion of the trunk of a tree or a large branch; often, but not always, darker colored than the outer sapwood.

Hemiacetal Groups Functional groups derived from carbonyl groups by addition of one molecule of an alcohol. The ring form of sugar molecules are hemiacetals.

Hemicellulose Polysaccharides associated with cellulose in wood fiber cell walls. They contain a variety of sugars, including galactose, xylose, mannose, arabinose, as well as glucose. The linkages between sugars in hemicellulose main chains are through carbons one and four and have a beta configuration. Most hemicelluloses have single sugar unit side chains. Hemicelluloses have lower molecular weights than cellulose, typically containing 50-150 sugar units per molecule. The four major types of hemicelluloses in papermaking woods are the glucuronoxylans, the arabinoglucuronoxylans, the glucomannans, and the galactoglucomannans.

Hemicellulose interact strongly with water and promote fiber swelling during refining. Surface hemicelluloses participate in interfiber bonding by contributing hydrogen bonds. Carboxyl groups on xylan molecules ionize and impart negative (anionic) electrostatic surface charge to wood fibers.

Hercules Size Test (HST) Common test for degree of sizing, or water resistance, where the time required for a decrease in light reflection of the opposite side of a paper specimen which has been covered with a given amount of ink is measured.

Heterocoagulation Aggregation of unlike particles by a charge neutralization or patching mechanism. Can refer to fiber-fines aggregation or the aggregation of sizing and/or filler particles with fibers and fines. The forces involved originate on the surfaces, and consequently the surface potential (zeta potential) is a decisive factor.

Heteroflocculation Bridging mechanism of filler and fines retention where a long-chain polymer acting as a retention aid provides a connection between particles and fibers by simultaneously adsorbing onto both components. The polymer and particle surface usually have opposite charges.

High Density Stock A designation that typically refers to stock in the 8-15% consistency range.

Holocellulose Sum of cellulose and hemicellulose in fibers.

Hydrate Substance that has had water added to its structure Refining is said to "hydrate" fibers. This refers to the fact that wood fibers absorb much water and swell when refined. Some of the water is held in the fiber pores and lumens and some of it is bound strongly to the fiber cellulose molecules by hydrogen bonds.

Hydration Incorporation of water into a complex molecule, as occurs in the gelatinization of starch and in the strong association of water with cellulose at a fiber surface.

Hydrogen Bond Attraction occurring between a covalently linked hydrogen atom and neighboring atom or group of atoms. The attraction between bound hydrogen atoms and adjacent bonded atoms is especially strong with respect to nitrogen, oxygen, and fluorine atoms, increasing in that order. Hydrogen bonds are responsible for holding cellulose molecules together in the fiber structure and for bonding fibers together in a sheet of paper. They are about one-tenth as strong as covalent bonds and, therefore, a large number of hydrogen bonds must exist for strong interactions to occur (cooperative bonding).

Hydrogen bonds are formed by water and alcohols. The positively polarized --OH hydrogen atoms from one molecule forms a weak bond to a negatively polarized oxygen atom on another molecule. This leads to the extensive association of molecules found in liquid water and alcohols. Although hydrogen bonds have a strength of only about 20 kJ/mol versus 430 kJ/mol for a typical covalent bond, the presence of many hydrogen bonds means that extra energy must be added to break them during the boiling process. Hence, alcohols and water have higher boiling points than would be expected based on molecular weight alone.

Hydrogen bonds are extremely important in papermaking. They are the primary bonds that exist between fibers at fiber crossing points in the web and, as such, are the chief source of paper bonding strength. The association of water and papermaking fibers involves hydrogen bonding. Many polymers adsorb onto fiber surfaces through hydrogen bonding interactions.

See also Polar Bonds.

Hydrolysis (1) Chemical decomposition or conversion of a complex molecule (e.g. a polymer) into two or more smaller molecules by reaction with water. A good example is the conversion of cellulose into glucose; one molecule of water is absorbed for each molecule of glucose formed. (2) Chemical reaction of dissolved substance with water to produce either weak acid or weak base.

Hydrolyzed Applied to chemical compounds that have incorporated water into their chemical structure, forming either weak acids or weak bases.

Hydrophilic	Having a strong affinity for binding and absorbing water, which results in swelling and formation of gels. All carbohydrates are hydrophilic, especially starches and gums.
Hydrophilic Colloid Systems	Colloidal systems composed of solutions of macromolecules whose dimensions lie in the colloidal size range or aggregates of small molecules (micelles), whose dimensions are colloidal in size.
Hydrophobic	Having no affinity for water, resistant to wetting and penetration by water. This property is characteristic of fats, oils and waxes.
Hydrophobic Colloid Systems	Colloid systems consisting of discrete particles suspended in water. The particles do not interact strongly with water. The particle surface charges determine the stability of the dispersion, which is thermodynamically unstable and tends to separate into two phases. Suspensions of fines, fillers, and sizing agents all have the properties of hydrophobic colloid systems.
Hydroxyl Group	Functional group, -OH, in organic chemistry. Alcohols are characterized by a hydroxyl group attached to an aliphatic carbon chain, while phenols are compounds that contain a hydroxyl group on a benzene ring. All carbohydrates contain multiple hydroxyl groups. The hydroxyl group is responsible for many of the unique properties of cellulosic fibers.
Hygroscopic	Having the ability to absorb water vapor from the surrounding atmosphere to an extent proportional to the relative humidity.
Inert	Incapable of participating in a chemical reaction.
Interfering Substances	See Anionic Trash.
Internal Bonding Strength.	The force with which fibers are held together within a sheet of paper or paperboard. The Scott Bond Test and z-direction tensile tests are often used to measure internal bond strength
Internal Sizing	Achieving resistance to penetration by fluids in a paper sheet through the addition of sizing agents to the furnish prior to sheet formation
Ion	Electrically charged atom or group of atoms. Ions in solution are due to ionization of a dissolved substance, and may be either positively or negatively charged.
Ion Exchange Reactions	Reactions between anionic surface charges on fibers and complexing cations. The strength of attraction of wood fibers for various cations occurs in the order monovalent < divalent < trivalent. Thus, Al^{+3} binds more strongly to fibers than Ca^{+2}, which binds more strongly than Na^{+1}. Studies have shown that the ion exchange capacity of fibers is related to their hemicellulose content.
Ionic Compound	Compound that ionizes or dissociates into ions in water solution.
Ionization	(1) Many molecules dissociate into two or more ions when dissolved in water. This process, called ionization, may be complete and essentially irreversible, or incomplete and reversible. When the latter is true, a dynamic equilibrium is established in solution whereby ions continuously reassociate to form molecules and molecules dissociate into ions. The ratio of the product of the ion concentrations to the concentration of the original molecule remains constant and is expressed by the Ionization Equilibrium Constant. (2) Any process by which a neutral atom or molecule gains or loses electrons, thereby acquiring a net charge. Ionization occurs by dissociation in water solution (e.g. NaCl => $Na^+ + Cl^-$.)
Irreversible Reactions	Chemical reactions that can proceed in one direction only.
Isoelectric Point	pH level of a colloid system at which the particles do not move in an electric field; i.e., the pH at which the net charge on the particles is zero.

Jet Cooking — Continuous starch cooking system in which a starch slurry is mixed with high pressure steam in a chamber where intense shear is developed. Starches processed in this way are referred to as "thermomechanical starches". This is a common method for preparing cationic starches.

Kajaani Fiber Length Analyzer — Automated device for scanning a pulp suspension which utilizes the ability of cellulose to depolarize a polarized light beam in order to measure the number of fibers in each of a large number of separate length categories.

Kaolin — Any of several clay minerals, derived from Kaolinite, with a two layer crystal in which silica and alumina sheets alternate. Kaolin clays are one of the most common filler pigments. Clays are also used as coating pigments. See Clay, Delaminated Clay.

Karaya Gum — Polysaccharide used as a deflocculating agent for long fibered stock dispersions.

Latewood — Relatively more dense and smaller celled wood produced in the annual ring during the later part of the growing season.

Latex — Colloidal suspensions or emulsion of natural or synthetic rubber or plastic in water.

Lignin — Natural binding substance of the cells of wood and plant stalks; a complex network polymer based on the phenylpropane units are linked through ether linkages and carbon-carbon single bonds. Hydroxyl and methoxyl groups are attached to the benzene carbon atoms. Lignin is synthesized by a random process that produces a three-dimensional network structure having no discernible structural regularity. It is intermingled intimately with cellulose and hemicellulose in the fiber wall and is the primary chemical component of the middle lamella -- where it bonds fibers together. Hemicellulose molecules are also thought to be covalently bonded to the lignin structure.

Lignin as it occurs in the plant is called "native lignin" or "protolignin" to distinguish it from the modified forms isolated by chemical means, e.g. Klason lignin or kraft lignin. Dissolved lignin is a major component of the interfering substances in the white water systems of many paper mills, particularly those employing mechanical pulp fibers.

Residual lignin in papermaking fibers adversely affects their abilty to participate in interfiber bonds; provides an inhibiting influence on the swelling of fibers in water; and contributes to the formation of a bulky, rapid-draining mat on the paper machine. The free phenolic groups in the structure contribute to the negative surface charge on fibers when they ionize at $pH > 10$.

Lignosulfonate Groups — SO_3^{-1} groups found on lignin in sulfite pulp fibers and chemithermomechanical pulp fibers. The sulfonate group is a strong acid under all papermaking pH conditions. Consequently, the group contributes a negative charge to the fiber surfaces.

Liquid Dye — Dye in concentrated liquid form.

Lumen — Central cavity of a wood fiber.

Lyophilic Colloid Systems — True solutions of very large molecules or aggregates of small molecules. There is no true interface between the medium and dissolved molecules or aggregates.

Lyophobic Colloid Systems — Particle suspensions that are unstable with respect to aggregation. The particle-suspending medium interface strongly affects system behavior.

Machine Direction of Paper — See Directionality of Paper.

Macromolecule — Synonym for polymer molecule.

Matrix — Intercellular substance. The wood matrix is made up of lignin, hemicellulose, and extractives.

Mechanical Properties of Paper Mechanical paper properties, or strength properties, determine the durability and resistance to applied forces exhibited by paper in manufacturing, converting, and end-use operations. Included in this category are stress-strain behavior, stress relaxation behavior, viscoelastic behavior, tensile breaking strength, elongation (stretch), tensile energy absorption, bursting strength, folding endurance, bending stiffness, internal tearing resistance, edge tearing resistance, surface strength, and internal bonding strength.

Melamine-formaldehyde Wet Strength Resin (M-F Resins) Wet strength resin produced by reacting melamine with formaldehyde to produce methylolamines which, when treated with hydrochloric acid, produces a highly cationic, colloidal suspension which is very substantive to papermaking fibers. Melamine-formaldehyde resins require high temperatures and acid pH values (ca. 4.5) for curing.

Micelles Aggregates of small molecules that are large enough to fall into the colloidal size range. The concentration at which micelle formation occurs spontaneously is called the "critical micelle concentration." Detergent molecules are an example.

Microbiological Analysis A measure of the number and types of microorganisms in a given amount of sample.

Microbiological Survey Systematic sampling and biological analysis of selected flows within the papermaking system so that a complete microbiological picture is obtained.

Microelectrophoresis A method of measuring zeta potential where a particle dispersion is placed in a specially designated cell. An electric field is placed across the cell and the migration velocity of the dispersed, charged particles is measured. The quantity directly derived from the measurement is the electrophoretic mobility, which is the migration velocity divided by the electric field strength. Zeta potential can be computed by incorporating the electrophoretic mobility value into the Helmholtz-Smoulochowski Equation. Microelectrophoresis is the most common way to measure zeta potential in the paper industry. Typical values of electrophoretic mobility range from -3 to +3, while zeta potential values range from -40 mV to +40 mV.

Microfibril Threadlike component of the cell wall structure composed of cellulose polymer chains and associated polysaccharides united through crystalline regions (crystallites) and amorphous regions. Microfibrils cannot be seen with a light microscope and must be observed with an electron microscope. See also Elementary Fibril.

Microorganism Microscopic or submicroscopic organism. In papermaking systems, the microorganisms that cause problems are certain bacteria and molds.

Microparticle Retention Systems Retention systems that combine polymers and/or starch with silica or bentonite particles. Such systems enhance drainage and formation as well as retention.

Middle Lamella General term for the lignin-rich cementing layer between cell walls in plants, usually referring to the intercellular layer, but sometimes including the primary walls of adjacent cells and even the outer layers of the secondary wall. The term, "true middle lamella," is sometimes applied to the compound layer or "intercellular matrix" between the secondary walls of contiguous cells. Consists primarily of lignin.

Middle Lamella Region between adjacent fibers in wood..

Mineral Acid Common inorganic acids, such as hydrochloric acid, sulfuric acid, or nitric acid.

Moisture Content Percentage of water in a material, determined by completely drying a representative sample to constant weight at 105°C. This procedure yields an oven-dry (OD) moisture content. Because of its hygroscopic nature, cellulosic materials may not attain a true moisture-free condition unless care is taken to dry the air which is introduced into the oven.

Mordant Substance that has the ability to fix or bind dye molecules to cellulosic fibers.

Morphological Properties of Fibers Pulp properties or characteristics that are dependent on the form and structure of the fibers. Basic morphological properties include average fiber length, cell wall thickness, cross sectional area, fibril angle, fiber coarseness, and specific surface.

Network Flocculation Flocculation whereby two or more polymers react together in a first step to produce a polymer network. The network then entraps particles mechanically or interacts specifically with sites on particle surfaces. Electrostatic interactions between polymer and particles play less important roles in network flocculation than in bridging flocculation.

Neutral Size Saponified rosin size which contains no free rosin.

Nonfibrous Additive Any component of the papermaking furnish that is nonfibrous in structure.

Nonionic Compound Compound that does not ionize in water nor take on an electrostatic charge when dissolved.

Opacity The property of a sheet that obstructs the passage of light and prevents seeing through the sheet objects on the opposite side. The property is especially important for printing papers.

Overall Retention Percentage of stock (or of any particular papermaking component) flowing into the paper machine system that is retained on the forming fabric.

Papermaking Additives Materials added to paper or to the papermaking furnish to modify or improve certain paper properties (functional additives) or to facilitate the papermaking process (processing aids). This classification encompasses all materials that enter the system except fiber and water. Common additives are alum, filler, sizing agents, dyes, wet strength resins, biocides, defoamers, retention aids, and dry strength agents. Syn. Nonfibrous Additives, Chemical Additives.

Paste Size Form of rosin size, usually containing approximately 5% free rosin, at a solids content of 70-80%. Mill preparation requires two stages of dilution in a suitable emulsification unit; hot water dilution down to 10-20% solids followed by cold-water dilution to a working concentration of 2-6%.

Patching Coagulation A charge neutralization phenomenon where high charge density, low-to-medium molecular weight polymers adsorb onto anionic fiber fines or filler particles in a flat conformation to form positive "patches" on the particle surface. When two particles approach one another, the positive patches are attracted to negative sites on the particles. The resulting attractive interaction leads to the formation of closely packed aggregates that promote good formation and drainage. Patch forming polyelectrolytes produce stronger aggregates than salts.

PEI (Polyethyleneimine) A polymer synthesized by a ring-opening, acid-catalyzed addition reaction of ethyleneimine that results in a branched structure having primary, secondary, and tertiary amine groups whose charge density is a function of pH. It has a very high charge density under acid pH conditions. PEI is used as a retention aid and as a charge neutralizing polymer.

PEO (Polyethylene Oxide) A polymer formed by polymerization of ethylene oxide. It has a low anionic charge and has been found to be effective as a fines control agent in groundwood containing papermaking systems.

Peptization Stabilization (i.e. prevention of aggregation) of hydrophobic colloidal particles within a water system by addition of a suitable electrolyte, called a "peptizing agent," which is adsorbed onto particle surfaces.

Peptizing Agent Substance which promotes colloidal dispersion.

Permanence The permanence of paper refers to the retention of significant use properties, particularly folding endurance and color, over prolonged periods. Permanence is affected by temperature, humidity, light, and the presence of chemical agents.

pH

Liquid water molecules dissociate to a small extent and the product of the concentrations of hydrogen ion and hydroxide ion are a constant, 1×10^{-14}. Since the concentration of these two ionic species is very important to the properties of many systems, it is convenient to express their concentration by a short-hand terminology that is simpler than scientific notation, known as "pH." pH is defined as the negative logarithm (base ten) of the hydrogen ion concentration.

Since most systems have hydrogen ion concentrations ranging from 10^{-1} Molar to 10^{-14} Molar, pH values range from 1 to 14. pH values less than 7 are referred to as *acid*, while pH values greater than 7 are said to be *alkaline*. pH values equal to 7 are called *neutral*. When an acid-base reaction occurs and the resulting hydrogen ion concentration is 10^{-7}, the process is called *neutralization*.

Phenolic Hydroxyl Groups

Acid OH groups on benzene rings. They are weaker acids than carboxyl groups and begin to ionize at pH 10 and become completely ionized at pH 12. As components of lignin, they contribute to fiber surface charge in unbleached pulps under high pH conditions found in the stock preparation area often.

Physico-Chemical Mechanism

Chemical reactions or interactions manifested by physical change. Examples in papermaking include flocculation, dispersion, solids retention, and sizing.

Pick Strength

See Surface Strength

Pigment

Finely divided solid organic or inorganic coloring material that is insoluble in the medium in which it is applied (as contrasted to a dye which is soluble). White pigments are commonly employed as fillers in papermaking furnishes and they constitute the principal ingredient in coating mixtures. See also Filler.

Pitch

Resinous material released from wood by mechanical and/or chemical processing. Pitch includes fatty acids, resin acids, their insoluble salts, and esters of fatty acids with glycerol and sterols, as well as other fats and waxes.

Pitch Control

Prevention of the formation of pitch deposits and agglomerates within the papermaking system. The usual control strategy is either to keep the material dispersed; or add a high specific-surface substance like talc, which has an affinity for pitch particles; or, to add a high charge density, low molecular weight, cationic polymer that promotes the retention of colloidal pitch particles in the produced paper.

Pitch Deposits

Complex resinous hydrophobic agglomerates containing organic and inorganic portions that deposit onto papermaking machinery and in the formed web of paper. Changes in temperature, pH, chemical conditions, and shear can induce the deposition of pitch during papermaking.

Plastic Pigments

Bulky organic pigment-like particles manufactured by an emulsion polymerization process, used mainly in coating formulations. A number of different polymers and copolymers can be used, but polystyrene is the usual choice from a cost/performance standpoint. Plastic pigments have found limited use in specialty grades. Syn. Hollow Sphere Pigment.

Plate Count

Determination of the number of colonies of bacteria that develop after plating appropriate dilutions of a sample under specified conditions.

Polar Bonds

Polar bonds are covalent bonds having unequal sharing of electrons between the two bonded atoms. For example, in the O--H bond of the water molecule the more electronegative oxygen atom attracts the electron pair to its vicinity to a greater extent than the hydrogen atom. The end result is that the oxygen takes on a slight negatively charged character, while the hydrogen becomes slightly positive. Hence, a dipolar bond situation is established. This situation leads to the tendency of water molecules to associate with themselves through hydrogen bonding networks.

Poly-DADMAC — This polymer is a quaternary cationic polyelectrolyte of medium-high molecular weight synthesized by polymerization of diallyl dimethyl ammonium chloride (DADMAC). High charge densities are achieved. The polymer is often used as a standard titrant in the colloid titration method, as a charge neutralizer, and as one of the polymer components in a dual polymer retention system.

Polyacrylamides — High molecular weight compounds based on the acrylamide monomer unit utilized as bridging flocculants and retention aids in papermaking suspensions. They can be cationic or anionic.

Polyaluminum Hydroxy Chloride (PAC) — Form of aluminum used to precipitate rosin size in neutral papermaking systems at pH 7.0. Chemically, the name refers to a series of polyhydroxide chlorides of aluminum with the general formula: $Al_n(OH)_mCl_{3n-m}$. PAC is used for a variety of applications, including sizing, charge neutralization, pitch control, and retention control. It accomplishes these tasks without introducing the sulfate anion into the system and under alkaline conditions.

Polyamide Amines — Polymers formed by condensation of adipic acid and diethylenetriamine, followed by reaction with a small amount of epichlorohydrin. These reactions produce a cationic polyamide-epichlorohydrin resin whose molecular weight is high enough to serve as a retention aid. These materials also find wide use as wet strength resins.

Polyamide-epichlorohydrin Wet Strength Resin (PAE Resins) — Wet strength resin produced by first preparing a polyamide by reacting a dibasic acid with a triamine to produce a polyamide intermediate. The polyamide is then treated with epichlorohydrin to alkylate secondary amine groups and produce tertiary aminochlorohydrin groups. These groups self-alkylate to form 3,4-hydroxylacetidinium groups, which are responsible for the cationic character and reactivity of the wet strength resin. PAE molecules undergo cross-linking reactions with themselves and reactions with carboxyl groups to produce wet strength in paper. High temperatures and pH values greater than 6 are required. PAE resins are the most common wet strength resins in use today.

Polyamines — Polyamines formed by the reaction of dimethylamine and epichlorohydrin. The molecule has its charged nitrogen in the molecular backbone chain rather than presiding on a pendant group.

Polydisperse System — System containing particles of different sizes, or in which polymer molecules vary in molecular weight.

Polydispersity — Distribution of molecular weights of a polymer within a sample. For cellulose, a full appreciation of its physical and chemical properties requires consideration of the fact that any given sample consists of a mixture of molecules of differing chain lengths. Natural cellulose has a rather narrow molecular weight distribution and a polydispersity close to 1.0. Polydispersity is increased by processes such as pulping and bleaching.

Polyelectrolyte — Long chain polymer having cationic and/or anionic functional groups bound to the polymer chain. Natural and synthetic polyelectrolytes are utilized in papermaking systems as either cationic or anionic retention aids, and for adjusting the zeta potential.

Polymer — High molecular weight compound consisting of long chains that may be open, closed, linear, branched, or cross-linked. The chains are composed of repeating units, called monomers, which may be either identical or different.

Polymer Adsorption — Reaction between polymers and fiber or fines surfaces where the polymer is attracted to the surface (usually by electrostatic attraction) and adheres to it by electrostatic interactions (ion exchange) and other bonding interactions, such as hydrogen bonding. Practically all polymer additives must adsorb onto fiber and fines surfaces in order to be effective.

Polymer Charge Density Refers to the amount of cationic or anionic charge associated with a polymer; expressed per unit weight, as a percentage of the average number of monomer units in the polymer molecule, or on some other similar basis.

Polymer Molecular Weight The weight, in grams, of one mol (6.023×10^{23}) of polymer molecules. Since the individual polymeric materials used in papermaking exhibit ranges of molecular weights, it is necessary to refer to the "average molecular weight" of a substance. Also, retention aids are often referred to as having "low," "medium," or "high" molecular weights, where the low and medium molecular weight materials act through charge neutralizing and patching mechanisms, while high molecular weight polymers promote bridging.

Polysaccharides Linear or branched polymers of more than ten monosaccharides that are linked by means of glycosidic bonds. A polymer of ten or less monosaccharides is called an oligosaccharide.

Porosity of Paper Porosity is the ratio of the pore volume, or void volume, of the web to its total volume. It is often evaluated by measuring the permeability of a paper to air or another fluid.

Post-Consumer Waste Paper Paper discarded from residences, institutions (e.g. schools, hospitals) and businesses that is collected and reused in papermaking.

Precipitate Solid substance formed from solution, usually by chemical action, when the solubility of the substance is exceeded.

Precipitated Calcium Carbonate (PCC) Calcium carbonate filler produced by a chemical process. The most common reaction used for this purpose is the reaction of slaked lime ($Ca(OH)_2$) with carbon dioxide gas (CO_2) to produce $CaCO_3$. This process is called the carbonation process. Many paper mills actually carry out this process on-site. Precipitated calcium carbonate can be produced in a variety of shapes and sizes, depending upon the carbonation reaction conditions.

Precipitating Agent In general chemical use, a substance that induces the production of a precipitate when added to a solution.

Pregelatinized Starch Starch which has been precooked and dried, and has the ability to hydrate and swell in cold water to form a solution.

Pressate Liquid which separates from solid/liquid mixture on being passed through a pressing operation.

Pressure Screen Pulp screening equipment which functions by the passage of accepted stock through a perforated plate at full line pressure. The plate is kept clean by means of a rotating hydrofoil or other rotating element designed to provide alternating pressure and vacuum pulses.

Primary Bonding Forces Covalent and ionic bonds that unite atoms into molecules or chemical compounds. High energy bonds having energies of 400-600 kJ/mol or higher. Bond distances of 0.1-0.2 nm are involved.

Primary Cleaners Centrifugal cleaners which handle the entire stock flow as well as certain recycle streams (e.g. secondary screen accepts). Generally, all accepted stock must pass through the primary cleaners.

Primary Wall Outermost, very thin, highly lignified layer of the cell wall.

Processing Aids Chemical additives used to promote aggregation interactions and other interactions among furnish components and enhance the performance of functional additives or promote the runnability and cleanliness of the paper machine. Defoamers, pitch dispersants, retention aids, drainage aids, formation aids, and microbiological control agents are all referred to as chemical processing aids.

Promoter Polymer additives that enhance the retention of sizing agents and dyes in papermaking.

Protective Colloid Material that surrounds colloidal particles and prevents their aggregation. In papermaking, the presence of oxidized starch, carboxymethylcellulose, or gelatin can act to keep a rosin size precipitate well dispersed.

PVSK A common anionic polymer used as a standard titrant in the colloid titration method. It is the potassium salt of polyvinyl sulfate.

Quaternary Starch Starch that has been substituted with cationic functional groups containing a quaternary amine group. Quaternary amine groups retain their cationic charge over the entire pH range used in papermaking.

Refining The mechanical treatment of fibers by a beater, conical refiner, or disk refiner to give them (the fibers) the characteristics required to be formed into sheets of paper or paperboard having the desired properties. Refining develops fiber-fiber bonding and strongly affects all properties that are related to fiber-fiber bonding. The mechanical properties, in particular, are affected.

Relative Bonded Area Ratio of the total bonded area of fibers within an assemblage to one-half of the total external surface area of component fibers.

Release Aid Blend of oils, solvents, and/or surfactants that is sprayed onto the paper surface or Yankee dryer surface to facilitate sheet release from the dryer surface.

Repulping Wet Strength Paper This refers to the processing of wet strength paper to reclaim the fibers for reuse. U-F and M-F papers are relatively easy to repulp, requiring mechanical action and high temperatures in a pulper. Acid treatment is also required in certain instances. PAE papers are more resistant to repulping and have to be treated with an oxidizing agent, such as sodium hypochlorite. Freshly made wet strength paper is more easily repulped because it is not fully cured.

Resin (1) Secretions of certain trees, consisting of mixtures of resin acids and esters insoluble in water, but soluble in ether, alcohol, and other neutral organic solvents. (2) Generally, any polymer that is a basic material for plastics; often designated as “synthetic resin” to differentiate from natural resin. The term is also used when referring to wet strength agents.

Resin Emulsions Non-elastomeric polymers used as stock additives to improve sheet strength, provide specific sizing effects, and improve other sheet properties. Examples are polystyrenes, polyalkylacrylate, and polyvinyl acetate.

Resistance Properties of Paper Resistance properties of paper, or sizing properties, refer to a paper’s ability to retard the rate at which a fluid, usually water, will penetrate into the web structure.

Retention Amount of any papermaking material which is retained in paper during the forming process, usually expressed as a percentage of that initially added. Exact percentages are often difficult to establish because of changes in the materials during the papermaking process or during analysis. See also Single-Pass Retention.

Retention Aid Chemical added to the papermaking furnish to increase the retention of fiber fines, mineral fillers, or other additives. Usually reserved to mean a high molecular weight polymer that promotes aggregation by bridging.

Retrogradation Formation of an insoluble precipitate in a starch solution. Usually associated with molecular association of the amylose fraction of starch.

Reverse Sizing Internal sizing with rosin in which the order of adding rosin and alum is reversed, i.e., alum is added to the papermaking stock before the rosin.

Reversible Reaction Chemical reaction that can proceed in either direction depending on conditions.

Rosin Emulsions Very high (95-100%) free rosin acid content rosin dispersions. Some types are stabilized with a protective colloid such as casein or soy protein.

Rosin Size — Forms of rosin acids used as a wet end sizing additive. Can be in a highly saponified form (sodium salts of resin acids), which is precipitated onto papermaking fibers by alum, or as dispersed, high free rosin acid material (>95%) that must be retained by cationic starch or size retention aid. The dispersed rosin reacts with alum on the papermaking fibers. Both materials increase the water resistance of paper.

Roughness of Paper — See Smoothness of Paper

Rutile — One of the two forms of titanium dioxide used as papermaking fillers.

S-Layers — The three layers of the secondary cell wall, designated as the S_1 for the outer layer, the S_2 for the central layer, and the S_3 for the inner layer. The last is sometimes called the tertiary wall.

Sap — Liquid within unseasoned wood or the living tree containing the nutrients and other chemicals in solution which are required for growth.

Sapwood — Lighter-colored, water-conducting outer portion of the tree stem with living ray tissue, as differentiated from the darker heartwood portion in the center of the stem.

Satin White — Bulky white pigment made from the interaction of slaked lime and aluminum sulfate. It can be described chemically as calcium sulfoaluminate complex.

Scalenohedral Calcium Carbonate — The most common form of precipitated calcium carbonate used as a filler. Scalenohedral particles have elongated, "football" shaped particles that form as clusters or "rosettes" of particles.

Scattering Power — Ability of fibers and filler particles in a sheet to scatter incident light. Scattered light contributes to opacity, brightness, gloss, and color. Scattering power is expressed in terms of the specific scattering factor for a material -- whose units are m^2/g or cm^2/g.

Secondary Bonding Forces — Bonds between atoms that have lower bonding energies (12-21 kJ/mol) than covalent or ionic bonds. Van der Waals bonds and hydrogen bonds fall into this category. They are basically nondirectional in nature and involve bonding distances of 0.2-0.3 nm. Strong bonding forces between molecules can be established with secondary bonding by combining many such bonds.

Secondary bonding forces (hydrogen bonding) produce fiber-fiber bonding in paper. All adsorption and aggregation phenomena in wet end chemistry involve secondary bonding forces.

Secondary Fibers — Fibers from paper that has been used by converters or consumers and then reclaimed for reuse in papermaking.

Secondary Wall — Portion of the cell wall that contains most of the fiber substance, formed in three distinct layers characterized by different fibril alignments.

Self Sizing — The gradual development of water resistance in a paper that does not have this property immediately after it is manufactured. Often a problem in papers like corrugating medium. Can also occur in tissues. Caused by redistribution of natural resinous materials throughout the sheet upon storage.

Shade of Paper — See Color of Paper.

Silica — Silicon dioxide (SiO_2) occurring naturally in many forms. A synthetically-prepared amorphous form is used as a papermaking filler. It has very small particle size and very high specific surface area and is used as an extender pigment.

Silicate — Any of a number of compounds containing silicon, oxygen, and one or more metals, with or without hydrogen. Certain synthetically-prepared amorphous silicates are used as papermaking fillers.

Single-Pass Retention — Percentage of the stock issuing from the headbox that is retained on the forming fabric during paper manufacture. Syn. First-Pass Retention.

Sintering Conversion of the rosin-alum precipitate, which is retained within the paper sheet during internal sizing, into an efficient hydrophobic material in the dryer section.

Size Any material used for internal or surface sizing, for example rosin with alum, starch, animal glue, gelatin, latex, AKD, ASA.

Size Hydrolyzate Product formed from the reaction between AKD or ASA size and water. In both instances, hydrolysis reduces the efficiency of the sizing agent. The ASA hydrolyzate also forms deposits that adversely affect paper machine runnability. Size hydrolysis is promoted by high temperatures and high pH.

Size Precipitant Chemical in the papermaking system which serves to precipitate soap rosin size onto fibers. Alum is commonly used for this purpose, but other aluminum compounds are suitable.

Sizing Treatment of paper to resist liquid penetration, either by means of wet end additives or surface application.

Slack Sized Papers Papers having intermediate levels of water resistance; e.g. offset printing papers, writing papers.

Slender Fibers Slender fibers have thinner cell walls and produce dense, smooth papers having higher tensile strength and lower tearing resistance. The opposite of slenderness is coarseness.

Slime Gelatinous, stringy, viscous, pasty or leathery masses of bacteria (anaerobic and aerobic) and fungi (yeasts and molds), often in symbiotic associations. See also Slime Deposits.

Slime Control Inhibition or elimination of slime formation.

Slime Deposits Deposits in the papermaking system characterized by some degree of microorganism activity, but also consisting of various combinations of fibers, fillers, and other materials.

Slimicide See Biocide.

Slow Stock Stock that does not release water readily in the free drainage zone of a paper forming section. Produces a low Canadian Standard Freeness test result.

Smoothness of Paper Smoothness relates to the magnitude of surface irregularities of paper. When a paper has small surface irregularities, it is said to be *smooth*. When it is has large surface irregularities, it is said to be *rough*.

Soap Size Rosin size composed of sodium salts of resin acids. Formed by neutralizing resin acids with NaOH or some other sodium base, such as Na_2CO_3. Different forms of soap sizes are available, including paste, extended, and dry sizes.

Sodium Aluminate ($Na_2Al_2O_4$) Chemical used in the sizing process as a source of alumina. Unlike alum solutions which are acidic, sodium aluminate solutions are alkaline. In most paper mill applications, sodium aluminate is used in combination with alum to achieve both the desired pH and alumina content.

Soft Flocs Aggregates that are not very shear resistance and break down under such conditions. They reform after the shear is removed. Hence, they are *reversible* in character. Soft flocs are formed by coagulation mechanisms (i.e. charge neutralization or patching).

Softening Agent Surfactant-type chemical which is added to papermaking stock to impart a softer feel to the paper product. Syn. Softener.

Softwood Wood produced by coniferous trees, i.e. nonporous wood. The term has no reference to the actual hardness of the wood. See also Gymnosperms.

Sol Solution of colloidal molecules or micelles. Solutions of starch, hemicellulose, and high molecular weight polymers are examples.

Solubility
Ability of one substance (solute) to blend uniformly with another substance (solvent). For example, the solubility of a dye in water is usually measured in terms of the highest concentration (g/L) at which the material is totally dissolved at a prescribed temperature.

Solubility Product
All apparently water-insoluble substances dissolve to some degree. A dynamic equilibrium is established between the insoluble material and the dissolved ions. The product of the respective ion concentrations is a constant called the solubility product.

Specific Surface Area
Surface area of a material per unit weight. Usually reported as m^2/g. Fibers have values in the 1-2 m^2/g range. Fiber fines lie in the 8-10 m^2/g. Fillers range from 8-80 m^2/g. The specific surface area of a substance is inversely related to its particle size.

Stabilizer
(1) Substance added to an emulsion or dispersion to prevent aggregation and separation. (2) Substance added to a formulation to protect against degradation during processing.

Starch
Polymer of glucose derived from various plants, principally corn, tapioca, potato, and wheat. It is widely used in papermaking as a dry strength additive, as the major component in surface sizing solutions, and as an adhesive in coating formulations. Many types and forms are available including the native product, called pearl starch, fractionated starches, and starches modified by chemical and thermal action.

Structurally, starch is characterized by an arrangement of repeating diglucose units bonded by α-1,4 glucosidic linkages -- which are called maltose units. Branches through C-6 also occur in one fraction of the material called *amylopectin*. The fraction that contains no such branches is called *amylose*. The starches used as wet end additives contain high percentages of amylopectin, which forms stable solutions that do not tend to retrograde.

In contrast to cellulose molecules which take on straight, ribbon-like shapes, starch molecules adopt helical and other less regular shapes. In addition, the amylopectin molecules have a built-in "herring bone" shape due to their primary structure.

Most wet end additive starch comes from corn or potato starch. In order to achieve high levels of starch retention, it is necessary to incorporate small quantities of cationically charged groups into its structure. Common substitution rates typically range from 1.5 cationic groups per 100 glucose units, on average, to 4.5 cationic groups per 100 glucose units. Cationic starch addition rates vary from a few pounds per ton to as high as 50-60 pounds per ton in most instances. Typical molecular weights of wet end starches range from 500,000 to 3,000,000.

The preparation of most wet end starches involves heating suspended starch granules (diameter 5-100 microns) to approximately 95°C in order to swell and rupture the starch granules and dissolve the individual starch molecules. The most common cooking practices include batch processing and continuous, steam "jet" cooking.

Cationic starch has several applications, including fines retention, synthetic size emulsion stabilization and retention, and dry strength enhancement. Of these, dry strength is by far the most important use and starch accounts for more than 95% of all dry strength additives.

Starch Conversion	Because of the high viscosity (due to very high molecular weight) and poor flow characteristics of naturally occurring starch solutions, some modification is usually required to facilitate their use as a sizing agent or in coating formulations. Chemical modifications by manufacturers include dextrinization (heating in powder form with acid or salt), acid thinning (treatment with mineral acid in aqueous slurry), oxidation (e.g reaction with hypochlorite in suspension), etherification, esterification, and cross-linking with reagents. Mill treatments of native starch include enzyme conversion (heating a solution with alpha amylase), thermal conversion (starch slurry subjected to high temperature steam and shear mixing in a pressurized heater), and thermochemical conversion (pressurized oxidation of starch solution at high temperature).
Steric Stabilization	Stabilization of a dispersion of colloidal particles by adsorbing long chain neutral or anionic polymers onto particle surfaces so that when two particles approach one another the highly solvated polymer chains interfere with one another by steric interactions and prevent the particles from approaching one another closely enough for aggregation to occur. Syn. Protective Colloid Effect.
Stickies	Deposits in secondary fiber papermaking systems created by such contraries as ink residuals, tars, latexes, and hot-melt materials. The term "tackies" refers to the same type of problem, perhaps less severe.
Stiffness	The ability to resist deformation under stress. Resistance to a force causing the specimen to bend is called bending or flexural stiffness.
Stock Preparation	That part of the papermaking process in which pulp is treated mechanically, and sometimes chemically by use of additives, to prepare it for forming into a sheet of paper or paperboard. Stock preparation functions include defibering, mixing, blending, consistency adjustment, refining, and contaminant removal.
Streaming Potential	In the streaming potential measurement, liquid is forced, under a pressure gradient, through a plug formed from fibers, fines and other furnish components. Electrodes are placed on either side of the plug to measure the streaming potential established across the plug by the flowing liquid. Zeta potential can be calculated from the streaming potential value. Currently, the most common use of streaming potential in the paper industry is for the on-line measurement of zeta potential.
Strength Properties of Paper	See Mechanical Paper Properties
Stretch of Paper	Stretch, or elongation, refers to the amount that a specimen will deform while being subjected to a tensile stress exerted in the plane of the paper. It is a dimensionless parameter, usually expressed as a percent of the original length of the specimen.
Structural Properties of Paper	Structural paper properties are those properties that describe how the components are arranged in a sheet of paper. Included in this category are grammage, basis weight, caliper, formation, directionality, two-sidedness, porosity, smoothness, and roughness.
Substantivity	Property of a paper dye which makes it capable of being adsorbed by cellulose fibers from an aqueous medium.
Surface Active Agent	Any chemical which spontaneously migrates to an interface (liquid/liquid, liquid/solid, or liquid/air) is said to be "surface active." Surface active agents usually comprise an organic molecule containing both a hydrocarbon portion (oil soluble; hydrophobic) and a polar portion (hydrophilic; water soluble). These chemicals can be further delineated according to application into categories such as detergents, dispersants, foaming agents, and defoamers. Syn. Surfactant.

Surface Activity Term associated with the spontaneous lowering of the surface tension when certain molecules are present, even at very low concentrations. Materials that exhibit surface activity are called "surfactants."

Surface Charge Electrostatic charge associated with the surface of papermaking particles. These charges in fibers are derived mainly from the ionization of surface hemicellulose carboxyl groups and phenolic hydroxyl groups. In other particles, such as fillers, surface charge can be derived from several different sources, such as the ionization of surface hydroxyl groups; isomorphous substitution of ions in the particle lattice structure; and the incorporation of charged groups in the pigment particle surface during its synthesis. Surface charge cannot be directly measured readily. It is most often measured indirectly by electrokinetic methods that determine zeta potential. See Zeta Potential.

Surface Sizing Application of a sizing additive solution to a formed and dried paper web by a size press, coater, or calender box. Surface sizing is done to enhance resistance properties and/or paper surface strength.

Surface Strength The property of paper that describes its ability to withstand rupture when subjected to a tensile stress perpendicular to the paper surface. Such stresses are often encountered in printing processes. It is important that the surface fibers or coating on the sheet be bonded to the paper strongly enough not to be pulled out, or picked, during the process.

Surface Tension Liquid surfaces are in a state of "tension" which causes them to contract to their minimum surface area. For example, liquid droplets adopt spherically shaped forms. Surface tension arises because surface molecules experience unbalanced interactions with bulk molecules below or within the surface of the liquid and vapor molecules above the liquid surface. Surface tension is expressed in units of force/distance (e.g. mN/m)) or energy per area (e.g. mJ/m^2)

Synthetic Size Sizing agent that has been developed as a wet end additive in a neutral/alkaline papermaking system where rosin has not been traditionally applied. Among the principal chemicals used for this purpose are alkyl ketene dimer (AKD) and alkenyl succinic anhydride (ASA).

Talc White pigment with a plate-like morphology, composed mainly of hydrated magnesium silicate [$Mg_3Si_4O_{10}(OH)_2$], occasionally used as a papermaking filler. Certain grades of talc with high specific surface can selectively adsorb pitch particles and are used to prevent their deposition within pulp and papermaking process equipment.

Tearing Resistance The force required to tear a specimen under standardized conditions. Two measurement methods are in common use: (1) Internal tearing resistance, wherein the edge of the specimen is cut before the actual test. The most common instrument used for this measurement is the Elmendorf tear tester. (2) Edge tearing resistance, wherein the specimen edge is not torn prior to the test. Both tests involve an out of plane tear.

Tensile Breaking Strength The maximum tensile stress developed in a specimen, in the plane of the paper, before rupture under prescribed conditions. It is usually expressed as force per unit width.

Tertiary Starch Starch that has been substituted with cationic functional groups containing a tertiary amine group. Tertiary amine groups have a high cationic charge only at pH values of less than 6. Increasing the pH above six causes the percent charge to decline to less than 50% at pH 9.

Thermal Conversion (of Starch) See Starch Conversion.

Thickening Water Removal Thickening occurs when the particles in a suspension are in contact with one another and are entangled or immobilized in a network. The structure collapses as drainage proceeds. There is no sharp boundary between the formed mat and the suspension, and water is removed from all parts of the network simultaneously and at equal rates.

Thickness of Paper See Caliper of Paper

Tinting Dyes Dyes used in small quantities to impart various colored tints to white papers.

Titanium Dioxide (TiO_2) White pigment used as a papermaking filler. Two molecular forms are available: anatase and rutile. Titanium dioxide fillers have spherical particle shapes and have average ultimate particle diameters of about 0.3 micron. It is not widely used anymore due to its relatively high cost. It has the highest refractive index of all papermaking fillers and, consequently, the highest light scattering coefficient.

Titration Laboratory technique of quantitative analysis whereby measured amounts of a standard solution are successively added and reacted with a sample substance in solution until a desired end point is reached as shown by a color change, potentiometric inflection point, or other indicator.

Toluidine Blue Indicator dye used to detect the endpoint in the colloid titration method. The dye is blue in cationic environments and pink in anionic environments in the presence of PVSK. See also Colloid Titration, PVSK.

Total Charge The charge associated with both the suspended and dissolved materials in a papermaking sample as determined by colloid titration.

Tracheid Botanical term for wood fibers. Tracheids have an elongated cell with bordered pits and imperforate ends. Relatively long tracheids (up to 7 mm) constitute the principal part of the cellular structure of softwoods, and shorter tracheids (seldom over 1.5 mm) are present in many hardwoods.

Tray Water Water passing through the forming fabric of a paper machine which contains various concentrations of fines and filler depending on location. Most of the water that passes through the forming section is recycled and reused in the system.

Two-Sidedness of Paper Two-sidedness refers to the difference in texture and composition of the two sides of a sheet of paper. It is most pronounced in papers formed on a single wire machine. The side of the web that lies against the forming fabric is called the *wire side*, and the side of the web away from the forming fabric is called the *felt side.* The wire side often has a cross-hatched pattern in its surface. The felt side contains more fines and is usually smoother in texture. The term *two-sidedness* is also used to refer to the tendency of dyes to be concentrated more in the felt side of a sheet than the wire side.

Ultrastructure Internal organization of the fiber wall, including the molecular structure of the cell wall constituents and their organization into various layers and structures. In order to observe fiber ultrastructure, it is necessary to employ the electron microscope and x-ray diffraction analysis.

Urea-Formaldehyde Wet Strength Resin (U-F Resins) Wet strength resin produced by reacting urea with formaldehyde to produce a cross-linkable dimethylol urea intermediate. Incorporation of a polyfunctional amine group imparts cationicity to the structure and makes it attractive to papermaking fibers during sheet formation. Urea-formaldehyde resins require high temperatures and acid pH values (ca. 4.5) for curing.

Vacuum Drainage Water removal by vacuum elements (wet and dry suction boxes, suction couch roll). Web consistency typically goes from 2-4% to 20-22% in the vacuum drainage zone of the paper machine.

Vessel Tube-like structures in connected at or near their ends by perforation plates. Vessel elements are responsible for the conduction of sap up and down the trunk of a hardwood. They exist in a wide variety of shapes and sizes. For example, birch wood contains vessel elements that are shaped like fibers and can be bonded into the paper web reasonably well. On the other hand, oak vessels are large, blocky structures that do not bond well in the sheet and pick out in printing presses.

Virgin Fiber Pulp used for papermaking that has not previously been used in any paper or board product.

Washburn Equation An equation that describes the penetration of liquids into fiber beds and packed powders. It includes terms for pore radius and pore length, time, contact angle, fluid viscosity, and fluid surface tension.

Water of Hydration Refers to the layer of water molecules that surround molecules when they are in solution. See also Hydration.

Water Retention Value (WRV) Ratio of water to dry fiber after centrifuging a wet pulp sample for a specified time at a specified force (usually 30 minutes at 900g). The WRV is used as an indication of how tightly free moisture is held within the fiber structure, and is a good approximation of the fiber saturation point.

Waterleaf Papers Very absorbent papers having no water resistance; e.g. toweling, tissue, corrugating medium.

Wet End Chemistry Surface and colloid chemistry of fines and additives and their interactions with fibers.

Wet Strength Broke Broke from wet strength papermaking operations. Because of its high wet strength, the broke is often difficult to repulp, and special techniques, such as acid, high temperatures, and bleaching agents, are sometimes employed, depending on the particular type of wet strength resin contained in the broke.

Wet Strength Curing Chemical reactions where wet strength resins retained in the paper web react to form intramolecular cross-links and produce a network structure within the paper web that protects fiber-fiber bonds against disruption by water. Also, chemical reactions between retained wet strength resin and fibers which lead to covalent links between the two. Again, the end result is the production of a water-resistant interfiber bonds. Wet strength resin curing required high temperatures and it is also sensitive to pH conditions. The most common resins in use today require pH values above 6 for the most efficient curing. Curing is a relatively slow reaction and continues after the paper is manufactured and removed from the paper machine.

Wet Strength Paper Paper whose tensile strength when saturated with water is more than 10-15% of its dry tensile strength.

Wet Strength Resins Chemicals added to the papermaking stock that serve to reinforce normal fiber-fiber bonds and provide strength retention in aqueous media. Common agents are urea-formaldehyde, melamine-formaldehyde, and polyamide resins; these are applied at an intermediate degree of polymerization, so that the final "cure" is obtained during and after drying the paper.

Wet Web Strength Strength of never-dried paper when it contains 20-45% solids content. Measured by a wet tensile test or a wet bursting strength test. Wet web strength is not to be confused with wet strength, where dry paper is saturated with water prior to testing. Wet web strength is related to paper machine runnability at the wet end and is most critical in paper grades that contain high mechanical fiber content, such as newsprint. See also Wet Strength Paper.

Wetting The interaction that occurs between a liquid and a solid surface when the liquid spreads spontaneously without forming a droplet having a finite contact angle.

Wetting Agent Material capable of lowering the surface tension of water solutions, and thus increasing their wetting power.

White Pitch Term used to describe agglomerated white stickies in repulped stock which are derived from the synthetic binders in coated broke.

White Water General term for any stock filtrate or process water that contains fiber fines. On the papermachine, white water is produced during the forming and dewatering of the web. "Rich" white water contains a high concentration of fiber fines, while "lean" white water contains a low concentration.

White Water Loop Paper machine process circuit in which white water that drains from the forming web on the paper machine is then reused for dilution of thick stock coming from the stock preparation section immediately prior to the approach flow system.

Zeta Potential Electrical potential at the interface between the dense layer of counterions that surround a charged particle suspended in a liquid and the counterions that are not closely associated with the particle. Zeta potential is measured by electrokinetic techniques where there is relative motion between the particle and its suspending medium. Most papermaking particles have natural negative charges. Zeta potential is very important in the aggregation of papermaking particles and in the adsorption of chemical additives by fibers. See Surface Charge.